Experimental and Thermodynamical Modeling of Ore-Forming Processes in Magmatic and Hydrothermal Systems

Experimental and Thermodynamical Modeling of Ore-Forming Processes in Magmatic and Hydrothermal Systems

Special Issue Editor

Galina Palyanova

MDPI • Basel • Beijing • Wuhan • Barcelona • Belgrade

MDPI

Special Issue Editor
Galina Palyanova
Sobolev Institute of Geology and Mineralogy
Siberian Branch
Russian Academy of Sciences
Russia

Editorial Office
MDPI
St. Alban-Anlage 66
4052 Basel, Switzerland

This is a reprint of articles from the Special Issue published online in the open access journal *Minerals* (ISSN 2075-163X) in 2018 (available at: https://www.mdpi.com/journal/minerals/special_issues/ Experimental_Thermodynamical_Modeling#)

For citation purposes, cite each article independently as indicated on the article page online and as indicated below:

LastName, A.A.; LastName, B.B.; LastName, C.C. Article Title. *Journal Name* **Year**, *Article Number*, Page Range.

ISBN 978-3-03897-515-1 (Pbk)
ISBN 978-3-03897-516-8 (PDF)

Contents

About the Special Issue Editor vii

Preface to "Experimental and Thermodynamical Modeling of Ore-Forming Processes in
Magmatic and Hydrothermal Systems" ix

Galina Palyanova
Editorial for the Special Issue: Experimental and Thermodynamic Modeling of Ore-Forming
Processes in Magmatic and Hydrothermal Systems
Reprinted from: *Minerals* 2018, *8*, 590, doi:10.3390/min8120590 1

Alexander V. Zotov, Nikolai N. Kuzmin, Vladimir L. Reukov and Boris R. Tagirov
Stability of $AuCl_2^-$ from 25 to 1000 °C at Pressures to 5000 bar and Consequences for
Hydrothermal Gold Mobilization
Reprinted from: *Minerals* 2018, *8*, 286, doi:10.3390/min8070286 6

Vladimir L. Tauson, Sergey V. Lipko, Nikolay V. Smagunov and Raisa G. Kravtsova
Trace Element Partitioning Dualism under Mineral–Fluid Interaction: Origin and
Geochemical Significance
Reprinted from: *Minerals* 2018, *8*, 282, doi:10.3390/min8070282 21

Yuliya Bataleva, Yuri Palyanov and Yuri Borzdov
Sulfide Formation as a Result of Sulfate Subduction into Silicate Mantle (Experimental
Modeling under High P,T-Parameters)
Reprinted from: *Minerals* 2018, *8*, 373, doi:10.3390/min8090373 47

Valery Murzin, Konstantin Chudnenko, Galina Palyanova, Aleksandr Kissin and
Dmitry Varlamov
Physicochemical Model of Formation of Gold-Bearing Magnetite-Chlorite-Carbonate Rocks at
the Karabash Ultramafic Massif (Southern Urals, Russia)
Reprinted from: *Minerals* 2018, *8*, 306, doi:10.3390/min8070306 62

Sasha Krneta, Cristiana L. Ciobanu, Nigel J. Cook and Kathy J. Ehrig
Numerical Modeling of REE Fractionation Patterns in Fluorapatite from the Olympic Dam
Deposit (South Australia)
Reprinted from: *Minerals* 2018, *8*, 342, doi:10.3390/min8080342 80

Francisco J. Testa, Lejun Zhang and David R. Cooke
Physicochemical Conditions of Formation for Bismuth Mineralization Hosted in a
Magmatic-Hydrothermal Breccia Complex: An Example from the Argentine Andes
Reprinted from: *Minerals* 2018, *8*, 486, doi:10.3390/min8110486 101

Ana Rita F. Sirqueira, Márcia A. Moura, Nilson F. Botelho and T. Kurt Kyser
Nature and Evolution of Paleoproterozoic Sn and Rare Metal Albitites from Central Brazil:
Constraints Based on Textural, Geochemical, Ar-Ar, and Oxygen Isotopes
Reprinted from: *Minerals* 2018, *8*, 396, doi:10.3390/min8090396 122

Jingya Cao, Qianhong Wu, Xiaoyong Yang, Hua Kong, Huan Li, Xiaoshuang Xi,
Qianfeng Huang and Biao Liu
Geochronology and Genesis of the Xitian W-Sn Polymetallic Deposit in Eastern Hunan
Province, South China: Evidence from Zircon U-Pb and Muscovite Ar-Ar Dating,
Petrochemistry, and Wolframite Sr-Nd-Pb Isotopes
Reprinted from: *Minerals* 2018, *8*, 111, doi:10.3390/min8030111 144

Lizhi Yang, Xiangbin Wu, Jingya Cao, Bin Hu, Xiaowen Zhang, Yushuang Gong and Weidong Liu
Geochronology, Petrology, and Genesis of Two Granitic Plutons of the Xianghualing Ore Field in South Hunan Province: Constraints from Zircon U–Pb Dating, Geochemistry, and Lu–Hf Isotopic Compositions
Reprinted from: *Minerals* **2018**, *8*, 213, doi:10.3390/min8050213 . . . . **167**

Hao Wei, Jiuhua Xu, Guorui Zhang, Xihui Cheng, Haixia Chu, Chunjing Bian and Zeyang Zhang
Hydrothermal Metasomatism and Gold Mineralization of Porphyritic Granite in the Dongping Deposit, North Hebei, China: Evidence from Zircon Dating
Reprinted from: *Minerals* **2018**, *8*, 363, doi:10.3390/min8090363 . . . . **187**

About the Special Issue Editor

Galina Palyanova, Ph.D., is the leading scientist in the V.S. Sobolev Institute of Geology and Mineralogy, of the Siberian Branch of the Russian Academy of Sciences. She graduated from the Novosibirsk State University with a degree in geochemistry. Her main area of research is ore mineralogy, as well as thermodynamic and experimental modeling of ore-forming processes. She has published over 70 research articles in international journals. Her monograph "Physical and chemical features of the behavior of gold and silver in the processes of hydrothermal ore formation" was awarded a medal and a diploma of the Russian Mineralogical Society. For her great contribution to the development of the mineral resource base of Russia, she was awarded medals of the Russian Ministry of Higher Education and the Russian Academy of Sciences and an honorary diploma of the Russian Ministry of Natural Resources and Energy Resources.

Preface to "Experimental and Thermodynamical Modeling of Ore-Forming Processes in Magmatic and Hydrothermal Systems"

This Special Issue book includes 10 original research articles that highlight, discuss and solve some problems of ore-forming processes in magmatic and hydrothermal systems. Several articles of this *Minerals* Special Issue are devoted to the study of the genesis, sources of ore matter, and geochronology of some ore occurrences in China (Cao et al., 2018; Yang et al., 2018; Wei et al., 2018), Brazil (Sirqueira et al., 2018), Australia (Krneta et al., 2018), Argentina (Testa et al., 2018), and Russia (Murzin et al., 2018). This Special Issue also contains articles that highlight recent experimental investigations on the solubility of gold in HCl/NaCl fluids (Zotov et al., 2018), the distribution behaviors of some precious metals and rare earth elements between Fe-rich minerals and chloride-based hydrothermal solutions (Tauson et al., 2018), and interactions of Mg,Fe,Ni-olivine with anhydrite and Mg-sulfate, as possible sources of oxidized S-rich fluid under lithospheric mantle P,T-parameters (Bataleva et al., 2018).

Ore-forming processes occur over a long period of time. What we know about them today is just one piece in the huge mosaic of the history of the development of the Earth. I hope that this special issue will be useful for scientists who work on the fundamental problems of ore-forming processes and genesis of the ore deposits, and will provide new ideas for future research.

Galina Palyanova
Special Issue Editor

minerals

MDPI

Editorial

Editorial for the Special Issue: Experimental and Thermodynamic Modeling of Ore-Forming Processes in Magmatic and Hydrothermal Systems

Galina Palyanova [1,2,3]

[1] V.S. Sobolev Institute of Geology and Mineralogy, Siberian Branch of the Russian Academy of Sciences, Akademika Koptyuga pr., 3, Novosibirsk 630090, Russia; palyan@igm.nsc.ru
[2] Novosibirsk State University, Pirogova str., 2, Novosibirsk 630090, Russia
[3] Diamond and Precious Metal Geology Institute, Siberian Branch of the Russian Academy of Sciences, pr. Lenina, 39, Yakutsk 677980, Russia

Received: 30 November 2018; Accepted: 11 December 2018; Published: 13 December 2018

A number of excellent books and articles on the ore-forming processes, magmatic and hydrothermal systems, physicochemical conditions of the ore-forming fluids, and thermodynamic modeling in the geosciences have been published over the past 50 years [1–13]. Fundamental knowledge about ore-forming processes has practical applications for mineral exploration.

Experimental investigations and thermodynamic modeling have been successfully used to solve many problems, in which ore-forming processes play an important role. The experimental study of simple systems with precisely controlled parameters provides the data necessary for obtaining reliable thermodynamic characteristics of minerals, aqueous species (ions, complex ions, and molecules), gas mixtures, and solid solutions. The main obstacles to the experiments are the multicomponent, multiphase and multiaggregate features of natural systems and the inevitability of their simplification. Thermodynamic modeling can be an alternative to complex model experiments. To construct quantitative genetic models of ore-forming processes, computer thermodynamic modeling on the basis of modern program complexes is used [14–18]. The reconstruction of the sources of ore components and fluids, physicochemical parameters, and the mechanisms of accumulation and separation of elements during the formation of ore deposits is an important task in addressing the fundamental problems of ore-forming processes. Some of the papers in this issue deal with experimental and thermodynamic modeling [19–25], while the others are devoted to the analytical geochemistry, geochronology, and genesis of some ore occurrences [26–28].

In the article by Zotov and coauthors [19], the solubility of gold was measured in HCl/NaCl fluids (NaCl concentration varied from 0.1 to 3 m) at 450 °C and pressures from 500 to 1500 bar. The obtained data for the stability constant of the $AuCl_2^-$ complex, together with values from the literature for temperatures from 25 to 1000 °C, were fitted to the simple equation of log Ks°. Fluid chlorinity, together with acidity and redox state, controls the concentration of many ore metals, which can migrate in the fluid phase in the form of chloride complexes. The speciation of gold in natural chloride-bearing fluids is discussed. The results of the study are important to the understanding of the formation process of hydrothermal gold deposits.

Tauson and coauthors [20] investigated the partitioning of precious metals (Au, Pt, and Pd) in the system of a "mineral–hydrothermal solution", where the minerals were pyrite, As–pyrite, magnetite, Mn–magnetite, and hematite and the fluids were ammonium chloride-based hydrothermal solutions at 450 and 500 °C and 100 MPa pressure. These authors study the cocrystallization (exchange) coefficients (D_e) of rare earth elements (Ce, Eu, Er, and Yb) and Fe in magnetite and hematite at 450 °C and 100 MPa. Trace element partitioning in the "mineral–hydrothermal solution" system was studied by the method of thermogradient crystal growth coupled with the internal sampling of a fluid phase. The analytical procedure used enables the evaluation of structurally bound and superficially bound modes of trace elements in crystals and the determination of corresponding dual partition coefficients. The obtained experimental results assume that the distribution of precious metals and rare earth elements is largely dependent on the composition of the superficial nonautonomous phases, which affects geochemical regularities.

Bataleva and coauthors [21] report the results of the experimental modelling of interactions of Mg, Fe, and Ni–olivine with anhydrite and Mg–sulfate, which are possible sources of oxidized S-rich fluid under lithospheric mantle P–T parameters. An experimental study was performed in the olivine–$CaSO_4$($MgSO_4$)–C systems at 6.3 GPa and temperatures of 1050 and 1450 °C using a multianvil high-pressure apparatus. Calcium and magnesium sulfates, being the most abundant sulfur-bearing minerals in the Earth's crust, are expected to play a principal role in the recycling of oxidized forms of sulfur into the deep mantle under subduction settings. It was experimentally shown that olivine–sulfate interaction can result in mantle sulfide formation and the generation of potential mantle metasomatic agents: S- and CO_2-dominated fluids and silicate–oxide or carbonate–silicate melts.

Murzin and coauthors [22] present a physicochemical model for the formation of magnetite–chlorite–carbonate rocks with copper–gold in the Karabash ultramafic massif (Southern Urals, Russia). This massif is located within a belt of ultramafic massifs stretching along the Main Ural fault zone in the Southern Urals. It was constructed based on the geotectonics of the Karabash ultramafic massif; features of the spatial distribution of metasomatically altered rocks, their geochemical characteristics, and mineral composition; data on the P–T parameters; and composition of the ore-forming fluids. This model included a four-reservoir calculation scheme. Thermodynamic modeling was performed using the Selektor-C software that employs the Gibbs energy minimization method, including minerals, aqueous solution components, and gases in the Na–K–Mg–Ca–Al–Si–Ti–Mn–Fe–Cu–Ag–Au–Hg–S–P–Cl–C–H–O system. The thermodynamic properties of various compounds were calculated using the Selektor-C database. The rocks (serpentinites, gabbro, harzburgite and limestones); deep magmatogenic fluids, mixed with metamorphogenic fluids released during the dehydration and deserpentinization of rocks in the lower crust; and meteoric waters were considered as the possible sources of the petrogenic and ore components in the model. It supports the involvement of sodium chloride–carbon dioxide fluids extracting ore components (Au, Ag, and Cu) from deep-seated rocks and is characterized by the ratio of ore elements corresponding to Clarke values in ultramafic rocks. The model calculations show that copper–gold can also be deposited during the serpentinization of deep-seated olivine-rich rocks and ore fluids risen by the tectonic flow to a higher hypsometric level. These results predict the copper–gold-rich ore occurrences in other ultramafic massifs.

Krneta and coauthors [23] investigated the REE fractionation in apatites from the Olympic Dam iron oxide–copper–gold deposit (South Australia) with the aim of constraining fluid evolution. The concentrations of trace elements and their variation within hydrothermal minerals can provide valuable information on the fluid parameters and conditions of ore deposition for assemblages. The behavior of the REE in hydrothermal fluids is affected by parameters such as pH, temperature, salinity, redox conditions, and fluid composition, thus allowing the REE to be used as a geochemical tracer in hydrothermal systems.

The REE signatures of three unique types of apatite from hydrothermal assemblages crystallized under partially constrained conditions were numerically modeled, and the partitioning coefficients between the apatite and fluid were calculated. Article by Krneta et al. [23] adds important information to the current knowledge on REE fractionation in apatites.

The article by Testa and coauthors [24] tried to resolve the evolution of hydrothermal fluids responsible for complex Bi–Cu–Pb–Zn–Mo–As–Fe–Ag–Au mineralization in the San Francisco de los Andes breccia-hosted deposit (Frontal Cordillera, Argentina). The authors provide well-documented insights into the complex mineralogy and mineral microtextures of the hydrothermal precipitates and reconstruct the physiochemical conditions of the ore-forming processes (temperature, fO_2, fS_2, fTe_2, and pH). They suggest that the mineralization formed due to the magmatic contribution of Te to the hydrothermal system. This study presents an interesting problem of genetic and (in part) practical importance related to the Bi and Te ores.

Sirqueira and coauthors [25] studied the genesis and evolution of Paleoproterozoic Sn and rare metal albitites in the Goiás Tin Province (central Brazil). Albitites are uncommon rocks with typically more than 70–80% of albite. Most of the known albitites worldwide have their origin attributed to the action of hydrothermal fluids on granites. More rarely, albitites are formed by direct crystallization from Na-rich magmas, generally related to specialized and rare-metal granites. Geological, petrological, and isotopic data from albitites and spatially associated granites were integrated to constrain the evolution history of the albitites. This study also intends to contribute to the petrogenesis and metallogenesis of evolved granitic systems rich in rare metals. These results extend the possibilities of a tin source in the Goiás Tin Province. They have implications for the province's economic potential and help understand the solubility and tin concentration in peraluminous granitic systems which are highly evolved and very rich in sodium.

Cao and coauthors [26] investigated the genesis and geochronology of the Xitian W–Sn polymetallic ore field (Hunan Province, South China). Analytical and calculated data obtained during the study of hydrothermal zircon U–Pb and muscovite $^{40}Ar/^{39}Ar$ dating allowed them to conclude: (1) there are two epochs of W–Sn mineralization formation: skarn-type at ca. 226 Ma and quartz vein/greisen-type at ca. 156 Ma; (2) the ore-forming fluids for the two metallogenic events are both characterized by enrichment in F and low oxygen fugacities; and (3) the ore components for the skarn-type and quartz vein/greisen-type W–Sn mineralization are both originated from a crust source. A good distinction is made between the skarn deposits, which are related to the earlier period of magmatic activity and the vein deposits, which are related to the second period of magmatic activity. This paper reports on some new geochemistry in the Xitian deposit with applications regarding the timing of mineralization and metal sources.

Yang and coauthors [27] studied the geochronology, petrology, and genesis of two small-sized granitic plutons (Laiziling and Jianfengling) of the Xianghualing Ore Field (South Hunan Province, South China). These granitic plutons have a close relationship with the super-large Xianghualing Sn deposit and large Dongshan W deposit, respectively, both in time and space. South Hunan, located in the central part of the Shi–Hang zone, is well known for its world-class W–Sn polymetallic deposits. The Shi–Hang zone, well known as the collision suture between the Yangtze Block and Cathaysia Block in the Neoproterozoic, is also an important granitic magmatic belt and polymetallic metallogenic belt.

New data of the zircon U–Pb dating, bulk-rock geochemical compositions, and zircon Lu–Hf isotopes of Laiziling and Jianfengling granites constrain the source and origin of the granitic magmas, discuss the tectonic setting, and clarify the relationship between these two plutons. Both of the Laiziling and Jianfengling granitic plutons are characterized by extremely similar elemental and Lu–Hf isotopic compositions. These features indicate that they both belong to highly fractionated A-type granites and were formed in an extensional setting and from the same magma chamber, which originated from the Paleoproterozoic metamorphic basement of South China with a certain amount of mantle-derived magma involving temperatures of ca. 730 °C and low oxygen fugacity.

Wei and coauthors [28] report the zircon U–Pb dating results of a porphyritic granite intrusion recently discovered in the Zhuanzhilian section and try to establish evidence of age and the relationship with the Au mineralization of the Dongping gold deposit (North Hebei, China). The gold deposits associated with alkaline rocks worldwide are widely distributed and have important economic value. The Dongping gold deposit, located in the middle northern part of the North China Craton, is the first giant gold deposit discovered among the alkaline complex-hosted in China in the 1980s. Mechanisms, possible sources, and tectonic environments of gold mineralization in the Dongping deposit are well discussed in this paper.

Funding: This work was supported by a state assignment project (No. 0330-2016-0001).

Acknowledgments: The author would like to express great appreciation for the editorial suggestion to be a guest editor for the Special Issue "Experimental and Thermodynamical Modeling of Ore-Forming Processes" for the journal Minerals. Special thanks go to the authors of the articles included in this Special Issue and the organizations that have financially supported research in the areas related to this topic. The author thanks the Editors-in-Chief, Editors, Assistant Editors, and Reviewers for their important comments and constructive suggestions, which helped the contributing authors to improve the quality of the manuscripts.

Conflicts of Interest: The author declares no conflicts of interest.

References

1. Garrels, R.M.; Christ, C.L. *Solutions, Minerals, and Equilibria*; Harper & Row: New York, NY, USA, 1965.
2. Barnes, H.L. *Geochemistry of Hydrothermal Ore Deposits*, 2nd ed.; Wiley-Interscience Inc.: New York, NY, USA, 1979.
3. Barnes, H.L. *Geochemistry of Hydrothermal Ore Deposits*, 3rd ed.; Wiley-Interscience Inc.: New York, NY, USA, 1997.
4. Barnes, H.L. Hydrothermal Processes. *Geochem. Perspect.* **2015**, *4*, 1–93. [CrossRef]
5. Robb, L. *Introduction to Ore-forming Processes*; Blackwell Publishing: Oxford, UK, 2005.
6. Pirajno, F. *Hydrothermal Processes and Mineral Systems*; Springer Science & Business Media: Berlin, Germany, 2009.
7. Anderson, G.M. *Thermodynamics of Natural Systems*; Cambridge University Press: New York, NY, USA, 2005.
8. Cemic, L. *Thermodynamics in Mineral Sciences*; Springer: Berlin, Germany, 2005.
9. Albarede, F. *Introduction in Geochemical Modeling*; Cambridge University Press: New York, NY, USA, 1995.
10. Carmichael, I.S.E.; Eugster, H.P. Thermodynamic Modeling of Geologic Materials: Minerals, Fluids, and Melts. *Rev. Mineral.* **1987**, *17*, 499.
11. Borisov, M.V. *Geochemical and Thermodynamic Models of Veined Hydrothermal Ore Formation*; Nauchnyi mir: Moscow, Russia, 2000. (In Russian)
12. Grichuk, D.V. *Thermodynamic Models of Submarine Hydrothermal Systems*; Nauchnyi mir: Moscow, Russia, 2000. (In Russian)
13. Ghiorso, M.S. Thermodynamic models of igneous processes. *Ann. Rev. Earth Planet. Sci.* **1997**, *25*, 221–241. [CrossRef]
14. Richet, P.; Ottonello, G. Thermodynamics of Phase Equilibria in Magma. *Elements* **2010**, *6*, 315–320. [CrossRef]
15. Karpov, I.K.; Chudnenko, K.V.; Kulik, D.A. Modeling chemical mass-transfer in geochemical processes: thermodynamic relations, conditions of equilibria and numerical algorithms. *Am. J. Sci.* **1997**, *297*, 767–806. [CrossRef]
16. Shvarov, Y.V. HCh: New potentialities for the thermodynamic simulation of geochemical systems offered by Windows. *Geochem. Int.* **2008**, *46*, 834–839. [CrossRef]
17. Chudnenko, K.V. *Thermodynamic Modeling in Geochemistry: Theory, Algorithms, Software, Application*; Academic Publishing House Geo: Novosibirsk, Russia, 2010; p. 287, ISBN 978-5-904682-18-7. (In Russian)
18. Kulik, D.A.; Wagner, T.; Dmytrieva, S.V.; Kosakowski, G.; Hingerl, F.F.; Chudnenko, K.V.; Berner, U.R. GEM-Selektor geochemical modeling package: Revised algorithm and GEMS3K numerical kernel for coupled simulation codes. *Comput. Geosci.* **2015**, *17*, 1–24. [CrossRef]
19. Zotov, A.V.; Kuzmin, N.N.; Reukov, V.L.; Tagirov, B.R. Stability of $AuCl_2^-$ from 25 to 1000 °C at Pressures to 5000 bar and Consequences for Hydrothermal Gold Mobilization. *Minerals* **2018**, *8*, 286. [CrossRef]
20. Tauson, V.L.; Lipko, S.V.; Smagunov, N.V.; Kravtsova, R.G. Trace Element Partitioning Dualism under Mineral–Fluid Interaction: Origin and Geochemical Significance. *Minerals* **2018**, *8*, 282. [CrossRef]
21. Bataleva, Y.; Palyanov, Y.; Borzdov, Y. Sulfide Formation as a Result of Sulfate Subduction into Silicate Mantle (Experimental Modeling under High P,T-Parameters). *Minerals* **2018**, *8*, 373. [CrossRef]
22. Murzin, V.; Chudnenko, K.; Palyanova, G.; Kissin, A.; Varlamov, D. Physicochemical Model of Formation of Gold-Bearing Magnetite-Chlorite-Carbonate Rocks at the Karabash Ultramafic Massif (Southern Urals, Russia). *Minerals* **2018**, *8*, 306. [CrossRef]
23. Krneta, S.; Ciobanu, C.L.; Cook, N.J.; Ehrig, K.J. Numerical Modeling of REE Fractionation Patterns in Fluorapatite from the Olympic Dam Deposit (South Australia). *Minerals* **2018**, *8*, 342. [CrossRef]
24. Testa, F.J.; Zhang, L.; Cooke, D.R. Physicochemical Conditions of Formation for Bismuth Mineralization Hosted in a Magmatic-Hydrothermal Breccia Complex: An Example from the Argentine Andes. *Minerals* **2018**, *8*, 486. [CrossRef]

25. Sirqueira, A.R.F.; Moura, M.A.; Botelho, N.F.; Kyser, T.K. Nature and Evolution of Paleoproterozoic Sn and Rare Metal Albitites from Central Brazil: Constraints Based on Textural, Geochemical, Ar-Ar, and Oxygen Isotopes. *Minerals* **2018**, *8*, 396. [CrossRef]

26. Cao, J.; Wu, Q.; Yang, X.; Kong, H.; Li, H.; Xi, X.; Huang, Q.; Liu, B. Geochronology and Genesis of the Xitian W-Sn Polymetallic Deposit in Eastern Hunan Province, South China: Evidence from Zircon U-Pb and Muscovite Ar-Ar Dating, Petrochemistry, and Wolframite Sr-Nd-Pb Isotopes. *Minerals* **2018**, *8*, 111. [CrossRef]

27. Yang, L.; Wu, X.; Cao, J.; Hu, B.; Zhang, X.; Gong, Y.; Liu, W. Geochronology, Petrology, and Genesis of Two Granitic Plutons of the Xianghualing Ore Field in South Hunan Province: Constraints from Zircon U–Pb Dating, Geochemistry, and Lu–Hf Isotopic Compositions. *Minerals* **2018**, *8*, 213. [CrossRef]

28. Wei, H.; Xu, J.; Zhang, G.; Cheng, X.; Chu, H.; Bian, C.; Zhang, Z. Hydrothermal Metasomatism and Gold Mineralization of Porphyritic Granite in the Dongping Deposit, North Hebei, China: Evidence from Zircon Dating. *Minerals* **2018**, *8*, 363. [CrossRef]

minerals

MDPI

Article

Stability of $AuCl_2^-$ from 25 to 1000 °C at Pressures to 5000 bar and Consequences for Hydrothermal Gold Mobilization

Alexander V. Zotov, Nikolai N. Kuzmin, Vladimir L. Reukov and Boris R. Tagirov *

Institute of Geology of Ore Deposits, Petrography, Mineralogy and Geochemistry (IGEM RAS),
35 Staromonetnyi per., 119017 Moscow, Russia; avzotov36@mail.ru (A.V.Z.); kolyanfclm@gmail.com (N.N.K.);
azotov@igem.ru (V.L.R.)
* Correspondence: boris1t@yandex.ru

Received: 10 May 2018; Accepted: 2 July 2018; Published: 4 July 2018

Abstract: Gold is transported in high-temperature chloride-bearing hydrothermal fluids in the form of $AuCl_2^-$. The stability of this complex has been extensively studied, but there is still considerable disagreement between available experimental data on the temperature region 300–500 °C. To solve this problem, we measured the solubility of gold in $HCl/NaCl$ fluids (NaCl concentration varied from 0.1 to 3 mol·(kg·H_2O)$^{-1}$) at 450 °C and pressures from 500 to 1500 bar (1 bar = 10^5 Pa). The experiments were performed using a batch autoclave method at contrasting redox conditions: in reduced experiments hydrogen was added to the autoclave, and in oxidized experiments the redox state was controlled by the aqueous SO_2/SO_3 buffer. Hydrogen pressure in the autoclaves was measured after the experiments in the reduced system. The gold solubility constant, $Au_{(cr)} + HCl°_{(aq)} + Cl^- = AuCl_2^- + 0.5 H_2°_{(aq)}$, was determined for the experimental *T-P* parameters as $\log K_s° = -4.77 \pm 0.07$ (500 bar), -5.11 ± 0.08 (1000 bar), and -5.43 ± 0.09 (1500 bar). These data, together with values from the literature for temperatures from 25 to 1000 °C, were fitted to the simple equation $\log K_s° = 4.302 - 7304 \cdot T(K)^{-1} - 4.77 \cdot \log d(w) + 11080 \cdot (\log d(w)) \cdot T(K)^{-1} - 6.94 \times 10^6 \cdot (\log d(w)) T(K)^{-2}$, where $d(w)$ is the pure water density. This equation can be used together with the extended Debye–Hückel equation for activity coefficients to calculate gold solubility at pressures up to 5000 bar at fluid chlorinities at least up to 30 wt %. The speciation of gold in natural chloride-bearing fluids is discussed.

Keywords: gold; solubility; hydrothermal solutions; chloride complex; experiment; thermodynamic modeling

---

1. Introduction

Chlorine is the most important component of natural fluids. Fluid chlorinity, together with acidity and redox state, controls the concentration of many ore metals, which can migrate in the fluid phase in the form of chloride complexes. The chlorinity of natural ore-forming fluids varies in a wide range. For example, solutions migrating in seafloor hydrothermal systems at temperatures from 250 to 400 °C are relatively diluted (~3 wt % NaCl eq., Seward et al. [1]), whereas the magmatic brines of porphyry systems may contain up to 50–60 wt % NaCl eq. at ~750 °C (Ulrich et al. [2]). In these fluids Au can be efficiently transported in the form of the Au(I)-Cl complexes [1]. The complexation of Au in chloride solutions has been extensively studied by potentiometry, solubility experiments, X-ray absorption spectroscopy, and ab initio molecular dynamics (AIMD). Nikolaeva et al. [3] determined the potential of Au electrode in 1 M (mol·L^{-1}) NaCl + Au(I) solution at 25–80 °C. The authors attributed the measured potentials to the reaction between the Au electrode and $AuCl_2^-$. The solubility of Au in chloride solutions was studied by Gammons and Willams-Jones [4] at 300 °C, Zotov et al. [5]

at 350–500 °C (see also references in this paper), and Stefánsson and Seward [6] at 300–600 °C. These authors [4–6] determined the formation constants of $AuCl_2^-$—the dominant Au complex in chloride-rich fluids—and calculated the thermodynamic properties of this complex. The Au solubility constants determined by Zotov et al. [5] are ~1.5 log units higher than those reported by Stefánsson and Seward [6] for similar *T-P* parameters. Au–Cl complexation at higher temperatures (750–1000 °C) was studied by Ryabchikov and Orlova [7], Zajacz et al. [8], and Guo et al. [9]. The dominant role of $AuCl_2^-$ in chloride-rich fluids was confirmed by X-ray absorption spectroscopic experiments performed at 250 °C (Pokrovski et al. [10]). Mei et al. [11] interpreted the results of AIMD simulations in terms of the formation of the $NaAuCl_2°$ ion associate, which was found to predominate over $AuCl_2^-$ at fluid densities $d < 0.7$ g·cm^{-3}. The solubility of Au in low-density vapor was determined by Archibald et al. [12].

The aim of our study is to obtain new experimental data on Au solubility, determine the stability of the dominant Au–Cl complex at an intermediate temperature of 450 °C, and resolve the disagreement between the key studies of Au–Cl complexation [5,6]. The uncertainty of redox potential is an important source of errors in the Au solubility constant. To eliminate this error, we performed solubility experiments under contrasting redox conditions: reduced (with the addition of hydrogen to the experimental system) and oxidized (with the sulfite (SO_2)/sulfate (SO_3) redox buffer). After all experiments under reduced conditions, hydrogen pressure was measured in the experimental system. Combining the two sets of data obtained in the reduced and oxidized systems, we obtained reliable values for the stability constant of the dominant Au complex, $AuCl_2^-$, free of error related to redox potential uncertainty. These new values were pooled with the literature data and fitted to a simple density model equation. The resulting equation enables accurate estimation of Au solubility in chloride fluids in a wide range of fluid chlorinity and over the whole range of *T-P* parameters characteristic of hydrothermal systems, up to temperatures of 1000 °C and pressures of 5000 bar. Based on this new model, the speciation and concentration of Au in natural chloride-bearing fluids are discussed.

2. Materials and Methods

2.1. Experimental

Solubility experiments were performed using two methods of redox control. To impose reduced conditions, hydrogen was added into the autoclave, and its pressure was measured after the experiment. The redox state of the oxidized system was controlled by the SO_2/SO_3 buffer.

2.1.1. Reduced Conditions

The solubility measurements were performed at 450 °C and pressures of 500 and 1000 bar using Ti autoclaves (VT-8 alloy) with a volume of 20 cm^3. A needle valve allowing gas pressure measurement was built in the cup of the autoclave (Kudrin [13]). Prior to the experiments the autoclaves were passivated with an HNO_3-H_2O (1:3 by volume) mixture at 400 °C and 500 bar. The total pressure in the autoclave was controlled by the degree of filling using the *PVT* properties of the H_2O-NaCl system (Driesner [14]; Driesner and Heinrich [15]). The uncertainty of total pressure was estimated as ±2%. Experimental solutions were prepared from distilled water, extra pure NaCl, and 0.1 M HCl fixanal. A piece of Au net (99.99%) with an outer surface of 1.5–3 cm^3 was fixed in the upper part of the autoclave. The autoclave was loaded in air. A weighed Al chip (99.9%) was used to produce hydrogen upon heating via the reaction: $Al + 2H_2O \rightarrow AlOOH + 1.5 H_{2(g)}$. The loaded autoclave was hermetically closed and placed in a gradient-free furnace. During the experiment, temperature was controlled within ±2 °C using a K-type thermocouple. The duration of the experiments varied from 5 to 12 days, which was sufficient for equilibration (Zotov and Baranova [16]). After the experiment, the autoclave was quenched in cold water. The hydrogen pressure in the autoclave was measured after quenching: the internal valve inside the autoclave cap was opened and hydrogen from the autoclave passed into an external cell with a built-in pressure piezometer. The volume of the cell was 6.7 cm^3. The piezometer was calibrated at atmospheric pressure and in vacuum (5×10^{-5} bar); its readings before the measurements were verified against an external reference pressure gauge in the pressure range 2–5 bar. The measured value was corrected for the pressure of atmospheric nitrogen (0.8 bar). We estimated the uncertainty of hydrogen pressure determination as ±10%. This uncertainty is accounted for mostly by the presence of nitrogen. After hydrogen pressure measurement, the autoclave was opened and the quench solution was extracted and diluted with an equal volume of aqua regia. Then the autoclave was filled with hot aqua regia and kept for 30 min at 70–80 °C, after which the washing solution was diluted with an equal volume of distilled H$_2$O. Finally the solution was diluted with 6 M HCl to match the concentration range suitable for analysis. The concentration of dissolved Au was analyzed by inductively coupled plasma mass spectrometry (ICP-MS) on an X Series 2 Thermo Scientific mass spectrometer, Waltham, MA, USA. The accuracy of the analysis was ±5% at the 95% confidence level, and the detection limit was 0.1 ppb. It was found that a considerable amount of dissolved Au (20–50%) was deposited on the autoclave walls during quenching.

The concentration of hydrogen in the experimental fluids calculated from the measured hydrogen pressure was lower than the initial value. The hydrogen loss varied from 20% to 90% at 500 bar and from 40% to 99% at 1000 bar. Hydrogen escaped through the needle valve and the autoclave seal. The hydrogen loss was not reproducible and, usually, increased after several experiments owing to the seal ring wear. The concentration of dissolved Au increased during the experiment. We believe that the kinetics of Au dissolution is fast enough to attain equilibrium Au concentration at any hydrogen pressure. This assumption is confirmed by the excellent agreement between Au solubility constants obtained from the Au solubility data under reduced and oxidized conditions (see next section). Therefore, the Au concentrations of experimental solutions determined after the experiments are considered identical to the equilibrium Au solubilities for given T-P-$f(H_2)$ parameters.

2.1.2. Oxidized Conditions

In the oxidized system, Au solubility was measured at 450 °C and pressures of 1000 and 1500 bar. Conventional batch Ti autoclaves (VT-8 alloy) with an internal volume of 20 cm^3 were used. The stock solutions were prepared from extra pure concentrated SO$_2$, H$_2$SO$_4$, HCl, and extra pure NaCl. The concentration of SO$_2$ was determined by iodometric titration; H$_2$SO$_4$, by densimetry; and HCl, by volumetric titration against Trizma® base (Sigma-Aldrich, St. Louis, MO, USA) with methyl red as an indicator. The procedures of autoclave loading, quenching, and solution preparation were the same as described for the reduced system. The duration of the experiments varied from 12 to 20 days.

2.2. Thermodynamic Calculations

The standard state of a pure solid phase and H$_2$O corresponds to a unit activity of the pure phase at a given temperature and pressure. The standard state adopted for the aqueous species is unit activity for a hypothetical one molal (m, mol·(kg·H$_2$O)$^{-1}$) ideal solution. The activity coefficients of charged aqueous species were calculated using the extended Debye–Hückel equation

$$\log \gamma_i = -\frac{A z_i^2 \sqrt{I}}{1 + B \overset{\circ}{a} \sqrt{I}} + \Gamma_\gamma \tag{1}$$

where the ion size parameter $\overset{\circ}{a}$ was taken to be 4.5 Å for all species, A and B are the Debye–Hückel activity coefficient parameters, I is the ionic strength, z_i is the charge of the species, and Γ_γ is the conversion factor from mole fraction to molality. For neutral species, it was assumed that $\log \gamma_n = \Gamma_\gamma = -\log (1 + 0.018 \cdot m^*)$, where m^* is the sum of the concentrations of all solute species. Speciation calculations were performed by means of the Gibbs computer code of the HCh software package (Shvarov [17]). The thermodynamic properties of the aqueous species Na$^+$, Cl$^-$, HSO$_4$$^-$, SO$_4$$^{2-}$, HSO$_3$$^-$, SO$_3$$^{2-}$, S$_2O_3$$^{2-}$, K$^+$, KCl°$_{(aq)}$ were taken from the SUPCRT92 database

(Johnson et al. [18]). The thermodynamic properties of H_2O, $HCl°$, $NaCl°$, and OH^- were adopted from Wagner and Pruss [19], Tagirov et al. [20], Ho et al. [21], and Bandura and Lvov [22], respectively; those of $H_2°_{(aq)}$, $O_2°_{(aq)}$, and $SO_2°_{(aq)}$, from Akinfiev and Diamond [23]. The values of Henry constants and dissociation constants of aqueous electrolytes calculated using the aforementioned thermodynamic data are presented in the Supplementary Material Section 1. The thermodynamic properties of $Au_{(cr)}$, hematite Fe_2O_3, magnetite Fe_3O_4, $Ni_{(cr)}$, and bunsenite NiO were taken from the SUPCRT92 database [18], and those of manganosite MnO and hausmannite Mn_3O_4, from Robie and Hemingway [24]. Thermodynamic data for quartz, muscovite, K-feldspar, and andalusite were adopted from Berman [25] with corrections of Sverjensky et al. [26].

The OptimA program of the HCh package (Shvarov [27]) was used to calculate the Gibbs free energies of Au aqueous complexes. The program computes the Gibbs free energies of aqueous complexes by the least squares minimization of the difference between the calculated and experimental values of activity or total solute concentration (i.e., $m(Au_{total})$ in this study). Finally, the optimized values of the Gibbs free energies of Au complexes were recalculated, together with the corresponding confidence intervals, to the logarithms of reaction constants as $\log K = -\Delta_r G/(2.303 \cdot RT)$.

3. Results

3.1. Experimental Au Solubility Determination

The results of experiments are given in Table 1. Figure 1a,b and Figure 2a–c show that the slopes of the solubility curves vs. $\log m(H_2°_{(aq)})$ and $\log m(Cl^-)$ are consistent, over the whole range of experimental temperature, pressure, and system composition, with the formation of $AuCl_2^-$ complex via the reaction

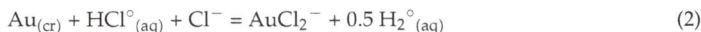

$$Au_{(cr)} + HCl°_{(aq)} + Cl^- = AuCl_2^- + 0.5\,H_2°_{(aq)} \tag{2}$$

for which

$$\log K_s° = \log m(AuCl_2^-) + 0.5 \log m(H_2°_{(aq)}) - \log m(Cl^-) - \log m(HCl°_{(aq)}) \tag{3}$$

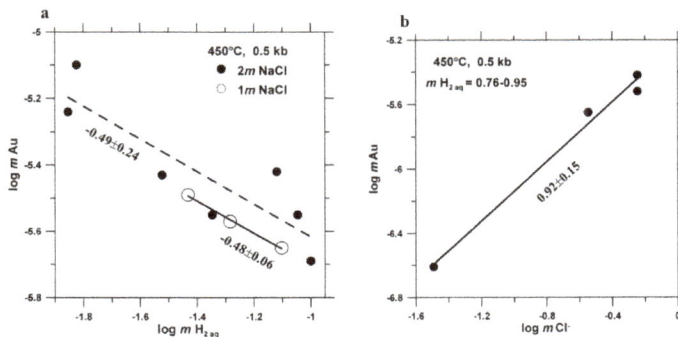

Figure 1. The concentration of Au (**a**) as a function of $H_2°_{(aq)}$ concentration at constant fluid salinities, and (**b**) as a function of Cl^- concentration at constant dissolved hydrogen concentration. The slopes of the solubility curves are indicated in the diagrams. Experiments were performed with hydrogen added to the system (reduced conditions).

It follows from Equations (2) and (3) that the dissolved Au concentration decreases as a square root of hydrogen fugacity (at constant fluid acidity and chlorinity). The dependence of the logarithm of dissolved Au concentration on the logarithm of hydrogen molality has a slope of -0.5 (Figures 1a and 2a,b). Note that the Au solubilities determined in the systems with contrasting redox states are in excellent agreement with the stoichiometry of Equation (2) (Figure 2b). As follows from Equations (2) and (3) and is demonstrated by Figures 1b and 2c, the concentration of Au increases proportionally to the fluid chlorinity (at constant concentrations of hydrogen and hydrochloric acid).

Table 1. Compositions of experimental solutions and results of Au solubility experiments (molal concentrations are given, mol·(kg·H$_2$O)$^{-1}$), t = 450 °C, P from 500 to 1500 bar.

m NaCl	m HCl	m H$_2$SO$_4$	m SO$_2$	m H$_{2(aq)}$	log m Au	Δlog m Au **
			500 bar			
2	0.1			0.014	−5.24	0.14
2	0.1			0.03	−5.43	0.17
2	0.1			0.076	−5.42	−0.05
2	0.1			0.09	−5.55	0.05
2	0.1			0.015	−5.1	−0.01
2	0.1			0.1	−5.69	0.16
2	0.1			0.045	−5.55	0.20
1	0.1			0.037	−5.49	−0.11
1	0.1			0.079	−5.65	−0.12
1	0.1			0.052	−5.57	−0.11
0.1	0.1			0.147	−6.79	−0.02
0.1	0.1			0.095	−6.64	−0.07
0.1	0.1			0.167	−6.95	0.11
			1000 bar			
2	0.1			0.0026	−5.16	0.20
2	0.1			0.057	−5.7	0.08
0.5	0.1			0.078	−5.86	−0.30
0.5	0.1			0.118	−6.34	0.09
0.5	0.1			0.0085	−5.61	−0.07
0.5	0.1			0.0082	−5.6	−0.07
0.1	0.1			0.0003	−5.28	−0.21
0.206	0.119	0.0478	0.0987	5.86 × 10^{-7} *	−3.69	−0.13
0.508	0.119	0.0478	0.0987	1.26 × 10^{-6} *	−3.62	0.01
1.018	0.119	0.0478	0.0987	2.22 × 10^{-6} *	−3.60	0.12
2.035	0.119	0.0478	0.0987	3.78 × 10^{-6} *	−3.44	0.09
3.043	0.119	0.0478	0.0987	5.07 × 10^{-6} *	−3.21	−0.07
			1500 bar			
0.218	0.119	0.0478	0.0987	8.05 × 10^{-7} *	−4.10	−0.001
0.497	0.119	0.0478	0.0987	1.62 × 10^{-6} *	−3.93	0.04
1.012	0.119	0.0478	0.0987	2.94 × 10^{-6} *	−3.68	−0.05
2.038	0.119	0.0478	0.0987	5.04 × 10^{-6} *	−3.42	−0.15
3.007	0.119	0.0478	0.0987	6.55 × 10^{-6} *	−3.64	0.15

* calculated values; ** Δlog m Au = log m Au(calc.) − log m Au(exp.).

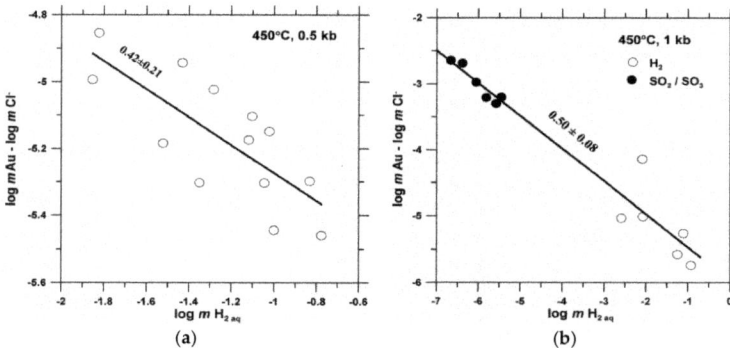

Figure 2. *Cont.*

(c)

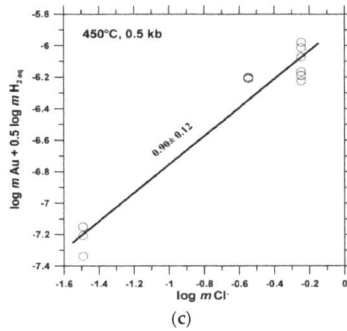

Figure 2. (**a**) The concentration of Au (corrected for Cl⁻ concentration according to Equation (2)) as a function of $H_2^\circ{}_{(aq)}$ concentration under reduced conditions. (**b**) The concentration of Au (corrected for Cl⁻ concentration) as a function of $H_2^\circ{}_{(aq)}$ concentration under reduced to oxidized conditions. (**c**) The concentration of Au (corrected for the of $H_2^\circ{}_{(aq)}$ concentration) as a function of Cl⁻ concentration under reduced conditions.

The differences between the experimental and calculated Au solubility values (last two columns of Table 1) do not exceed 0.2 log units and are independent of NaCl concentration in the wide range of fluid salinities (from 0.1 to 3 mol·(kg·H₂O)⁻¹), HCl concentrations, and redox conditions (Figure 3). This confirms the high accuracy of our method of the calculation of activity coefficients (in particular, the constant value of the ion size parameter $\overset{\circ}{a}$ = 4.5 Å), including the activity coefficient of $H_2^\circ{}_{(aq)}$, which was calculated ignoring the salting-out effect even in concentrated NaCl solutions.

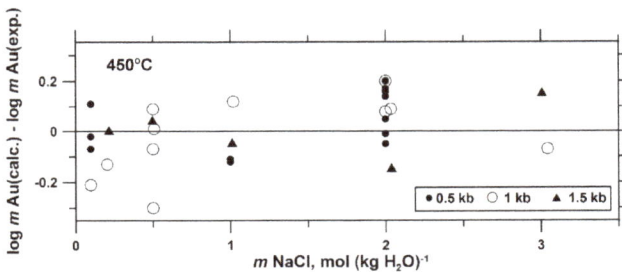

Figure 3. Difference between Au solubilities calculated using Equation (5) and observed in the experiments vs. NaCl molality.

3.2. Au Solubility Constant at 25–1000 °C and Pressures up to 5 kbar

The Au solubility constant values, log K_s°, determined in this study are given in Table 2 together with values obtained from the literature data. The log K_s° values are plotted as a function of temperature and pressure in Figure 4. Nikolaeva et al. [3] determined the electromotive force (e.m.f.) of the reaction

$$AuCl_2^- + 0.5H_{2(g)} = Au + 2Cl^- + H^+ \tag{4}$$

at 25–80 °C and saturated vapor pressure ($P_{sat.}$). The ionic strength of experimental solutions was fixed at 1 M. In the present study, the data of [3] were corrected for ionic strength for the calculation of the thermodynamic value of Au solubility constant (Supplementary Material Section 2). Experimental Au solubility data reported by Ryabchikov and Orlova [7] (750 °C, 1500 bar, Supplementary Material Section 3.1), Guo et al. [9] (800 °C, 2000 bar, Supplementary Material Section 3.2), and Zajacz et al. [8]

11

(1000 °C, 1500 bar, Supplementary Material Section 3.3) were evaluated using the OptimA computer code for the calculation of log $K_s°$. The results of regression analysis are summarized in the Supplementary Materials together with the original experimental data. In the calculation we used the experimental data of Guo et al. [9] obtained under redox conditions reliably controlled by the Ni-NiO, Fe_3O_4-Fe_2O_3, MnO-Mn_3O_4 buffers (five experiments in total). Experimental data of Zajacz et al. [8] for sulfur-free NaCl-HCl-H_2 fluids were used in our calculations, because thermodynamic data for sulfur-bearing species are highly uncertain at the experimental *T-P* parameters. The Au solubility constant reported by Gammons and Willams-Jones [4] for 300 °C and $P_{sat.}$ was recalculated to satisfy Equation (3) as described in Supplementary Material Section 3.4.

The Au solubility constants obtained in the present study and values calculated from the literature data were fitted to a simple density model (Anderson et al. [28]),

$$\log K_s° = 4.302 - 7304 \cdot T(K)^{-1} - 4.77 \cdot \log d(w) + 11080 \cdot (\log d(w)) \cdot T(K)^{-1} - 6.94 \times 10^6 \cdot (\log d(w)) \cdot T(K)^{-2} \quad (5)$$

where $d(w)$ is the pure water density. The calculated values of log $K_s°$ are listed in Table 3, compared with the experimental data in Figure 4, and with the literature theoretical models in Figures S1 and S2 of the Supplementary Materials. The values of Gibbs free energy of $AuCl_2^-$, calculated using Equation (5), are tabulated in Table S14 of the Supplementary Materials. The values of Gibbs free energy of aqueous species—components of the experimental systems, for which we used equations of state other than HKF (Helgeson, Kirkham, Flowers) are listed in Table S15 ($NaCl°$), Table S16 ($SO_2°$), Table S17 ($O_2°$), and Table S18 ($H_2°$) of the Supplementary Materials. During the fit of log $K_s°$ to the density model, the statistical weights of log $K_s°$ values were set to 2 for the 25 °C data (Nikolaeva et al. [3]) and 0.5 for $t \geq 750$ °C, because of the higher uncertainty of the thermodynamic properties of aqueous ions and electrolytes at high temperatures. The fit quality, expressed as the standard deviation of the difference between the calculated and experimental log $K_s°$ values, is equal to 0.12, which is close to the experimental uncertainty (Table 2). As can be seen in Figure 4, the dependence of log $K_s°$ on the reciprocal absolute temperature is close to linear. This supports the plausibility of the Au solubility reaction (Equation (2)), which can be considered as isocoulombic ($\Delta_r C_p° \sim 0$).

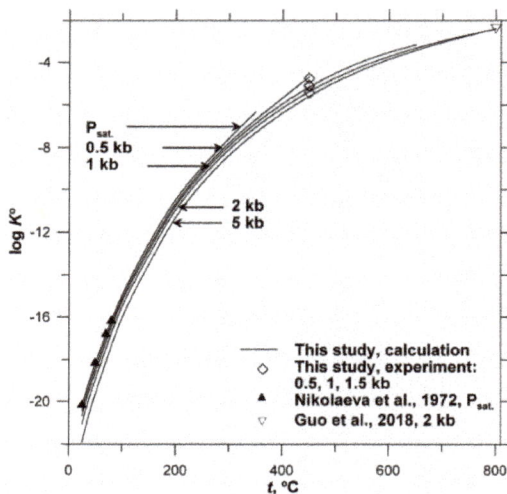

Figure 4. *Cont.*

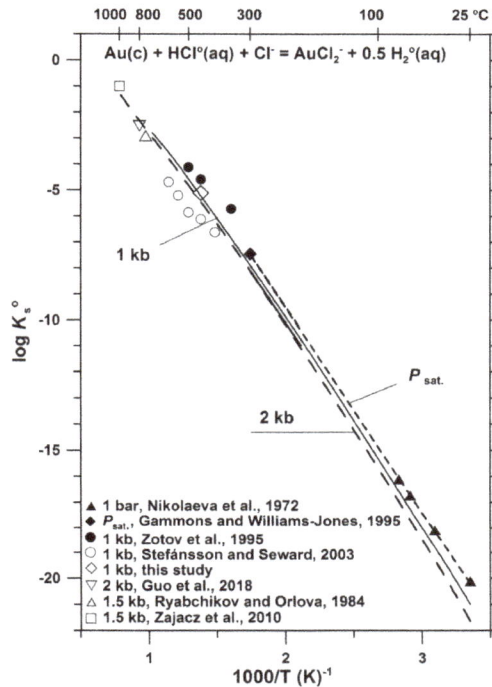

Figure 4. The Au solubility constant (Equation (3)) as a function of pressure and reciprocal temperature. Symbols are experimental data, and lines were calculated using Equation (5). The effect of pressure up to 5 kbar is shown in the insert.

Table 2. Comparison of experimental values of the Au solubility constant (Equation (3)) with model values calculated using Equation (5): log K_s° = 4.302 − 7304·T(K)$^{-1}$ − 4.77·log d(w) + 11080·(log d(w))·T(K)$^{-1}$ − 6.94 × 10^6·(log d(w))·T(K)$^{-2}$, where d(w) is the pure water density.

t, °C	P, bar	log K_s° (exp.)	log K_s° (calc.)	Δlog K_s° (calc. − exp.)	Reference
450	500	−4.77 ± 0.07	−4.72	0.05	This study
450	1000	−5.10 ± 0.09 *	−5.22	−0.12	*Ibid.*
450	1000	−5.12 ± 0.06 **	−5.22	−0.10	*Ibid.*
450	1500	−5.43 ± 0.09	−5.36	0.07	*Ibid.*
25	1	−20.12 ± 0.08	−20.14	−0.02	Nikolaeva et al. [3]
50	1	−18.09 ± 0.08	−18.11	−0.02	*Ibid.*
70	1	−16.74 ± 0.07	−16.68	0.06	*Ibid.*
80	1	−16.12 ± 0.07	−16.02	0.10	*Ibid.*
300	$P_{sat.}$	−7.46 ± 0.40	−7.47	−0.01	Gammons and Williams-Jones [4]
750	1500	−2.91 ± 0.30	−2.59	0.32	Ryabchikov and Orlova [7]
800	2000	−2.54 ± 0.20	−2.33	0.21	Guo et al. [9]
1000	1500	−0.99 ± 0.20	−1.23	−0.24	Zajacz et al. [8]

* Au-NaCl-HCl-H$_2$-H$_2$O system; ** Au-NaCl-HCl-H$_2$SO$_4$-SO$_2$-H$_2$O system.

The experiments reported here fall within the *T-P* range of the key studies of Au solubility in chloride fluids performed by Zotov et al. [5] and Stefánsson and Seward [6] (Supplementary Material Section 3.5). Our Au solubility data are 0.5 log units lower than those of [5]. At 350 °C the difference between the log K_s° values predicted by Equation (5) and data of [5] increases to one log unit. This disagreement can be explained by partial loss of hydrogen, whose concentration was calculated, but not measured in [5]. In contrast, the log $K_s°$ values reported by Stefánsson and Seward [6] are one log unit lower compared with our results. The reason for this disagreement is unclear.

Table 3. The Au solubility constant of the reaction $Au_{(cr)} + HCl°_{(aq)} + Cl^- = AuCl_2^- + 0.5H_2°_{(aq)}$ as a function of temperature and pressure. Calculations were performed using Equation (5) obtained in the present study.

t, °C	Pressure, bar					
	$P_{sat.}$	500	1000	1500	2000	5000
25	−20.14	−20.56	−20.93	−21.27	−21.57	−22.96
100	−14.83	−15.06	−15.27	−15.46	−15.63	−16.41
200	−10.33	−10.55	−10.71	−10.84	−10.95	−11.42
250	−8.79	−8.99	−9.15	−9.27	−9.37	−9.76
300	−7.47	−7.72	−7.89	−8.00	−8.09	−8.43
350	−6.25	−6.65	−6.85	−6.97	−7.05	−7.34
400		−5.69	−5.97	−6.10	−6.18	−6.44
450		−4.72	−5.22	−5.36	−5.45	−5.68
500		−3.94	−4.58	−4.73	−4.82	−5.02
550			−4.02	−4.19	−4.27	−4.45
600			−3.56	−3.71	−3.79	−3.95
650			−3.16	−3.30	−3.37	−3.51
700			−2.81	−2.93	−2.99	−3.11
750				−2.59	−2.64	−2.76
800				−2.28	−2.33	−2.43
850				−2.00	−2.04	−2.13
900				−1.73	−1.77	−1.85
950				−1.47	−1.51	−1.60
1000				−1.23	−1.27	−1.36

4. Discussion

4.1. Effect of Temperature, pH, and Chlorinity on Au Solubility

The range of chlorinity of aqueous fluids used to derive Equation (5) extends from pure HCl to 4.4m NaCl (20 wt % NaCl). In our recent study (Tagirov et al. [29]), the speciation of Au was studied by means of X-ray absorption spectroscopy and ab initio molecular dynamics up to 4m HCl/7m KCl (34 wt % KCl). Moreover, the experimental spectra were acquired and theoretical calculations were performed for dry (anhydrous) chloride melts. It was demonstrated that at high temperatures the microscopic state (local atomic environments) of the system is identical over the whole range of chloride concentrations, from dilute fluids to dry melts. This means that $AuCl_2^-$ is the main chloride complex at any chloride concentration at fluid density $d > 0.3$ g·cm^{-3} (this density corresponds to the experimental T-P parameters of the study of Zajacz et al. [8]). The study [29] and the present work clearly demonstrate that there is no need to invoke ionic associates, such as $NaAuCl_2°$ and $HAuCl_2°$, to describe Au solubility. The solubility of Au in high temperature fluids and brines can be accurately calculated, as described in Section 2.2, using the equilibrium constant of reaction (2) and the extended Debye–Hückel equation for the activity coefficients of aqueous species (Equation (1)). At lower densities ($d < 0.3$ g·cm^{-3}), the simple neutral complex $AuCl°_{(aq)}$ predominates Au speciation (Archibald et al. [12]).

The excellent agreement between Au solubilities calculated by Equation (5) and experimental data in the wide range of T-P parameters (from 25 to 1000 °C and from $P_{sat.}$ to 2000 bar) enables accurate prediction of Au mobilization in most natural environments where ore metals are transported and deposited by chloride-rich fluids. The effect of temperature, pH, and fluid salinity on Au solubility is shown in Figure 5. This diagram was constructed for a redox state buffered by the Ni–NiO equilibrium and salinities of (i) 1m NaCl, which is typical, for example, of relatively low-temperature (300–400 °C) seafloor hydrothermal systems (Bortnikov et al. [30]; Hannington et al. [31]) and high-temperature magmatic fluids (800 °C) separated from arc dacite magma of porphyry-generating intrusion (Blundy et al. [32]); and (ii) $7m$ NaCl which is typical of magmatic brine inclusions and products of the separation of single-phase magmatic fluids (Seward et al. [1], Aranovich et al. [33]). At a low temperature of 200 °C, hydrosulfide complexes predominate at any pH and chlorinity (Seward et al. [1], Trigub et al. [34]). The maximum solubility is observed in near-neutral solutions in the $Au(HS)_2^-$ predominance region with pH ~pK_{H2S} (that is, at similar concentrations of $H_2S^\circ_{(aq)}$ and HS^-, which are necessary for the formation of the Au complex). As temperature increases, the Au concentration of acidic solution starts to increase, because the Au solubility is controlled in this region by $AuCl_2^-$ (Equation (2)). In the $AuCl_2^-$ predominance region, the Au solubility increases systematically with increasing temperature and fluid acidity. The Au concentration in $7m$ NaCl at neutral pH can reach a few hundred ppb at 500 °C and 1 kbar and increases by three orders of magnitude at 800 °C and 2 kbar. We note that the Au concentration at the minima of the solubility curves is weakly dependent on *T*, *P*, and fluid chemistry and is close to 10 ppb at the given redox state (Ni-NiO) and H_2S molality (0.01m). In this region, $AuHS^\circ_{(aq)}$ dominates at $t \le 400$ °C, and $Au(OH)^\circ_{(aq)}$ becomes the main Au complex at higher temperatures.

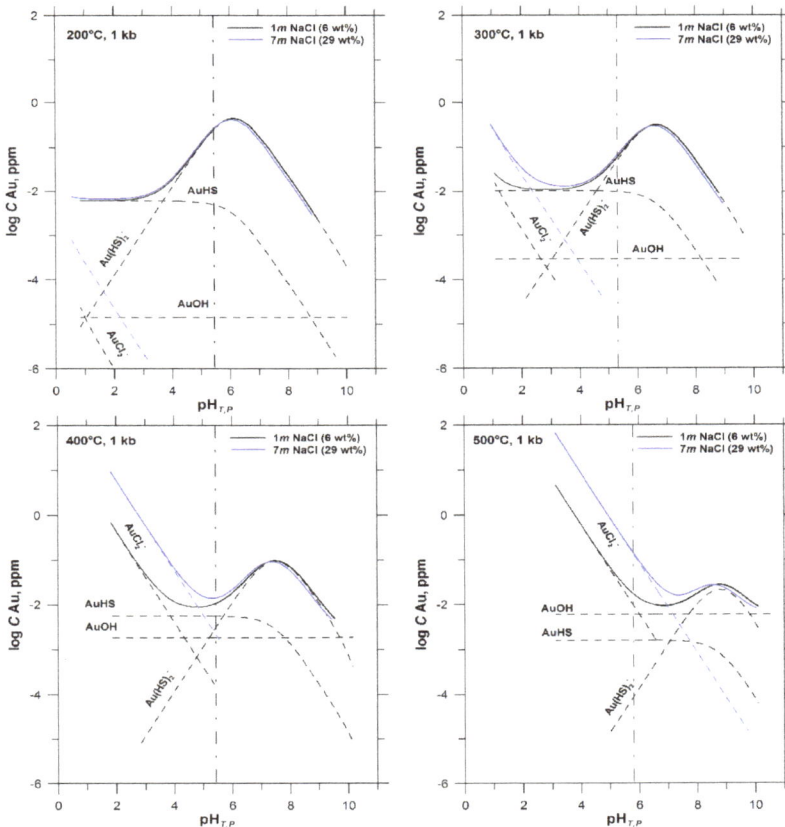

Figure 5. *Cont.*

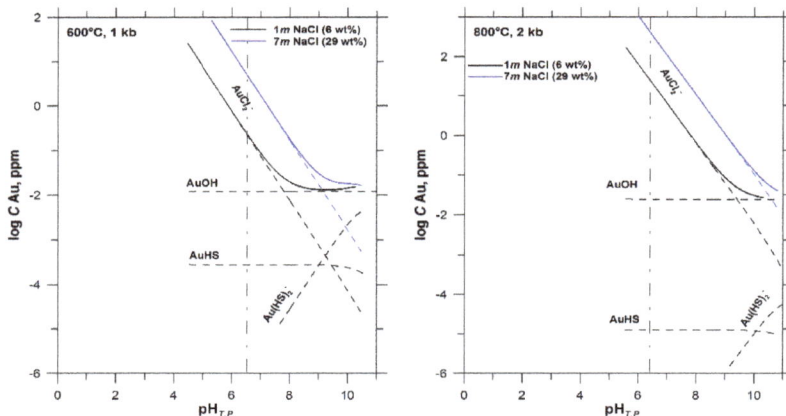

Figure 5. Gold solubility as a function of $pH_{T,P}$. The concentration of dissolved sulfur is 0.01 m, the redox conditions are controlled by the Ni/NiO buffer. Lines are calculated using the results of the present study: solid lines—total dissolved Au, and dashed lines—individual aqueous complexes. The vertical dashed-dotted line corresponds to neutral pH. Thermodynamic data for Au–OH and Au–HS complexes were adopted from Akinfiev and Zotov [35].

4.2. Gold Concentration in Natural Fluids

Figure 6 compares the Au concentrations calculated using Equation (5) with the data of Ulrich et al. [2] for high-temperature fluid inclusions of the Bajo de la Alumbrera porphyry Cu–Au deposit (Argentina). The highest-temperature primary brine inclusions from this deposit (650–770 °C, 50–60 wt % NaCl eq.) contain 0.55 ppm Au. The hydrothermal fluids of porphyry systems are usually oxidized, and the redox state can be considered close to the hematite-magnetite buffer. The calculated Au solubility in the chloride brine of the Bajo de la Alumbrera deposit is $\sim 10^3$ times higher than the measured Au concentration (0.55 ppm). This implies that the high-temperature ore fluids are strongly undersaturated with respect to native Au. Therefore, at high temperatures during the early stages of porphyry ore formation, Au can be deposited only in an "invisible" (or refractory) state, either in the form of nanoscale particles or as a component of solid solutions (e.g., Tagirov et al. [36], Trigub et al. [37]). A decrease in temperature results in a drastic decrease in Au solubility. The Au saturation limit of 0.55 ppm is attained at a temperature of slightly above 400 °C at 1 kbar under the hematite-magnetite buffer redox conditions. At these conditions, the deposition of native gold becomes possible, which is consistent with estimates for ore precipitation at the Bajo de la Alumbrera deposit.

Figure 7 shows Au solubility as a function of fluid chlorinity at temperatures from 300 to 1000 °C. The pH values are controlled by silicate mineral buffers, and the redox state, by the hematite–magnetite buffer. At the given KCl concentration, the pH values range from near-neutral to weakly acidic. Another feature of fluid systems buffered by mineral assemblages is that the muscovite–andalusite–quartz buffer yields more acidic pH values (black lines in Figure 7) than the K-feldspar–muscovite–quartz buffer (blue lines). At subcritical temperatures and near-neutral pH values imposed by fluid interaction with silicate mineral assemblages, the hydroxide complex AuOH° dominates over the whole range of chlorinity up to 60 wt % NaCl eq. An increase in temperature results in an increase in dissolved Au concentration. At low Cl concentrations in the AuOH° predominance region, an increase in temperature from 400 to 1000 °C results in an increase in Au solubility from 10 to 400 ppb. At higher chlorinities in the $AuCl_2^-$ predominance region, an increase in Cl concentration results in an increase in Au solubility, which can reach at 10 wt % NaCl eq. ~1 ppm at 500 °C and ~1000 ppm at 1000 °C. We note again that the pH values are near-neutral at both temperatures: pH ~5.3 at 500 °C/1 kbar and ~8 at 1000 °C/2 kbar (neutral pH is 5.8 and 7.4, respectively). Therefore, despite the neutral pH conditions, the Au solubility in high-temperature fluids is sufficiently high, owing to the high stability of $AuCl_2^-$, for efficient Au transport even in Au-undersaturated fluids.

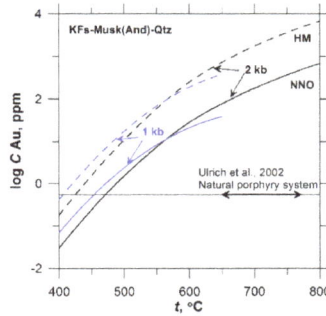

Figure 6. The solubility of Au as a function of temperature at 1 and 2 kbar (blue and black lines, respectively). The pH value is controlled by the K-feldspar–muscovite–quartz buffer at $t < 550\,°C$ at 1 kbar and $t < 600\,°C$ at 2 kbar and the K-feldspar–andalusite–quartz buffer at higher temperatures. The redox state of the system corresponds to the Fe_2O_3/Fe_3O_4 (HM) and Ni/NiO (NNO) buffers. The horizontal line shows the concentration of Au (0.55 ppm) determined in the highest-temperature fluid inclusions from the Bajo de la Alumbrera porphyry Cu–Au deposit (Argentina, Ulrich et al. [2]).

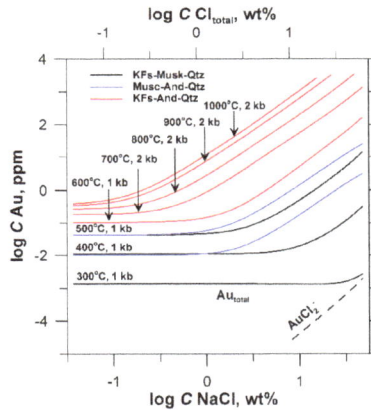

Figure 7. The solubility of Au as a function of fluid chlorinity for S-free system. The concentration of KCl is fixed as $m(KCl) = 1/5\ m(NaCl)$. The pH value is controlled by the K-feldspar–muscovite–quartz, muscovite–andalusite–quartz, and K-feldspar–andalusite–quartz buffers. Horizontal segments of the solubility curves correspond to the $AuOH°$ predominance fields. Thermodynamic data for Au–OH complexes are adopted from Akinfiev and Zotov [35].

5. Conclusions

The solubility of Au was measured at 450 °C and pressures of 500, 1000, and 1500 bar in a wide range of fluid chlorinity, from $0.1m$ to $3m$ NaCl. The redox state of the experimental system was controlled either by hydrogen added to the autoclave (reduced conditions), or by the SO_2/SO_3 buffer (oxidized conditions). The measured Au solubilities are best described by the formation of the $AuCl_2^-$ complex. The equilibrium Au solubility constants calculated for the contrasting redox states are in perfect agreement with each other and fall between the values reported by Zotov et al. [5] and

Stefánsson and Seward [6]. The experimental values of the Au solubility constant were fitted together with Au solubility data available in the literature for temperatures from 25 to 1000 °C and pressures up to 2 kbar to a simple density model equation. The equation describes the experimental data with an accuracy approaching the experimental uncertainties in the whole region of *T-P* parameters and can be used to predict Au solubility at pressures up to 5 kbar. The results of our study demonstrate that the simple complex $AuCl_2^-$ predominates Au speciation in natural fluids in a wide range of salinity at fluid densities $d > 0.3$ g·cm^{-3}. The solubility of Au can be accurately calculated at solute concentrations up to at least $5m$ NaCl eq. using the equilibrium constant of the reaction of $AuCl_2^-$ formation and the extended Debye–Hückel equation for activity coefficients. Thermodynamic calculations performed using the obtained results showed that the formation of $AuCl_2^-$ can account for efficient hydrothermal transport of Au at $t > 400$ °C and any chlorinity of hydrothermal fluid whose density (d) is > 0.3 g·cm^{-3}. At $t < 400$ °C hydroxide and hydrosulfide complexes become the dominant Au species. At $d < 0.3$ g·cm^{-3} the role of $AuCl_2^-$ is masked by the formation of $AuCl°$.

Supplementary Materials: The following are available online at http://www.mdpi.com/2075-163X/8/7/286/s1, Table S1: The log $K°$ values of the reaction $Na^+ + Cl^- = NaCl°_{(aq)}$, Table S2: The log $K°$ values of the reaction $HCl°_{(aq)} = H^+ + Cl^-$, Table S3: The log $K°$ values of the reaction $H_{2(g)} = H_2°_{(aq)}$, Table S4: The log $K°$ values of the reaction $O_{2(g)} = O_2°_{(aq)}$, Table S5: The log $K°$ values of the reaction $SO_{2(g)} = SO_2°_{(aq)}$, Table S6: The log $K°$ values of the reaction $H_2O_{(l)} = H^+ + OH^-$, Table S7: Calculation of Gibbs free energy of $AuCl_2^-$ using data of Nikolaeva et al., Table S8: Calculation of equilibrium constant of Au solubility reaction using data of Nikolaeva et al., Table S9: Treatment of experimental data of Ryabchikov and Orlova, Table S10: Treatment of experimental data of Guo et al., Table S11: Treatment of experimental data of Zajacz et al., Table S12: Recalculation of data of Gammons and Williams-Jones into the Au solubility constant, Table S13: Original and recalculated values of Au solubility constant, log $K_s°$, based on data of Zotov et al. and Stefánsson and Seward, Table S14: Gibbs free energy of $AuCl_2^-$, Table S15: Gibbs free energy of $NaCl°_{(aq)}$, Table S16: Gibbs free energy of $SO_2°_{(aq)}$, Table S17: Gibbs free energy of $O_2°_{(aq)}$, Table S18: Gibbs free energy of $H_2°_{(aq)}$.

Author Contributions: A.V.Z. and B.R.T. designed and performed the experiments and thermodynamic calculations, and wrote the paper; N.N.K. participated in the experimental work and data treatment; V.L.R. performed technical operations and participated in the experimental work; all the authors participated in the manuscript preparation.

Funding: This study was supported by the Russian Science Foundation, grant No. 17-17-01220 (experimental work and data treatment) and the Base Theme of IGEM RAS (compilation of literature data).

Acknowledgments: We thank Dmitry Petrenko for determination of Au concentrations and Ludmila Koroleva for determination of SO$_2$ concentrations. Nikolai Akinfiev is acknowledged for useful discussions during the manuscript preparation. We are grateful to Andrey Girnis for English grammar correction. The authors thank three anonymous reviewers and Galina Palyanova for helpful comments and suggestions. Chemical analyses were carried out at the "IGEM-Analytica" Center for collective use.

Conflicts of Interest: The authors declare no conflict of interest.

References

1. Seward, T.M.; Williams-Jones, A.E.; Migdisov, A.A. The chemistry of metal transport and deposition by ore-forming hydrothermal fluids. In *Treatise on Geochemistry*, 2nd ed.; Turekian, K., Holland, H., Eds.; Elsevier: New York, NY, USA, 2014; Volume 13, pp. 29–57.

2. Ulrich, T.; Günther, D.; Heinrich, C.A. The Evolution of a porphyry Cu-Au deposit, based on LA-ICP-MS analysis of fluid inclusions: Bajo de la Alumbrera, Argentina. *Econ. Geol.* **2001**, *96*, 1743–1774. [CrossRef]

3. Nikolaeva, N.M.; Erenburg, A.M.; Antipina, V.A. About the temperature dependence of standard potentials of halide complexes of gold. *Izvestiya Sibirskogo Otdeleniya Akademii Nauk SSSR Seriya Khimicheskih Nauk* **1972**, *4*, 126–128. (In Russian)

4. Gammons, C.H.; Williams-Jones, A.E. The solubility of Au-Ag alloy + AgCl in HCl/NaCl solutions at 300 °C: New data on the stability of Au(I) chloride complexes in hydrothermal fluids. *Geochim. Cosmochim. Acta* **1995**, *59*, 3453–3468. [CrossRef]

5. Zotov, A.V.; Kudrin, A.V.; Levin, K.A.; Shikina, N.D.; Var'yash, L.N. Experimental studies of the solubility and complexing of selected ore elements (Au, Ag, Cu, Mo, As, Sb, Hg) in aqueous solutions. In *Fluids in the Crust. Equilibrium and Transport Properties*; Shmulovich, K.I., Yardley, B.W.D., Gonchar, G.G., Eds.; Springer: Dordrecht, The Netherlands, 1995; pp. 95–137.
6. Stefánsson, A.; Seward, T.M. Stability of chloridogold(I) complexes in aqueous solutions from 300 to 600 °C and from 500 to 1800 bar. *Geochim. Cosmochim. Acta* **2003**, *67*, 4559–4576. [CrossRef]
7. Ryabchikov, I.D.; Orlova, G.P. Gold in magmatic fluids. In *Phizico-khimicheskie Modeli Petrogeneza i Rudoobrazovaniya*; Nauka Press: Novosibirsk, Russia, 1984; pp. 103–111. (In Russian)
8. Zajacz, Z.; Seo, J.H.; Candela, P.A.; Piccoli, P.M.; Heinrich, C.A.; Guillong, M. Alkali metals control the release of gold from volatile-rich magmas. *Earth Planet. Sci. Lett.* **2010**, *297*, 50–56. [CrossRef]
9. Guo, H.; Audétat, A.; Dolejš, D. Solubility of gold in oxidized, sulfur-bearing fluids at 500–850 °C and 200–230 MPa: A synthetic fluid inclusion study. *Geochim. Cosmochim. Acta* **2018**, *222*, 655–670. [CrossRef]
10. Pokrovski, G.S.; Tagirov, B.R.; Schott, J.; Bazarkina, E.F.; Hazemann, J.-L.; Proux, O. An in situ X-ray absorption spectroscopy study of gold-chloride complexing in hydrothermal fluids. *Chem. Geol.* **2009**, *259*, 17–29. [CrossRef]
11. Mei, Y.; Liu, W.; Sherman, D.M.; Brugger, J. Metal complexing and ion hydration in low density hydrothermal fluids: Ab initio molecular dynamics simulation of Cu(I) and Au(I) in chloride solutions (25–1000 °C, 1–5000 bar). *Geochim. Cosmochim. Acta* **2014**, *131*, 196–212. [CrossRef]
12. Archibald, S.M.; Migdisov, A.A.; Williams-Jones, A.E. The stability of Au-chloride complexes in water vapor at elevated temperatures and pressures. *Geochim. Cosmochim. Acta* **2001**, *65*, 4413–4423. [CrossRef]
13. Kudrin, A.V. Behaviour of Mo in aqueous NaCl and KCl solutions at 300–450 °C. *Geochem. Int.* **1989**, *26*, 87–99.
14. Driesner, T. The system H_2O-NaCl. II. Correlations for molar volume, enthalpy, and isobaric heat capacity from 0 to 1000 °C, 1 to 5000 bar, and 0 to 1 X_{NaCl}. *Geochim. Cosmochim. Acta* **2007**, *71*, 4902–4919. [CrossRef]
15. Driesner, T.; Heinrich, C.A. The system H_2O-NaCl. I. Correlation formulae for phase relations in temperature-pressure-composition space from 0 to 1000 °C, 0 to 5000 bar, and 0 to 1 X_{NaCl}. *Geochim. Cosmochim. Acta* **2007**, *71*, 4880–4901. [CrossRef]
16. Zotov, A.V.; Baranova, N.N. Thermodynamic properties of the aurochloride solute complex $AuCl_2{}^-$ at temperatures of 350–500 °C and pressures of 500–1500 bars. *Sciences Géologiques Bulletins Mémoires* **1989**, *42*, 335–342. [CrossRef]
17. Shvarov, Y.V. HCh: New potentialities for the thermodynamic simulation of geochemical systems offered by Windows. *Geochem. Int.* **2008**, *46*, 834–839. [CrossRef]
18. Johnson, J.W.; Oelkers, E.H.; Helgeson, H.C. SUPCRT92: A software package for calculating the standard molal thermodynamic properties of minerals, gases, aqueous species, and reactions from 1 to 5000 bar and 0 to 1000 °C. *Comput. Geosci.* **1992**, *18*, 899–947. [CrossRef]
19. Wagner, W.; Pruss, A. The IAPWS formulation 1995 for the thermodynamic properties of ordinary water substance for general and scientific use. *J. Phys. Chem. Ref. Data* **2002**, *31*, 387–535. [CrossRef]
20. Tagirov, B.R.; Zotov, A.V.; Akinfiev, N.N. Experimental study of dissociation of HCl from 350 to 500 °C and from 500 to 2500 bars: Thermodynamic properties of $HCl_{(aq)}$. *Geochim. Cosmochim. Acta* **1997**, *61*, 4267–4280. [CrossRef]
21. Ho, P.C.; Palmer, D.A.; Mesmer, R.E. Electrical conductivity measurements of aqueous sodium chloride solutions to 600 °C and 300 MPa. *J. Solut. Chem.* **1994**, *23*, 997–1018. [CrossRef]
22. Bandura, A.V.; Lvov, S.N. The ionization constant of water over wide ranges of temperature and density. *J Phys. Chem. Ref. Data* **2006**, *35*, 15–30. [CrossRef]
23. Akinfiev, N.N.; Diamond, L.W. Thermodynamic description of aqueous nonelectrolytes at infinite dilution over a wide range of state parameters. *Geochim. Cosmochim. Acta* **2003**, *67*, 613–627. [CrossRef]
24. Robie, R.A.; Hemingway, B.S. *Thermodynamic Properties of Minerals and Related Substances at 298.15 K and 1 bar (10^5 Pascals) Pressure and at Higher Temperatures, U.S. Geological Survey Bulletin 2131*; U.S. Government Printing Office: Washington, DC, USA, 1995.
25. Berman, R.G. Internally consistent thermodynamic data for minerals in the system Na_2O-K_2O-CaO-MgO-FeO-Fe_2O_3-Al_2O_3-SiO_2-TiO_2-H_2O-CO_2. *J. Petrol.* **1988**, *29*, 445–522. [CrossRef]
26. Sverjensky, D.A.; Hemley, J.J.; D'Angelo, M.D. Thermodynamic assessment of hydrothermal alkali feldspar-mica-aluminosilicate equilibria. *Geochim. Cosmochim. Acta* **1991**, *55*, 989–1004. [CrossRef]

27. Shvarov, Y. A suite of programs, OptimA, OptimB, OptimC, and OptimS compatible with the Unitherm database, for deriving the thermodynamic properties of aqueous species from solubility, potentiometry and spectroscopy measurements. *Appl. Geochem.* **2015**, *55*, 17–27. [CrossRef]

28. Anderson, G.M.; Castet, S.; Schott, J.; Mesmer, R.E. The density model for estimation of thermodynamic parameters of reactions at high temperatures and pressures. *Geochim. Cosmochim. Acta* **1991**, *55*, 1769–1779. [CrossRef]

29. Tagirov, B.R.; Trigub, A.L.; Filimonova, O.N.; Kvashnina, K.O.; Nickolsky, M.S.; Lafuerza, S.; Chareev, D.A. Gold speciation and solubility in chloride fluids and melts: Insights from X-ray absorption spectroscopy, Ab initio molecular dynamics and thermodynamic modeling. *Geochim. Cosmochim. Acta* **2018**. under review.

30. Bortnikov, N.S.; Simonov, V.A.; Bogdanov, Y.A. Fluid inclusions in minerals from modern sulfide edifices: Physicochemical conditions of formation and evolution of fluids. *Geol. Ore Depos.* **2004**, *46*, 64–75.

31. Hannington, M.D.; de Ronde, C.D.J.; Petersen, S. Sea-floor tectonics and submarine hydrothermal systems. In *Economic Geology 100th Anniversary Volume*; Henderquist, J.W., Thompson, J.F.H., Goldfarb, R.J., Richards, J.P., Eds.; Society of Economic Geologists: Littelton, CO, USA, 2005; pp. 111–141.

32. Blundy, J.; Mavrogenes, J.; Tattitch, B.; Sparks, S.; Gilmer, A. Generation of porphyry copper deposits by gas–brine reaction in volcanic arcs. *Nat. Geosci.* **2015**, *8*, 235–240. [CrossRef]

33. Aranovich, L.Y.; Newton, R.C.; Manning, C.E. Brine-assisted anatexis: Experimental melting in the system haplogranite–H_2O–NaCl–KCl at deep-crustal conditions. *Earth Plan. Sci. Lett.* **2013**, *374*, 111–120. [CrossRef]

34. Trigub, A.L.; Tagirov, B.R.; Kvashnina, K.O.; Lafuerza, S.; Filimonova, O.N.; Nickolsky, M.S. Experimental determination of gold speciation in sulfide-rich hydrothermal fluids under a wide range of redox conditions. *Chem. Geol.* **2017**, *471*, 52–64. [CrossRef]

35. Akinfiev, N.N.; Zotov, A.V. Thermodynamic description of aqueous species in the system Cu-Ag-Au-S-O-H at temperatures of 0–600 °C and pressures of 1–3000 bar. *Geochim. Int.* **2010**, *48*, 714–720. [CrossRef]

36. Tagirov, B.R.; Trigub, A.L.; Kvashnina, K.O.; Shiryaev, A.A.; Chareev, D.A.; Nickolsky, M.S.; Abramova, V.D.; Kovalchuk, E.V. Covellite CuS as a matrix for "invisible" gold: X-ray spectroscopic study of the chemical state of Cu and Au in synthetic minerals. *Geochim. Cosmochim. Acta* **2016**, *191*, 58–69. [CrossRef]

37. Trigub, A.L.; Tagirov, B.R.; Kvashnina, K.O.; Chareev, D.A.; Nickolsky, M.S.; Shiryaev, A.A.; Baranova, N.N.; Kovalchuk, E.V.; Mokhov, A.V. X-ray spectroscopy study of the chemical state of "invisible" Au in synthetic minerals in the Fe-As-S system. *Am. Mineral.* **2017**, *102*, 1057–1065.

minerals

MDPI

Article

Trace Element Partitioning Dualism under Mineral–Fluid Interaction: Origin and Geochemical Significance

Vladimir L. Tauson, Sergey V. Lipko *, Nikolay V. Smagunov and Raisa G. Kravtsova

A.P.Vinogradov Institute of Geochemistry, Siberian Branch of Russian Academy of Sciences,
Irkutsk 664033, Russia; vltauson@igc.irk.ru (V.L.T.); nicksm@igc.irk.ru (N.V.S.); krg@igc.irk.ru (R.G.K.)
* Correspondence: slipko@yandex.ru; Tel.: +7-3952-42-99-67

Received: 3 May 2018; Accepted: 26 June 2018; Published: 30 June 2018

Abstract: Trace element (TE) partitioning in the system "mineral-hydrothermal solution" is studied by the method of thermo-gradient crystal growth coupled with internal sampling of a fluid phase. The analytical procedure used enables evaluating of structurally bound and superficially bound modes of TE in crystals and determining corresponding dual partition coefficients. The case of precious metals (PM—Au, Pt, Pd) at 450 and 500 °C and 100 MPa pressure is considered. The minerals are pyrite, As-pyrite, magnetite, Mn-magnetite and hematite and fluids are ammonium chloride-based hydrothermal solutions. The partition coefficients for structural and surficial modes, D_p^{str} and D_p^{sur}, are found to be unexpectedly high (except for Au in pyrite). High concentrations of PM are attributed to superficial nonautonomous phases (NAPs), which can be considered as primary concentrators of PM. We also have studied the co-crystallization (exchange) coefficients (D_e) of REE (Ce, Eu, Er, Yb) and Fe in magnetite and hematite at 450 °C and 100 MPa. D_e^{sur} is elevated to two orders of magnitude as compared to D_e^{str}. It is shown that not only physicochemical parameters affect REE distribution in hydrothermal systems, but also NAP presence and its composition. The crystal growth mechanism specified by the agency of NAP is suggested. The study of PM distribution in natural pyrite of gold-ore deposits supported the importance of differentiating between structurally and superficially bound TE modes for correct use of experimental D values to determining element concentrations in ore-forming fluids.

Keywords: trace elements; partitioning; precious metals; rare earth elements; hydrothermal experiment; partition and co-crystallization coefficients; structurally and superficially bound modes; pyrite; magnetite; hematite; single crystals; AAS-GF; ICP-MS; XPS

1. Introduction

Study of mineral crystals as real structurally imperfect objects suggested that the distribution of trace elements (TE) in the reactions with their participation depends heavily on the admixture interaction with defects and may fail to be in accordance with the known thermodynamic ratios defined by Nernst's and Henry's laws [1–4]. For this reason, at the end of the last century microelements distribution faced a crisis, which was described by J. Jones in a rather emotional manner: "Elements entering crystals at the ppm level (or worse, at the ppb level) might partition into defects, rather than into well-defined crystallographic sites. This was discouraging" [5] (p. 161).

The surface is the most common defect of a real structure of a crystal. Its contribution to the distribution of elements is not hard to evaluate in principle; however, for reasons we do not fully understand, such studies are not popular in geological and mineralogical works. Perhaps this is because many authors published their results without taking into account this factor and explaining anomalies in the distribution of TE by other reasons. More and more of the latter are found with the upgrade of micro-analytical equipment. For example, the STEM/EDX study of high-angle grain boundaries in olivines revealed chemical segregation of some admixture elements [6]. The effect was qualitatively explained by thermodynamic and non-kinetic factors, believing that these elements are better compatible with the boundary than with the bulk structure; their presence at the boundary is associated with a lower strain energy in comparison with that when entering the crystal volume [6]. If TE in "closed" boundaries behaves in such a way, then free surfaces of crystal faces provide much greater opportunities for accommodation of TE, particularly when these elements are incompatible and are practically very scarcely represented in crystal structures. However, studies of quantitative separation of surface and bulk constituents of gold and other incompatible elements are extremely rare in the geological literature. Usually, this separation is not carried out and various concentration modes are associated with different generations of mineral in the sample and not with different grain (crystal) sizes of the same sample.

The accumulation of admixture elements is normally attributed to surface adsorption phenomenon, in particular, chemisorption [7]. However, being an exothermal process, adsorption is ineffective at high temperatures and under the impact of reactive fluid, causing, in contrast, desorption. It has been previously shown [8,9], that minor components such as Hg and Cd in sulfides can be part of the superficial non-autonomous phase (NAP) that is the product of chemical modification and structural reconstruction of the growing crystal surface layer bordering the hydrothermal fluid [10,11]. A later more extensive study of gold distribution in pyrite and other hydrothermal minerals in deposits of different genetic types confirmed the active role of the surface in the absorption of Au, an element, which is incompatible within the structures of most minerals [12]. There are highly determinate exponential dependences of the average content of evenly distributed Au on a specific surface area of an average crystal in the size fraction. The determination coefficients (R^2) often reached 0.9–0.99, and it is extremely difficult to reconcile with the existing patterns of Au uptake by mineral surfaces by an adsorption-reduction mechanism [13–15]. Incompatible elements are prone to camouflage, that is, to enter minerals, as mentioned above, they opt not for regular positions of atoms in a perfect crystal structure, but for various types of defects and their complexes [16]. It is very difficult to explain how highly deterministic dependencies on crystal specific surface areas appear. With $R^2 = 0.99$, it is difficult to talk about random distribution, which could have been the case with proper (autonomous) mineral forms emerging on the active centers associated with superficial defects [17,18]. This is because the number of randomly occurring defects, on which Au ions are reduced, depends on crystal surface area rather than on crystal specific area. Such clear dependence on the specific surface area will only take place if the element-absorbing NAP is homothetic to crystal (replicates its geometrical shape) and covers a significant portion of its surface.

In hydrothermal systems, the interaction of phase surfaces with the solution is usually faster than between solid phases, and a more soluble phase does not persist in such a system. Nevertheless, in the natural environments, including gold ore deposits [19], associations of such mineral phases are not uncommon. These are difficult to understand, except in one case—if the surface layers of these phases, and not their volumes, are in equilibrium with the solution. A.I. Rusanov [20] considered such layers as non-autonomous virtual phases, the number of which should be included in the extended Gibbs' phase rule. The detailed study of mineral surfaces by the methods of electron spectroscopy, scanning electron microscopy and scanning probe microscopy allowed us to establish that similar phases do exist, and they are real, because they have certain (albeit variable subject to circumstances) composition and thickness, element valence states differing from the volume [10,21]. They share a common feature with non-autonomous phases introduced by Rusanov—they cannot exist in isolation, without interaction with the matrix crystal. Therefore, they should evolve not by themselves, but together with the growing crystal, which means that NAP takes part in the growth process. Atomic force microscopic images of such phases forming on the surface of synthetic and natural crystals were presented in our numerous previous publications [10,11,21,22].

The emergence of NAP has a number of important implications, two of which are dealt with in this work: dualism of TE distribution coefficient and formation of hidden forms of metal content; the so-called invisible modes of metal occurrence. When considering these interrelated phenomena, we will refer to the results of earlier pilot studies [21,23–25], as well as to the data of studies of natural ore minerals with admixtures of precious metals.

2. Background

The philosophy behind our approach is to identify the part of the element concentration, which obeys the regularities for structurally bound impurity in the crystal and conforms to Henry's law, with minimal and relatively constant value and lower dispersion of distribution coefficient (compared with the dispersion of its total value). A coexisting phase can be either the ambient solution or the so-called reference mineral containing this element as a structurally bound admixture [23]. This approach is primarily aimed at distinguishing between volume and surface-related components of the element content, which is necessary to assess true and apparent (that is dual) distribution coefficient. There are special techniques designed to determine the speciation of the element [9].

Partitioning of TE can be described both by partition coefficient, D_p, and by co-crystallization (exchange) coefficient, D_e. D_p is a simple ratio of TE contents in solid phase (crystal, C^{cr}) and fluid (solution, C^{aq}), whereas D_e assumes an exchange reaction between TE and crystal matrix element (ME): $ME^{cr} + TE^{aq} = ME^{aq} + TE^{cr}$, $D_e = (C_{TE}/C_{ME})^{cr}/(C_{TE}/C_{ME})^{aq}$.

In an experiment, there is usually determined the total distribution coefficient, D^{tot}, which contains structural and superficial, nonstructural, component:

$$D_{p,TE}^{tot} = \left(f^V \cdot C_{TE}^{str} + f_h^S \cdot C_{TE}^{sur}\right)/C_{TE}^{aq} \tag{1}$$

where f^V and f_h^S are mass fractions ($f^V + f_h^S = 1$) of substance in the volume (core) of crystal and surface layer S with the thickness h, depending on the crystal size and shape [23]. Expression (1) allows us to calculate C_{TE}^{sur} and the corresponding partition coefficient from data on $D_{p,TE}^{tot}$ and C^{tot} for each size fraction with the crystal size r, if the concentration of TE structural component is known:

$$C_{TE}^{sur} = \left(C_{TE,r}^{tot} - f^V \cdot C_{TE}^{str}\right)/f_h^S \tag{2}$$

The manifestation of D_p duality is identified by D_p^{tot} increase in relation to the values that are defined by the law of distribution for isomorphic admixtures. The specific implementation of the approach and method of evaluating of C_{TE}^{str} and C_{TE}^{sur} contributions into TE total concentration for acquiring respective coefficients of dual distribution D^{str} and D^{sur} will be discussed below (Section 3.2.2).

3. Methods

3.1. Experimental Procedure

The methods of D determination in two series of experiments are described below. The distribution of precious metals (PM) was previously discussed [21,23–25] and is only supplemented here with fresh data. The distribution of rare earth elements (REE) in the systems magnetite (hematite) fluid has not been previously considered.

The crystals were obtained by standard techniques of hydrothermal thermogradient synthesis in stainless steel autoclaves of about 200 cm^3 [26]. Autoclaves were equipped with titanium alloy (VT–6, VT–8) inserts, with the volume of about 50 cm^3 and passivated surface. Ammonium chloride based solutions having demonstrated high efficiency in the course of synthesis of various mineral crystals were used. To obtain data on the composition of high-temperature fluid we applied internal sampling using traps attached to the insert plug (Figure 1).

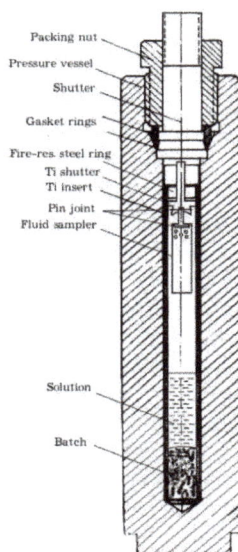

Figure 1. Schematic of the experiment. Crystals grow on the Ti shutter surface and insert wall.

The temperature in the zone of crystal growth was either 450 or 500 °C, pressure 100 MPa (1 kbar). The experiments were conducted in two stages. At the first stage, lasting 4 days at 450 and 3 days at 500 °C, the isothermal conditions were maintained to homogenize the batch material and ensure close to equilibrium conditions during subsequent thermogradient recrystallization held with 15 °C temperature drop (on the outer wall of the autoclave). The duration of the gradient stage is from 9 to 25 days depending on the temperature of the experiment and the desired size of the crystalline product. The experiments were terminated by autoclave quenching in cold running water at a rate of ~5 °C/s. The elements contained in low concentration in the solution do not persist in it under cooling to room temperature [9]. The method of capturing fluid portions (supercritical hydrothermal solution) by a trap is not strictly isothermal one. However, if the difference between the temperature of the experiment and the critical point of the aqueous-salt solution is not very big, and the cooling rate at quenching is high enough, it is reasonable to believe that the bulk mass of the fluid was captured at the highest density of the fluid, i.e., in the proximity of the experimental temperature [23]. Upon completion of the synthesis, the solution was extracted from the sampler which was rinsed with aqua regia to dissolve the precipitate. The cleaning solution was subsequently combined with the directly extracted one. Thereafter, the special chemical medium was created to determine the elements by appropriate analytical methods.

The batch was made up from domestically produced reagents of high chemical purity (pure reagent-grade). In the course of synthesis of magnetite, manganmagnetite and hematite the initial components were Fe, FeO, Fe_2O_3, Cr_2O_3, Mn_2O_3 and the batch mass was 3 or 5 g. NH_4Cl concentration was 8 or 10 wt %, in some experiments 2 wt % HCl or NaOH were added, as well as small quantities of potassium dichromate to improve magnetite growth. In the course of pyrite synthesis the batch weighing 6 g consisted of element substances—Fe, S and As. Solutions were prepared on the basis of NH_4Cl, adding in some cases 0.5–1 wt % of Na_2S, HCl or NaOH. In the experiments with precious metals, the batch was replenished with 1 wt % of Au, Pd and Pt in various combinations. In experiments with REE, Ce, Eu, Er and Yb were added in the form of oxides for 1 percent each of the batch mass of 5 g. The mineralizer concentration (NH_4Cl solution, "high purity" grade) amounted to 5 or 10 wt %.

Nitrogen is considered to be a common component of hydrothermal fluids forming the orogenic gold deposits [27,28]. Pyrite crystals from black shale-hosted gold deposits constantly exhibit nitrogen peaks in Auger spectra obtained from chips and natural crystal facets [29,30]. The ammonium ion is involved in ore formation processes; it is used in precious and non-ferrous metals exploration [31]. However, the main reason for the use of NH_4Cl in our experiments is related to its excellent transport properties rather than its geochemical significance.

3.2. Analytical Methods

The bulk of the analytical information on noble metals and other elements (Fe, Mn, As) was obtained by atomic absorption spectrometry (AAS). The measurements were performed on Perkin-Elmer (Model 503 and Analyst 800, The Perkin Elmer corp., Norwalk, CT, USA) devices in the Center of Collective Use (CCU) of isotope-geochemical research of the Institute of Geochemistry, Siberian Branch of Russian Academy of Sciences (SB RAS). REE were determined by inductively coupled plasma-mass spectrometry (ICP-MS) method on Agilent 7500ce unit manufactured by Agilent Technologies (Agilent Tech., Santa Clara, CA, USA) in the CCU "Ul'tramikroanaliz" of the Limnological Institute, SB RAS (Irkutsk, Russia).

3.2.1. Analysis of Precious Metals

When analyzing crystals, we used AAS method with the element electrothermal atomization in graphite furnace (AAS-GF). The analytical data selections for single crystals (ADSSC) version of data processing was applied (see below). If for gold the direct determination from the solution after crystal acid decomposition with minimum detection limit (MDL) of 0.3 ppb and an accuracy of 12% was reliable, Pd, Pt and Ru were determined after preliminary extraction concentrating and separation from the matrix [32]. Tristyrylphosphine ($C_6H_5CH-CH)_3P$ was used as an extracting agent. Extraction was performed in weak (0.5 M) hydrochloric solutions. The extracting agent concentration was 0.05 M (in toluene) and contact time of phases was 30 min. The proportion of aqueous and organic phases was 2:1 (by volume). Extraction was carried out in static mode at room temperature and without labilizing additives. Organic phase was used to measure element concentrations. Measurements were accurate to $\pm10\%$ with MDL of 5 µg/L (5 ppb) for Pd and 50 µg/L for Pt and Ru.

3.2.2. Analytical Data Selections for Single Crystals (ADSSC)

Experimental observations and the simulation of TE behavior allowed us to propose an approach based on rank-scaled statistical sampling of analytical data for individual TE-containing crystals. As the starting point, we consider the properties of TE uptake by mineral crystals [9]. Although TE can be contained in various binding forms, there exists a finite probability that any representative assemblage of individual (single) crystals involves several that are free from active sites or defects responsible for the non-structure element uptake from slightly oversaturated or undersaturated solutions. It is desirable that such crystals should be regularly facetted and have exclusively perfect faces, and thus they contain only two indiscernible element species: superficially bound and structurally bound. Other forms obviously enhance the total content of the element, so the task is to obtain a statistical sample of single crystals that contain it in minor but significant (above the detection limit) amounts. Let us take for example the gold determination by atomic absorption spectrometry with a graphite furnace (AAS-GF). The procedure of statistical data processing is the following [33]. The initial assemblage N usually contains 40–100 individual crystals. The more analyses, the more reliable the result can be obtained, if their number does not increase at the expense of quality of crystals. Only the reliable values N_1 that exceed at least three times MDL are taken for further consideration. The Au detection limit for AAS-GF determination was estimated with a special standard sample as 0.3 µg/L in solution or 0.3 ppb in solid [23]. In processing the data, the whole database of Au concentrations in individual single crystals is subdivided into ranges according to crystal masses in such a manner that all ranges contain roughly equal numbers of crystals and no less than 15. Strictly speaking, the crystal mass m_i

should be nearly equal for individual crystals subsets but this is an unrealistic requirement. Therefore, we follow the rule of minimum Δm in each subset taking into account both the quality of internal statistics and representatives of size fractions. Then an average concentration is determined for each range, the root-mean-square deviation (s) is evaluated, and values >1 s are rejected. The crystals showing negative deviations (<-1 s) remain in the sample. According to the concept adopted here, in principle, a crystal can contain only the structurally bound mode of TE and any additional forms can only increase the concentration of the element above the value for the structural (isomorphous) form. Because of this, low concentrations should be pre-treated as possible if they are not accounted for by the analytical errors or <3 MDL. Thus, the action described above allows the purification of the subsets of sample N_1 from crystals, which contain binding forms distributed non-uniformly between the crystals, first, fine inclusions of native gold. Thereafter for the subsets, we determine the mean $\overline{C}_{Au}$ and the standard error of the mean $\pm\sigma$. Finally, we use a procedure to separate structurally bound gold from modes bound to the surface. For this purpose, we introduce the criterion by which every C_{Au} in the size (mass) subset must be rejected if it is higher than $\overline{C}_{Au} + 30\% \cdot \overline{C}_{Au}$. This is a condition for structural mode: the variation coefficient having a maximum value of 20% (without random error) plus in round 10% of analytical uncertainty of AAS-GF [26]. Thereafter, we determine the average concentration of an evenly distributed Au in each subset ($\overline{C}_{Au}$), the average mass of crystal ($\overline{m}$) in the subset and then calculate the average crystal size and the specific surface area of the average crystal:

$$\overline{S}_{sp} = k \cdot \overline{r}^2 / \overline{m} \tag{3}$$

The crystal shape is approximated by a true polyhedron with an edge $\overline{r}$ and form coefficient k. If an adequate number of size fractions (subsets) is available (≥ 4), we get the possibility to evaluate the structurally bound constituent by the extrapolation of evenly distributed gold content to a zero-specific surface, i.e., to a very large crystal containing only bulk and no surface atoms.

Of course, we cannot exclude some special cases, for instance, gold nanoparticles [34] distributed in the same manner as structurally bound gold (Au^{str}). If this form is related to the surface (as is often the case for fine Au^0 particles), it can be separated from Au^{str}.

The validity of this procedure has been tested using the data for pyrite crystals of different size fractions [35]. The $\overline{C}_{Au}$ dependence upon $\overline{S}_{sp}$ indicated that the amount of structural gold in pyrite is equal to 2.3 ppm, in reasonable agreement with the value of gold solubility in low-arsenic pyrite at the same conditions (3 ± 1 ppm) [26]. However, it is not a simple matter to evaluate the uncertainty of the value Au^{str} obtained by extrapolation. The comparison with X-ray photoelectron spectroscopic (XPS) data within the system $Ag_2S{:}Au$ as well as with the data on other elements and methods showed that structurally bound constituent can be determined precise to $\pm 30\%$ [35,36].

There are two approaches to the evaluation of the surface-associated component of TE. In the works [21,23] calculation performed in respect of NAP, assuming true cubic (pyrite, Py) or octahedral (magnetite, Mt)) crystals uniformly covered by NAP of ~500 and ~330 nm thick, respectively (according to the XPS with ion etching of the surface and atomic force microscopy). Then, accounting for the apparent condition $r \gg h$, where r is the edge of the cube or octahedron, you can define f_h^S value in the Formula (2):

$$f_h^S(Py) = \frac{3r^2 h d^s}{(r - 2h)^3 d^V} \tag{4}$$

$$f_h^S(Mt) = \frac{6\sqrt{3/2} r^2 h d^s}{(r - 2h)^3 d^V} \tag{5}$$

where d^S and d^V are the densities of the surface and bulk phases, respectively.

Substituting (4) or (5) into (2) in the light of the values found through extrapolation C_{TE}^{str} we acquire evaluation of TE content in NAP in every size fraction, which is then averaged across all the fractions. Formulas, similar to (4) and (5) can be obtained for other shapes of crystals. For example, the pyrite crystals in experiments with As better correlate with the hemispherical shape because of the abundance of vicinal faces.

Thus obtained values $\overline{C}_{TE}^{NAP}$, first, are significantly dispersed due to highly simplified NAP model and r variations in size fractions, and, second, they are not directly comparable with C_{TE}^{str}, since the latter refers to the whole crystal, while C^{NAP} refers to its surface layer. So, papers [21,23] offer a different way of $\overline{C}_{TE}^{sur}$ evaluation, where TE content in the superficially bound mode characterizes an average crystal among all size samples; that is, the surface-related excess concentration of element:

$$\overline{C}_{TE}^{sur} = \frac{\sum \left(\overline{C}_{TE}^{sm} - C_{TE}^{str} \right) n^{sm} \overline{m}^{sm}}{\sum n^{sm} \overline{m}^{sm}} \qquad (6)$$

where $\overline{C}_{TE}^{sm}$ is the average content of TE evenly distributed in each size sample sm with the number of crystals n^{sm} and average crystal mass $\overline{m}^{sm}$. The surficial distribution coefficient D_p^{sur} is attributed to that value, and not to $\overline{C}_{TE}^{NAP}$, which is more practical and instrumental in comparing contributions of surface and bulk related modes.

3.2.3. Analysis of Rare Earth Elements

In the experiments with REE, Ce and Eu represented "light" rare earth elements (LREE), Er and Yb "heavy" rare earth elements (HREE). The fluid samples from the traps were analyzed by ICP-MS method on the Agilent 7500ce unit manufactured by Agilent Technologies with quadrupol mass analyzer (Agilent Tech., Santa Clara, CA, USA). Measurements were performed under the following optimal conditions: plasma power 1550 W, reflected power 1 W, the carrying gas flow rate 0.8 L/min, auxiliary gas flow—0.13 L/min, integration time 0.1 s, points on mass—3, oxides (CeO/Ce, 156/140) = 0.3%. The device was tuned by Tuning solution with the concentration of 10 µg/L (10 ppb) Li, Co, Y, Ce, Tl. The calculation of elements concentrations was carried out by external calibration method using multi-element certified solutions 68A, 68A–B (High-purity Standards) containing all of the elements from Li to U. The calibration solutions were diluted with 2% HNO_3 down to the concentration 0.5–50 µg/L of each element. The drift tracking of the device was carried out by internal standard In (10 ppb). Calculation of the concentrations was based on the most abundant isotopes, giving signals, free of overlapping or subject to minimum isobar and polyatomic disturbances. Elements detection limits are at the level of 1 ppt, determination error is 5%.

To determine REE in magnetite and hematite crystals the Laser Ablation platform of New Wave Research UP-213 was used. Parameters of the LA-ICP-MS experiment: plasma power 1400 W, carrier gas flow rate 1.18 L/min, plasma forming gas flow rate 15 L/min, cooling gas 1 L/min, laser energy 0.16–0.19 mJ, frequency 10 Hz, laser spot diameter 55 µm. Accumulation time per channel—0.15 s; acquisition time 27 s. Calculation of the concentrations was based on the standard sample NIST 612. The correctness was verified by using the in-house standard sample of hematite with elevated Ce content estimated by EMPA. It is known that synthetic hematite can incorporate up to 1.7 wt % Ce as a structurally bound impurity [37]. Standard deviations are consistent with counting statistic uncertainties at around 20–30 per cent. REE detection limits are estimated at ~0.1 ppm. Error analysis is rated as 30%.

3.2.4. X-ray Photoelectron Spectroscopy (XPS)

The X-ray photoelectron spectra of the samples were obtained on photoelectron spectrometer SPECS (SPECS, Berlin, Germany) equipped with a PHOIBOS 150 MCD 9 energy analyzer (Krasnoyarsk regional Centre for Collective Use, SB RAS). The spectra were acquired at excitation initiated by the radiation of Mg anode of an X-ray tube (Mg Kα = 1253.6 eV), the power was 180 W, voltage on the tube—12.5 kV. The survey spectrum was recorded with a step of 0.5 eV at transmission energy of the energy analyzer 20 eV. The high-resolution spectra of individual elements (narrow scans) were recorded usually with 0.05 eV interval and transmission energy of 8 eV. The line C 1s (285.0 eV) of hydrocarbon contaminants was used as an internal standard to account for electrostatic charging. Ion etching was conducted using raster ion gun PU-IQE112/38, working with accelerating voltage 2.5 kV and ion current of 30 µA. The etching rate was approximately 6 nm/min. The deconvolution of the spectra was performed using the program CasaXPS after subtracting the nonlinear background by Shirley's method, peak shape was approximated by symmetrical Gauss-Lorentz function. To interpret the spectra, we used NIST XPS database [38], the handbook [39] and publications [21,40]. In addition, some of the work on defining the chemical state of PM was performed on the spectrometer LAS-3000 manufactured by "Riber" (Riber, Bezons, France). Atoms of sample surface were excited by aluminum anode non-monochromatized radiation (Al Kα = 1486.6 eV) emitted by an electron beam accelerated at 10 kV with a current of 20 mA. Vacuum pressure in the analysis chamber typically was 6.7×10^{-10} mbar. The spectrometer was equipped with an OPX-150 hemisphere detector and Auger electron analyzer of the cylindrical mirror type OPC-200.

4. Natural Materials

Pyrite crystals to be analyzed were selected from the four ore sites: Degdekan, Natalkinskoe, Zolotaya Rechka (North-East of Russia) and Krasnoye (Eastern Siberia) (Figure 2). All of them belong to orogenic gold deposits [41]. The native gold in the deposits of ~740‰ to ~920‰ fineness is mainly coarse-grained and disseminated in vein quartz and sulfide mineral aggregates. Another reason for choosing these objects is that ore mineralization is located in carboniferous sediments. Such ore objects within black shale strata that are challenging for diagnostics of Pt occurrence modes. Study of gold deposits platinum content in the rocks of the carbonaceous formations, determination of Pt content in ores and minerals, identification of Pt occurrence forms, hosting entities and concentrators of this element occupy a special place in the investigations and have a significant theoretical as well as practical value. Presence of this noble metal in ores and the possibility of its extraction can significantly complement the range of known platinum-bearing ore formations and considerably increase the value of extracted gold at the fields, where platinoids accompany gold mineralization. The geological setting and mineralogical features of these objects (except for Zolotaya Rechka, see below) are given in detail in [42–47].

Pyrite is one of the most common sulfide minerals in all the gold deposits we studied. We studied pyrites from the ores with the highest gold grade selected in surface and underground mining workings (quarries, adits) and holes. Large mineralogical–geochemical samples weighing up to 10 kg were used for the study. The samples were crushed, sieved into fractions, washed in distilled water. Pyrite crystals were hand-picked under binocular microscope from fractions of 0.25–0.5, 0.5–1 and 1–2 mm.

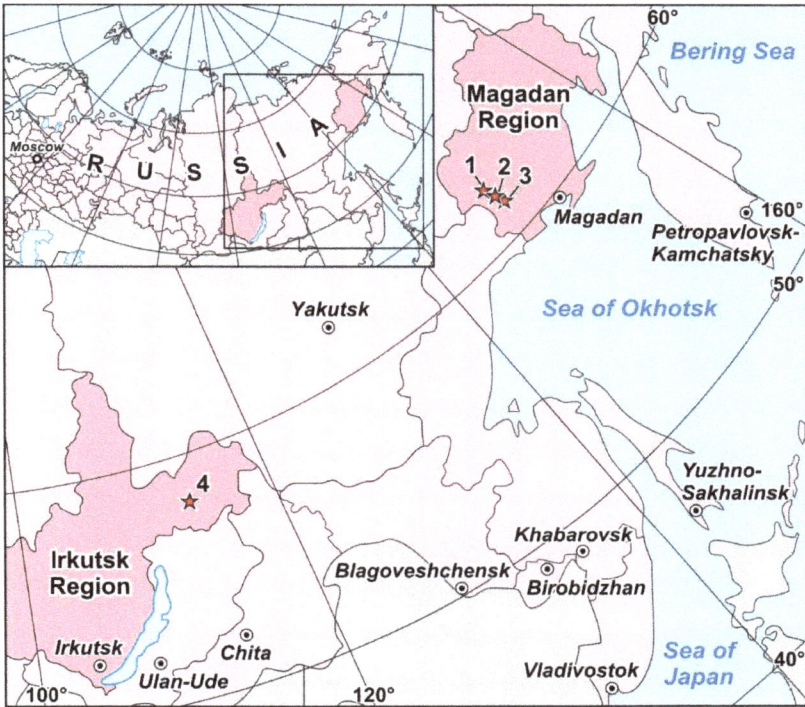

Figure 2. Location of the Degdekan (1), the Natalkinskoe (2), the Zolotaya Rechka (3) and the Krasnoe (4) deposits within the territories of Magadan and Irkutsk regions of the Russian Federation.

4.1. Degdekan, Natalkinskoe and Zolotaya Rechka Deposits

These objects are located in the North-East of Russia (Magadan Region, Tenkinsky ore area) and are within the gold-bearing Yano–Kolyma fold belt being confined to its orogenic zones formed by collision. Their metallogenic characteristics are similar and they are classed as arsenopyrite type of gold-quartz low sulfide formation. The deposits are characterized by complex and long-continued character of development, and, according to most researchers, by metamorphogenic-hydrothermal genesis. For all the deposits, the prevailing non-metallic mineral is quartz, while arsenopyrite and pyrite are most widely spread among ore minerals. Of subordinate significance are pyrrhotite, galena, sphalerite, chalcopyrite, gersdorffite, native gold. Silver as a proper mineral form is extremely rare. It is mainly present in the form of micro-inclusions of argentite (acanthite) and native silver in arsenopyrite and pyrite, freibergite in chalcopyrite. Among the samples considered here, only a few crystals of pyrite selected in the Zolotaya Rechka deposit were found to contain finely dispersed inclusions of argentite or acanthite (microprobe analysis). As regards platinum, with the contents ranging from 12.6 to 61.7 ppm in the pyrites we studied, no proper mineral forms have been identified [42–45].

4.1.1. Degdekan Deposit

The Degdekan gold deposit is an example of localization of vein zones and veinlet-disseminated sulfide ores in stratified Permian sediments. Ore-hosting sections are largely composed of carbonaceous siltstones and shales with a large number of interlayers of graphitized host rocks. They form a granite-metamorphic dome associated with the North-Western zone of the Tenkinsky deep fault. The time interval of the formation of the dome hosting the deposit amounts to 130–142 million years (m.y.). The absolute age of the gold mineralization by U-Pb SHRIMP is estimated at 133–137 m.y., by the Ar-Ar method—at 137 m.y., that is, according to isotopic geochronology ore formation dates back to the early Cretaceous period. The main vein mineral is quartz. The host rocks, besides silification, demonstrate albitization, sericitization, there are found carbonates (ankerite, calcite) and chlorite. The principal ore minerals are pyrite and its subordinate arsenopyrite. These sulfides amount on average about 3%. Pyrite occurs in the form of small anhedral grains of 1–2 mm in size and large, over 2 mm, euhedral crystals of hydrothermal genesis. In addition to these minerals, there are found sphalerite, galena; less frequent are pyrrhotite, chalcopyrite, there are rare occurrences of gersdorffite. Gold is mainly in native form and <1 mm in size. Its fineness varies from 740 to 800‰. Extremely rare are electrum and kustelite [42,43]. Gold-arsenopyrite-polymetallic (with pyrite) mineral association is identified as productive [42]. The temperature range of its formation (according to the study of fluid inclusions) is estimated at 200–230 °C, the pressure is about 1 kbar. The solutions were weakly mineralized (25 g/L) and had mostly sodium bicarbonate composition, which points to their amagmatogenic (metamorphogenic-hydrothermal) origin. A large part of the pyrite at the Degdekan deposit was selected from the veinlet disseminated ores with gold content from 1.4 to 15.2 ppm.

4.1.2. Natalkinskoe Deposit

Ore mineralization of the unique giant Natalkinskoe ore deposit, despite all the diversity, forms a uniform (by internal constitution) ore lode consisting of quartz, quartz-carbonate, quartz-sulphide veins and veinlets surrounded by a wide halo of sulfidized rocks [44,45]. The geological position of the deposit is mainly due to its closeness to the marginal part of the alleged granite pluton in the central zone of the Tenkinsky deep fault. Overall, the deposit is traced along strike for 5 km, with a width of 1 km. Host rocks are classed as late Permian sediments represented by clay rocks, siltstones, clayey shales with increased carbon content. According to K/Ar dating, the age of gold mineralization was estimated from 135–130 m.y. up to 110–100 m.y., i.e., according to isotope geochronology, this gold mineralization might date back to early Cretaceous period. Such non-metallic minerals as quartz, carbonates, feldspars, sericite and chlorite were identified at the deposit. Ore minerals, along with pyrite and arsenopyrite (amounting to 4–7 wt %), are presented by galena, sphalerite, chalcopyrite, native gold and rutile (less than 1 wt %). Native gold with the fineness from 750 to 900‰, less frequent electrum, mostly coarse-grained, occurs mainly in native state in vein quartz and conglomerates with sulfide minerals. Pyrite-arsenopyrite mineral association with gold and galena is found to be the most productive [44]. Industrial ore formed in the course of interaction of host rocks with low- and medium salty water-bicarbonate fluids within the salinity interval 3–12 wt % NaCl eq., at temperatures of 360–280 °C and approximate pressure of 2.4–1.1 kbar [45]. Pyrites from the richest (1.5–30.2 ppm Au) vein and veinlet-vein ores were studied.

4.1.3. Zolotaya Rechka Deposit

The Zolotaya Rechka deposit has not been sufficiently studied. There are no data on this deposit in the available published sources. Geological description and mineralogy of the deposit are presented here on the basis of unpublished materials of geological foundations; the material composition is supplemented with data provided by the authors of this article. The Zolotaya Rechka deposit is located in the anticline arch made up of weakly metamorphosed and intensely dislocated terrigenous sediments of the Upper Permian. It is dark grey, to black, siltstone clayey carbonaceous shales with streaks and lenses of sandstone and sand tuff. The deposit, as well as Natalkinskoe, is confined to the central zone of the Tenkinsk deep fault, but is located slightly further to the South. In metallogenic relation, it is of the same type as the previously considered ore objects. By structural and morphological characteristics the deposit falls under the category of vein-veinlet and veinlet-mineralized zones. Ore bodies are a combination of mineralized shatter zones and veinlet zones. Veinlet and disseminated sulfide mineralization (no more than 3 wt %) is presented by pyrite and arsenopyrite, with rare occurrences of chalcopyrite. Native gold occurs in vein quartz and conglomerates with sulfide minerals. Native gold with the fineness of 750–800‰, largely fine (up to 0.8 mm), and is mostly in quartz and conglomerates with sulfide minerals. Pyrite-arsenopyrite mineral association with gold is identified as productive. Pyrites were selected mainly from veinlet-mineralized zones with gold content from 1.2 to 15.2 ppm.

4.2. Krasnoye Deposit

The Krasnoye deposit is located in the territory of Eastern Siberia (Irkutsk region, Bodaibo ore area) (Figure 2). Bodaibo ore area as one of the largest gold reserves in Russia is a classic example of orogenic gold deposits in black-shale formations of Proterozoic age. The Krasnoye deposit is situated in the south of the area (Artyomovsk ore node). Structurally, the gold ore mineralization is confined to the core parts of anticlines and fault zones. The genesis is traditionally considered to be metamorphogenic-hydrothermal, with hydrogenic processes being assigned an essential role by a number of researchers [46,47].

The regressive phase of metamorphism, which is associated with gold ore mineralization, is estimated at 330–300 m.y. and is connected with the formation of the Angara-Vitim granitoid batholite. Ore mineralization was formed by moderately saline fluids of sodium bicarbonate composition with salinity range of 7.5–13.0 wt % NaCl eq., at temperatures of 140–300 °C and a pressure of 0.9–1.8 kbar.

The mineral composition of the Krasnoye ores is similar to that of the ore sites of the North East of Russia we considered. Non-metallic minerals are mainly presented by quartz, the primary mineral among metallic minerals being pyrite with the secondary and rare ones presented by arsenopyrite, chalcopyrite, galena, sphalerite, pyrrhotite, fahl ores of tennantite-tetrahedrite group, gersdorffite, gold. No proper mineral forms of silver have been identified [46,47].

Pyrite in ores is presented by two varieties: large crystalline pyrite, and fine dissemination of idiomorphic pyrite. Large crystals of pyrite largely develop in association with chalcopyrite, galena, pyrrhotite, gold and gray copper ores. Chalcopyrite is often closely associated with pyrrhotite and galena. Native gold with a fineness of 860–870‰ is associated with cracks in pyrite, develops as independent occurrences or in association with galena. Gold is mostly fine (<0.1 mm). Gold-pyritic mineral association with polymetals is most productive. Large pyrite crystals were selected from quartz-pyrite veins with gold content of 1.2 ppm up to 11.2 ppm.

4.3. Analysis Procedure

To study pyrite with ADSSC technology, idiomorphic crystals of different sizes with clean surfaces and overwhelming prevalence of cubic facets {100} were selected from ore samples. Each crystal was weighed on the analytical microbalance and transferred into beakers for subsequent dissolution (see analysis methods above). We used only the crystals with the mass exceeding 0.1 mg. According to our long-term experience of work with this method, acquisition of reliable data requires the starting sample to contain at least 70–80 crystals, their screening in the course of data processing being on the order of one third. This enables selection of 4–6 size fractions in the final sample, which provides extrapolation towards zero specific surface area of the average crystal.

For all samples analyzed, we acquired highly deterministic dependences of average sample PM contents on specific surface of the average crystal in size fraction. A structural component of impurity is identified with pre-exponential factor; the surface-bound component is calculated according to the Formula (6).

5. Results

5.1. Distribution of Precious Metals

Table 1 presents the procedure for calculation of PM (Au and Pd) contents in structurally and superficially bound modes by ADSSC method while studying their distribution between hematite crystals and hydrothermal solution.

Table 1. Example of data acquisition by the ADSSC method (TE = Au and Pd in hematite crystallized at 450 °C and 1 kbar in 10% NH₄Cl).

Number of Cryst. (Starting-Final Sample)	Characteristics of the Final Sample						TE Contents (ppm) *			
	Num. of Cryst.	Range of Mass. (mg)	$\bar{m}$ (mg)	$\bar{r}$ (mm)	$\bar{S}_{sp}$ (mm²/mg)	$\bar{C}_{TE} \pm 1\sigma$ (ppm)	$\bar{C}^{tot}$	$\bar{C}^{ev}$	C^{str}	$\bar{C}^{sur}$
				Gold						
43–26	7	0.09–0.12	0.11	0.360	4.081	85 ± 66	110	30	2.0	29.0
	7	0.14–0.20	0.17	0.416	3.526	41 ± 24				
	6	0.21–0.38	0.29	0.497	2.951	34 ± 20				
	6	0.39–0.83	0.61	0.637	2.303	15 ± 8				
				Palladium						
41–31	9	0.09–0.12	0.11	0.360	4.081	254 ± 51	111	78	4.7	74.8
	7	0.14–0.21	0.20	0.439	3.338	122 ± 28				
	7	0.23–0.38	0.33	0.519	2.828	78 ± 9				
	8	0.39–1.48	0.71	0.670	2.190	39 ± 8				

* Average total, evenly distributed and structurally and superficially bound forms concentrations.

Given the small size interval of the obtained crystals ($r \sim 0.3$–0.7 mm) it was possible to identify only four size fractions. This is the minimum number of fractions, with which the method of separation of structural and surface modes is efficient. The number of crystals in the final samples is not quite optimal either; it is advisable to have no less than 10, but this is not always easily achievable without compromising the quality of the data because it is influenced by the characteristics of individual crystals, perfect habitus forms and pure face surfaces. The data of Table 1 are graphically presented in Figure 3.

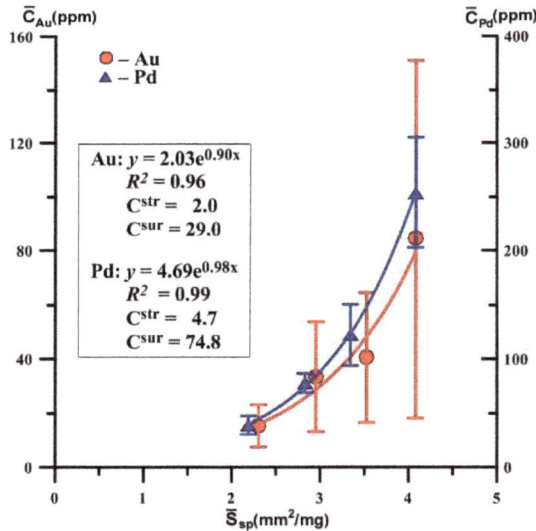

Figure 3. Dependences of the average concentrations of evenly distributed precious metals on the specific surface area of an average crystal in size fractions. Hematite crystals were obtained at 450 °C and 1 kbar in 10% NH₄Cl. The expressions for approximate curves and concentrations of structurally and superficially bound modes are shown (see Table 1 for details).

Despite the large amount of deviations from average values, especially for Au, the coefficients of determination of the dependencies presented here are quite high (0.96 and 0.99). However, error estimation shows that structural Au is determined with great inaccuracy. For Pd, the dependence is more correct and described by the equation $\overline{C}_{Pd} = (4.7 \pm 1.4) \times e^{[(1 \pm 0.1) \cdot \overline{S}_{sp}]}$, i.e., a structural component of Pd admixture in hematite is 4.7 ± 1.4 ppm. The assessment accuracy is 30%, typical for this method (see Section 3.2.2). It should be also noted that the concentration of elements in the superficially bound mode exceeds by more than an order of magnitude the content in structural mode. Hence, it is clear that the total distribution coefficient will be imposed mainly by surface mode. Table 2 summarizes previously acquired data [21,23–25] and the data of the current research on partition coefficients of PM (Au, Pt, Pd) in the systems "mineral-hydrothermal solution".

The experiments were conducted with magnetite, its manganese variety (manganomagnetite), hematite, pyrite and As-pyrite at a temperature of 450 and 500 °C in solutions based on ammonium chloride (see Section 3.1). Table 2 demonstrates that the surficial partition coefficient overwhelmingly excels the structural one by an order of magnitude or more. The only exception is D_p Au in pyrite in the absence of As, for which surficial coefficient is only double of the structural one.

The data in Table 2 also indicate one rather unexpected circumstance: except for Au in pyrite, PMs are compatible elements in magnetite, pyrite and hematite, and taking into account superficial component, they are highly coherent elements.

Calculated by Equations (2), (4) and (5) contents of Au in NAP amounted to ~(3–5) × 10^3 in pyrite, ~(1.5–2) × 10^3 in As-pyrite and ~(1.5–3) × 10^3 ppm in magnetite [21,23]. Despite these relatively high values, satisfactory Au 4f XPS spectra were not obtained. However, Pt 4f and Pd 3d spectra, fairly conditioned, though of low intensity, were obtained on magnetite samples (Figure 4). They confirm the presence in the surface of two PM forms; oxidized (likely bivalent state) and elemental metal ones. Their relationship together with other characteristics of the spectral lines are presented in Table 3.

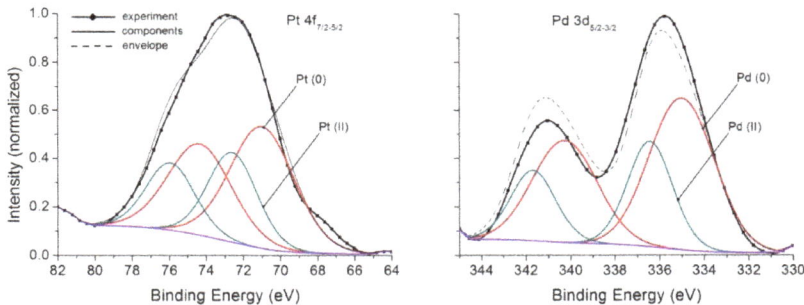

Figure 4. X-ray photoelectron spectra Pt 4f and Pd 3d of magnetite crystals. See Table 3 for parameters of spectral lines.

Table 2. Dual partition coefficients of precious metals in the system "mineral–fluid" determined in thermo-gradient hydrothermal experiments at 450 and 500 °C and 1 kbar.

Metal	Mineral	Experimental Conditions			D_p^{str}	D_p^{sur} **	Ref.
		T (°C)	Batch	Solution *			
Gold	Magnetite	450	Fe$_2$O$_3$ + Fe(FeO)	a.c., a.c. + HCl	1.0 ± 0.3	9.5 ± 5.5	[21]
		500	Fe$_2$O$_3$ + FeO	a.c.	0.5	10	[24]
	Mangan-magnetite	450	Fe$_2$O$_3$ + Mn$_2$O$_3$ ± FeO	a.c. + HCl	1.6	9.4	[24]
		500	FeO + Mn$_2$O$_3$	a.c.	0.8 ± 0.5	17 ± 9	[24]
	Pyrite	450	Fe + S	a.c., a.c. + Na$_2$S, a.c. + HCl	0.14 ± 0.04	0.3 ± 0.2	[23]
	As-pyrite	450	Fe + S + As	same	0.05 ± 0.03	0.4 ± 0.5	[23]
	Hematite	450	Fe$_2$O$_3$ + FeO + Cr$_2$O$_3$	a.c.	4.3	63	This work
Platinum	Magnetite	500	Fe$_2$O$_3$ + FeO	same	46	350	same
	Mangan-magnetite	500	FeO + Mn$_2$O$_3$	same	27	394	same
	Pyrite	500	Fe + S + As	a.c., a.c. + Na$_2$S, a.c. + HCl	21 ± 7	210 ± 80	[25]
Palladium	Magnetite	450	Fe$_2$O$_3$ + FeO + Mn$_2$O$_3$	a.c. + HCl	2.6	53	[24]
		500	Fe$_2$O$_3$ + FeO	a.c.	3.4	84	[24]
	Mangan-magnetite	450	Fe$_2$O$_3$ + FeO + Mn$_2$O$_3$	a.c. + HCl	3.0	71	[24]
		500	FeO + Mn$_2$O$_3$	a.c.	2.9 ± 0.7	52 ± 3	[24]
	Hematite	450	Fe$_2$O$_3$ + FeO + Cr$_2$O$_3$	a.c.	10.9	174	This work

* a.c.—ammonium chloride of 8–10 wt % concentration. ** Estimated with Equation (6).

Table 3. XPS data for magnetite crystals containing Pd and Pt.

Sample No.	Photo-Electron Peak	Lorentz [a]	Binding Energy (eV)	FWHM [b] (eV)	MPE [c]	MF [d]
1	$3d_{5/2}$	0.7	335.0	3.4		
	$3d_{3/2}$	0.7	340.3	3.4	Pd(0)	0.67
	$3d_{5/2}$	1.0	336.4	2.4	Pd(II)	0.33
	$3d_{3/2}$	1.0	341.7	2.4		
2	$4f_{7/2}$	0.7	71.0	4.0		
	$4f_{5/2}$	0.7	74.4	4.0	Pt(0)	0.64
	$4f_{7/2}$	0.7	72.6	3.1	Pt(II)	0.36
	$4f_{5/2}$	0.7	75.9	3.1		

[a] Lorentz contribution to mixed function "Lorentzian + Gaussian". [b] Full width at half maximum of peak height.
[c] Most probable entities from BE values. [d] Mole (atomic) fraction (total element = 1) calculated from areas of doublets.

Thus, in contrast to the widely held view about PM reduction on mineral surfaces to neutral state [13–15], in hydrothermal growth systems presence of two forms of each of the elements, PM (0) and PM (II) is registered. This shows that there is not only a reductive adsorption, as at low, close to normal conditions, P,T parameters, but also the uptake of these elements by NAP in chemically bound form. In respect of magnetite, an important fact should be mentioned. Adding Mn to the system and causing crystallization of solid solutions (Fe, Mn) [Fe$_2$O$_4$], it is possible to prevent the formation of NAP, resulting in flattening the curves $\overline{C}_{PM} - \overline{S}_{SP}$ [21,24]. This is accounted for by the enrichment of the surface with manganese confirmed by layer-by-layer analysis based on LA-ICP-MS method [24]. That is, on the surface there is too little iron to form Fe oxyhydroxide (similar to hydromagnetite) surficial phase. Thus, the development of NAP depends on physicochemical conditions and chemical composition of crystallization medium.

Considering the experimental data presented above, it is necessary to pay attention to one important fact. Spontaneous deposition of substances onto crystals under quenching definitely takes place in the interval from T of experiment to ~380 °C critical point, below which the crystals are no longer in contact with the solution. Simple balance calculations show that precipitation of the substance cannot provide the observed concentrations of trace elements in the surface layer. This may be exemplified by Pt in pyrite. In the previously described experiment [25] under the number D23–4, the value established in the solution from the trap equaled 0.42 mg/kg Pt. Platinum content in the surface form calculated for average crystal mass in the sample was 100 mg/kg. The mass of the obtained crystals was about 900 mg, which yields 0.09 mg of Pt. Even if we assume that the surface of grown crystals (and only it) during quenching uptakes Pt of the entire fluid occupied the insert volume of ~50 mL, with the content of 0.42 mg/kg Pt, we get only 0.02 mg of Pt, i.e., 4.5 times less. If we take into account the real ratio of the surfaces of crystals, insert and trap walls, which are also readily available for deposition of Pt, the difference will attain several orders of magnitude. Therefore, the effect of trace elements enrichment cannot result from the deposition of substances during quenching. In this case, the content of the element would be proportional to the crystal contact area with fluid phase, rather than its specific surface area. Indeed, various sectors to a different degree are available for precipitated substances during quenching, with some part of the crystal surface inside conglomerates being closed from fluid at the time of quenching. Other differences between the NAP (and the products of their evolution) and autonomous quenching phases lie within the presence of ordered surface structures and regular change of the composition with depth, for example, reducing oxygen content in pyrite with crystal depth, difference of morphology and composition of segregations on the faces that belong to different simple forms of crystals [10,22,25].

5.2. Distribution of REE

Rare earth elements are currently viewed as universal indicators of rocks and minerals genesis. The Ce anomaly in REE series is a redox-sensor of oxidative state of magmatic systems [48], the proportion Ce^{4+}/Ce^{3+} is regarded as a sensitive indicator of the oxygen fugacity [49]. However,

REE distribution coefficients of mineral-fluid in hydrothermal systems are still poorly understood. Described below are the results of determination of REE and Fe co-crystallization coefficients ($D_e = D_{REE/Fe}$) and modes of their occurrence in magnetite and hematite. In the experiments, these minerals were synthesized with the addition of Ce and Eu oxides as representatives of light REE (LREE), Er, and Yb, as representatives of heavy REE (HREE). NH_4Cl solutions were used as growth medium. The information on elements contents in minerals and coexisting fluids is presented in Table 4. Two concentrations of mineralizing solution NH_4Cl were used 5 and 10 wt %.

Table 4. REE contents in the bulk of magnetite and hematite crystals and in coexisting fluids at 450 °C and 1 kbar by ICP-MS (solution) and LA-ICP-MS (crystal interior).

Exper. No.	Mineral	Solution-Mineralizer	REE Concentrations, ppm								
			In Crystals				In Solution from Sampler				
			Ce	Eu	Er	Yb	Ce	Eu	Er	Yb	Fe
1	Magnetite	5% NH_4Cl	0.45	0.36	0.23	0.22	600	234	30.3	19.3	9880
2	Magnetite	10% NH_4Cl	0.26	0.30	0.83	1.23	407	210	136	77.7	11,290
3	Hematite	5% NH_4Cl	3.79	2.67	1.51	2.59	296	54.8	6.7	3.7	2340
4	Hematite	10% NH_4Cl	0.99	2.20	2.25	4.20	1190	135	11.5	7.3	9380

Table 5 presents the D_e co-crystallization coefficients, which reveal a proper accordance in both solutions used for magnetite (Mt) and a slightly worse accordance for hematite (Hm). Two approaches were used for evaluation of these parameters. The first approach included LA-ICP-MS analysis on crystal sections, pressed in epoxy cartridge and polished, i.e., the data obtained referred to REE volume contents (Table 5). These data show that D_e values are low, at the level of $(1–2) \times 10^{-5}$–$(1–2) \times 10^{-4}$ for Mt and $(1–4) \times 10^{-5}$ –$(2–8) \times 10^{-3}$ for Hm, and there is a clear tendency towards their increase for HREE, approximately one order of magnitude for Mt and two orders for Hm.

Table 5. Co-crystallization coefficients of REE and Fe in Mt and Hm at 450 °C and 1 kbar in ammonium chloride solutions.

Exper. No. *	Mineral	$D_e^{str} = (C_{REE}/C_{Fe})^{cr}/(C_{REE}/C_{Fe})^{aq}$			
		Ce	Eu	Er	Yb
1	Magnetite	10^{-5}	2.1×10^{-5}	10^{-4}	1.6×10^{-4}
2	Magnetite	10^{-5}	2.2×10^{-5}	9.5×10^{-5}	2.5×10^{-4}
3	Hematite	4.3×10^{-5}	1.6×10^{-4}	7.5×10^{-4}	2.3×10^{-3}
4	Hematite	1.1×10^{-5}	2.2×10^{-4}	2.6×10^{-3}	7.7×10^{-3}

* See Table 4.

In the second case, native crystal faces were analyzed, which was possible because the obtained crystals were large enough (up to 2 mm). The analysis showed a strong enrichment of surface in REE in relation to the crystal core and, consequently, a significant increase in D_e^{sur} as compared to D_e^{str} (confer Tables 5 and 6). The measurements of depth of the hole following laser evaporation of the material by microscopic methods (light microscopy, scanning electron microscopy (SEM)) showed an average of ~20 microns, so the laser-removed material may be presumed to be significantly diluted with the bulk material. However, superficial enrichment even in this version is ~2 orders of magnitude, with D_e increasing by the same value (Table 6).

Unfortunately, of all the studied REE, we managed to fix XPS peaks, which are reliably superior to background, only for Ce. The survey spectra analysis showed that its content in the surface layer could reach several atomic percent. The narrow spectra indicate two forms of the element, III and IV, with a noticeable predominance of the first one (Figure 5).

Table 6. REE contents in near-surface regions (up to ~20 μm depth) of Mt and Hm crystals from LA-ICP-MS data, and surficial co-crystallization coefficients REE/Fe.

Exper. No. *	Mineral	C^{sur}, ppm				D_e^{sur}			
		Ce	Eu	Er	Yb	Ce	Eu	Er	Yb
1	Magnetite	24	26	55	44	5.3×10^{-4}	1.5×10^{-3}	2.4×10^{-2}	3.2×10^{-2}
2	Magnetite	33	64	93	74	1.3×10^{-3}	4.7×10^{-3}	1.1×10^{-2}	1.5×10^{-2}
3	Hematite	433	60	21	24	4.9×10^{-3}	3.5×10^{-3}	10^{-2}	2.1×10^{-2}
4	Hematite	412	64	43	55	4.6×10^{-3}	6.4×10^{-3}	4.9×10^{-2}	10^{-2}

* See Table 4.

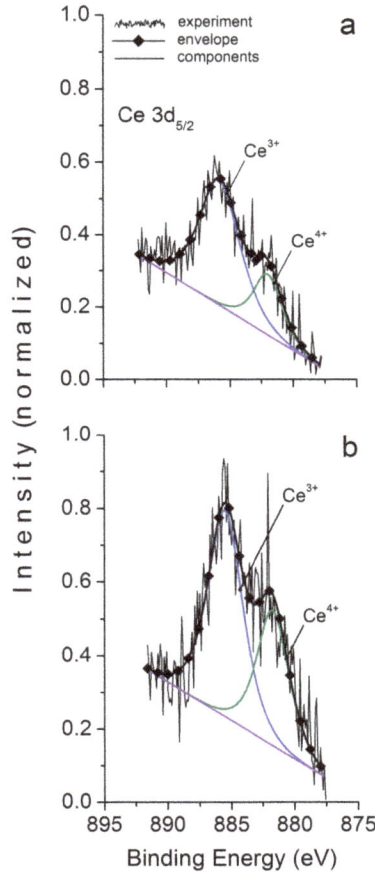

Figure 5. X-ray photoelectron spectra Ce $3d_{5/2}$ in magnetite. (**a**) initial sample, (**b**) after Ar^+ ion etching. See Table 7, sample 2.

Table 7. XPS data on chemical state of cerium in synthesized magnetite and hematite crystals.

Sample *	Photo-Electron Peak	Binding Energy (eV)	FWHM (eV) **	MPE **	MF **
2	$3d_{5/2}$	882.0	2.8	Ce^{4+}	28
	$3d_{5/2}$	885.8	3.6	Ce^{3+}	72
2, Ar^+	$3d_{5/2}$	881.7	3.0	Ce^{4+}	38
	$3d_{5/2}$	885.5	3.3	Ce^{3+}	62
3	$3d_{5/2}$	882.0	3.0	Ce^{4+}	30
	$3d_{3/2}$	900.6	3.0		
	$3d_{5/2}$	885.8	3.5	Ce^{3+}	70
	$3d_{3/2}$	904.4	3.5		
3, Ar^+	$3d_{5/2}$	881.2	3.0	Ce^{4+}	31
	$3d_{3/2}$	899.8	3.0		
	$3d_{5/2}$	885.4	4.1	Ce^{3+}	69
	$3d_{3/2}$	904.0	4.1		

* Correspond to experiment numbers in Table 4. Ar^+ ion etching during 10 min (~60 nm in depth). ** See Table 3.

In hematite, Ce content in the surface proved sufficient to articulate well both spin-orbital 3d doublet peaks, 5/2 and 3/2 (Figure 6).

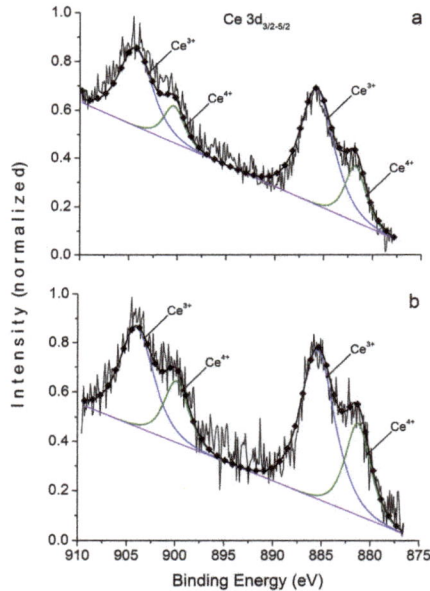

Figure 6. XPS doublets Ce 3d in hematite crystals. (**a**) initial sample, (**b**) after Ar^+ ion etching. See Table 7, sample 3.

It is important to note that etching with Ar^+ to ~60 nm has little effect on the ratio of Ce valence forms of magnetite and absolutely does not affect it in hematite (see Table 7). Therefore, the presence of two Ce forms is not an artifact; the effect of surface enrichment in REE is not caused by the formation of diffusion zone or normal adsorption, and Ce valence state is not as rigidly associated with fO_2 as is commonly believed, because the ratio of Ce^{3+}/Ce^{4+} is approximately the same for magnetite and hematite. The composition of Mt surface is consistent with the XPS data [24]. The ratio of Fe^{3+}/Fe^{2+} is approximately 1.1 (Table 8), in the O 1s spectrum there are identified O^{2-} and OH^- in the ratio 60:40 for magnetite and 80:20 for hematite. The data obtained show that

NAP compositions on Mt and Hm are similar to each other and correspond with oxyhydroxide composition [24]: $Fe^{3+}_{1+x}Fe^{2+}_{1-x}\left[O^{2-}_{1+y}(OH)^-_{1-y}\right]V^-_{2-y+x}$, where V denotes cation vacancies.

Table 8. Binding energies (eV) of Fe $2p_{3/2}$ multiple peaks and ratios of iron valence forms on the surface of magnetite and hematite crystals after 5-min Ar^+ etching.

Sample *	High-Spin Multiplet Fe³⁺, Peak Number **				High-Spin Multiplet Fe²⁺, Peak Number **			Fe³⁺/Fe²⁺
	1	2	3	4	1	2	3	
1	710.6	711.8	713.0	714.2	708.3	709.5	710.7	1.1
2	710.6	711.8	713.0	714.2	708.4	709.6	710.8	1.2
3	710.0	711.2	712.4	713.6	708.5	709.7	710.9	1.1

* Correspond to experiment numbers in Table 4. ** FWHM values (see Table 3) are taken 2.0 eV for Fe³⁺ peaks and 1.6 eV for Fe²⁺ peaks. Shake-up satellites are not shown.

5.3. Gold and PGE in Natural Pyrite and Ore-Forming Fluids

Figure 7 exemplifies the dependencies $\overline{C}_{PM} - \overline{S}_{sp}$ for gold and platinum group elements (PGE—Pt, Pd and Ru) obtained by ADSSC method for pyrite of the black-shale hosted gold ore deposits (namely, Degdekan deposit, see Section 4). In terms of quantity, the superficial mode exceeds by approximately one order the structural one in the content of all the studied elements, and therefore, determines mainly the total content of evenly distributed element. The data obtained (Figures 3 and 7) justify the geochemical role of NAP in TE distribution both in experimental and natural systems.

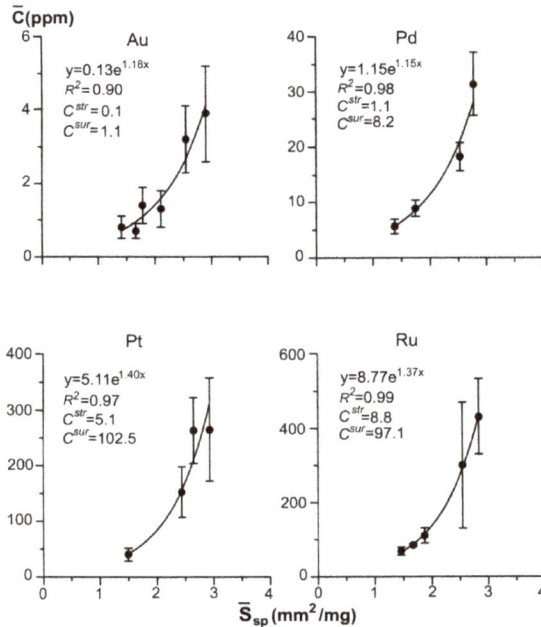

Figure 7. Dependences of the average concentrations of evenly distributed precious metals on the specific surface area of an average crystal in size fractions. Pyrite from the pyrite-arsenopyrite association of orogenic gold deposit Degdekan. Note high coefficients of determination of approximative exponents and predominance of superficially bound mode.

Using D_p^{str} data in Table 2, it is possible to assess the contents of Au and Pt in the fluid forming pyrite and pyrite-arsenopyrite assemblages (Table 9). For gold, we have accepted the D_p^{str} value of 0.1 averaging the data for pure and As-containing pyrite crystals.

Table 9. Estimates of the precious metal content in pyrite and ore-forming fluids using experimental distribution coefficients D_p^{str} (Py/aq).

Sample No.	Deposit	Number of Crystals (Starting–Final Sample)	Element	Element Concentration in Py (ppm)		C^{aq} (ppm)	$(Au/Pt)^{aq}$
				Structural Mode	Surficial Mode		
M-163/10	Degdekan	95–59	Au	0.21	0.53	2.1	21
		92–71	Pt	2.0	7.1	0.1	
DG-10/14 *	same	67–44	Au	0.13	1.1	1.3	5.4
		34–22	Pt	5.1	102	0.24	
Kr-9	Krasnoye	80–56	Au	0.04	0.36	0.4	4.0
		78–53	Pt	2.2	16	0.1	
Kr-39	same	86–55	Au	0.11	0.25	1.1	7.3
		72–61	Pt	3.2	37	0.15	
UV-3/13 *	Natalkinskoe	80–52	Au	0.29	1.37	2.9	1.6
		79–52	Pt	37	189	1.8	
ZR-10/13 *	Zolotaya Rechka	66–41	Au	3.4	6.3	34	56.7
		65–47	Pt	12.0	259	0.6	

* Pyrite from the pyrite-arsenopyrite association.

Such estimates are certainly approximate, as they do not take into account the details of the hydrothermal fluid chemical composition and differences in P, T conditions of the experiment and natural mineral formation. Au/Pt ratios reflect the situation more objectively, as for each sample they refer to the same parameters of the process of crystal formation, and fluid composition.

According to Table 9, Au content in fluids is within the range of 0.4–2.9 ppm, being substantially higher only for the Zolotaya Rechka (up to 34 ppm). However, elevated Ag content characterizing pyrite from this deposit might give rise to the additional to the structurally bound form intrinsic gold concentrator (Ag_2S nanoparticles, containing gold) [25]. Pt content generally varies from 0.1 up to 0.6 ppm and at the Natalkinskoe deposit it reaches 1.8 ppm, where the fluid is definitely enriched with both elements as compared to the other samples. Table 9 also demonstrates that the bulk concentration of the element cannot be used to assess its content in the fluid, as the surface component of the element content in most cases considerably exceeds the structural one, thus overestimating the result. However, it is important to note that if we use the data for the surface-associated mode and relevant distribution coefficients D_p^{sur} (which, as mentioned above, are largely responsible for total values); we will acquire the PM concentrations in fluid close, in certain cases, to those presented in Table 9. This presumes an important circumstance, possible "camouflage" of the true PM distribution by surficial modes of elements. This might prompt an illusion that separation of structural and surficial modes of the element fails to provide new knowledge about the fluid composition. Nevertheless, bearing in mind that NAP characteristics and properties depend on the growth conditions, one may assume a situation where the results of solving the inverse problem of PM content determination in the fluid will vary significantly for bulk and structural PM contents.

It is also important to note that in the course of the hydrothermal experiment, the same reaction vessel may not contain different crystal generations, which are distinguished by PM concentration, since these elements (Au, Pt, Pd) are stored in the batch throughout the run, i.e., their chemical potentials in hydrothermal solution are constant and determined by their solubility. As far as natural crystals are concerned, the issue of generations is not so obvious, but the determination coefficients of size dependencies at the level of 0.9–0.99 (Figure 7) exclude the possibility of crystal formation time being the only affecting factor, since various mineral generations contain crystals of different sizes as well. Such clear dependencies on the specific surface area will only take place if the NAP absorbing the element is homothetic to the crystal (that is, geometrically repeats its shape) and covers a significant portion of its surface. For the same reason, the effect of enrichment with microelements cannot be due to the deposition of a substance in the course of phase nucleation following heterogeneous mechanism; in this case, the content of the element would be proportional to the contact area of the crystal with fluid phase, rather than its specific surface area.

6. Discussion

The peculiarities discussed in the Introduction could be reconciled with the obtained experimental results, assuming that the crystal surface layer contains NAP participates in the growth process. Determining the nature of this participation, we proceed from the following considerations. In the course of crystallization, it is the surface layer of the crystal and not the crystal core that is in equilibrium with the solution. This provision is used in models of crystal zoning formation during fractional crystallization, when "crystallization of minerals in surface equilibrium with the surrounding liquid" is presumed ([50], p. 2143). The equilibrium is to be understood in the sense of equality of chemical potentials, taking into account the difference of chemical potentials $\Delta\mu$, which is necessary for growth that is as forced equilibrium under the action of extrinsic forcing factor [4]. This layer is structurally reconstructed and chemically modified into NAP, inseparable from the crystal matrix. An altered state of the layer is accounted for by the fact that the solid phase in equilibrium with the oversaturated solution is characterized by a higher chemical potential value than bulk solid phase in a saturated solution. The excess of μ can be realized both in changing the structure of the surface phase (for example, the degree of ordering or type of polymorphic modification), and in the change of its chemical composition (for example, absorption of admixture, which is incompatible with bulk structure). If several phases are jointly formed in the system, it is NAP rather than core sections that are in local phase equilibrium. The relative stability of such systems is due to the principle of continuity of phase formation on coexisting mineral surfaces, according to which NAP at different bulk phases are able to adapt to each other with the change of physicochemical conditions of growth [51]. Another important property of NAP is, as we saw above, the ability to absorb incompatible elements with distribution and co-crystallization coefficients greatly exceeding the equilibrium amounts for the bulk of crystals.

With NAP layer thickness increasing, there comes a time when adjacent to the matrix surface, sections lose their diffusive connection with oversaturated solution, and then part of the layer on the border is transformed into a matrix crystal structure following coherent or semi-coherent solid phase transformation. In the course of this process, structurally incompatible with the crystal matrix admixtures are isolated in the form of nano- and micro-inclusions prone to defects, pores and other imperfections of the transition zone. NAP has additional opportunities for accommodation of incompatible elements that is why the admixture partially remains in it, but its "excessive" part is withdrawn from the boundary layer on the surface, forming its own phases, including PM native forms (Figure 4). However, the admixture does not always reach the surface, so submicron and nano-size particles of the admixture form at the boundary. In any case, the distribution coefficient will be greatly overestimated for both the transition layer and NAP in total, as compared to crystal core. A similar effect is known in physical chemistry, it takes place during the separation of isotopes of simple ions of well-soluble salts, when there seems to be no obstacles to equalizing the composition by diffusion processes in the surface layer [52]. Nevertheless, it produces maximum effect on poorly soluble substances and incompatible admixtures, as shown above on the example of precious metals and rare earth elements in iron sulfides and oxides. It is important to realize that this phenomenon is thermodynamic and not kinetic in nature [52] and surface enrichment in the admixture should not be interpreted as diffusion zones, even if the admixture element forms a wide isomorphic mixture in the matrix crystal. This is a fundamentally important fact explaining the analogy of $\overline{C}_{PM} - \overline{S}_{sp}$ dependencies (Figures 3 and 7) in laboratory and natural systems. A traditional kinetic interpretation of admixture segregation and micro-zonality ("oscillatory zoning") of mineral crystals, based on

Peclet's growth number [53,54], unusable in case of incompatible elements such as PM, which may be partially reduced in the course of NAP evolution during aggregation of its nano-size fragments or decomposition caused by changing environment.

For pyrite, with its high nucleation energy, thermodynamically more preferable is growth through NAP represented, depending on the conditions (T, $\Delta\mu$, fS_2, fO_2), by the phase close to marcasite, pyrrhotite or oxysulfide containing trivalent iron and sulfoxy-anions [10,55].

With a pyrite crystal growth rate in our experiments of about 5×10^{-10} m/s, ~500 nm thick NAP layer is formed for about 10^3 s. During this time, in the surface layer equilibrium may be established with oversaturated solution, since the diffusion mobility in surficial areas of crystal is higher than in crystal core. For instance, for pyrite at 500 °C "bulk" diffusion coefficients of Fe, S and sulfur vacancy V_S are 3.7×10^{-19} [56], 2.1×10^{-23} [57] and 1.8×10^{-19} m^2/s [58], which correspond to the maximum diffusion lengths l of 27, 0.2 and 19 nm respectively. Calculations in density functional theory [58] provide for near-surface vacancies of sulfur the value of 1.5×10^{-16} m^2/s, which corresponds to $l \sim 550$ nm, i.e., comparable with a NAP layer thickness [10]. The diffusion coefficients in the surface layer may be assumed higher for both Fe and S as well, which will ensure a state of chemical equilibrium of surficial NAP layer with oversaturated solution. However, a much lower diffusion coefficient of sulfur is probably the reason for its deficit in the layer and hence "pyrrhotite" stoichiometry of NAP under moderate fS_2 values [10]. As regards magnetite, for stoichiometric crystal Fe diffusion coefficient extrapolated from high T to 500 °C is $\sim 2 \times 10^{-21}$ m^2/s, which corresponds to $l \sim 2$ nm. However, with the deviation from stoichiometry it increases by several orders of magnitude, both with the rise, and fall of fO_2 [59]. In defective oxyhydroxide NAP, it should be even higher to ensure that the ~330 nm thick surface layer establishes equilibrium with oversaturated solution [21].

The distribution of PM and REE is largely dependent on the composition of the NAP, which affects geochemical regularities. We saw above how partition and co-crystallization coefficients increase with superficially bound mode taken into account. On the other hand, the camouflage is possible, which creates an illusion of permanence of total distribution coefficients, since surface constituent mostly determines them. This is important in practical aspect, since higher D^{sur} as compared to D^{str} determine PM content in surface layers of crystals growing according to the mechanism involving NAP participation. This mechanism appears to be the same for both experimental and natural ore formation systems. In the latter case, we deal with "hidden" metal content, which does not show in distinct forms—presence of proper PM phases and increased contents of PM in bulk samples. The gold and PGE Clarks in the Earth's rocks are very low, and it is extremely difficult to explain their concentration in the deposits, unless we acknowledge that the mechanism of the process involves, firstly, the metal concentration within NAP, and, secondly, the subsequent NAP transformation with separation of nano-and micro-size PM forms. Thus, NAP are treated here as intermediate modes of concentration of rare and trace elements in geochemical systems.

The accumulation of PGE by NAP can explain the lack of their mineral (visible) forms together with the elevated content in black shale-hosted gold-ore deposits. This is practically important because gives motivation for PGE recovery accompanying gold ore processing.

The aggregation of virtually two-dimensional NAP, as well as their transformation to the bulk crystal structure, is accompanied by removal of the admixtures, which are incompatible with the matrix and preclude this transformation. They largely remain in intermediate meso-crystalline phases of submicron and micron-size [25]. We believe that the effect of surface enrichment in incompatible elements is largely accounted for by these phases. On the other hand, this means that microinclusions in minerals and, especially, on their surface may not reflect adequately the composition of mother liquor due to higher TE partition and co-crystallization coefficients for NAP, as described above.

The NAP sensitivity to the conditions of crystal formation is, on the one hand, an instrumental feature in determining these conditions. For instance, the pyrrhotite-like NAP almost does not occur on the surface of pyrite at high fS_2 [10]. The oxyhydroxide NAP is also weakly developed at high Mn content at the surface of magnetite crystals. On the other hand, the observed similarities in surface composition and element valence states on magnetite and hematite reduce indicating capacities of this mineral pair in fO_2 assessment, for example, by chemical state of Ce (Table 7). This can be an important consideration when interpreting Ce and Eu anomalies in determining the redox state and hydrothermal fluxes in the ancient oceans [60].

7. Conclusions

We suggest that the crystal growth mechanism specified by the agency of superficial nonautonomous phase operates both in experimental and natural hydrothermal systems because it has a thermodynamic rather than a kinetic nature. The important, although not exclusive, consequences are the dualism of partition and co-crystallization coefficients in the system "mineral–fluid" and so-called "hidden" metal content in hydrothermally grown ore minerals. The surficial D values are much greater than the structural ones for both the PM (Au, Pt and Pd) in pyrite, As-pyrite, hematite, magnetite, Mn-magnetite and REE (Ce, Eu, Er, Yb) in magnetite and hematite. Unexpectedly, except for Au in pyrite, the precious metals are found to be compatible in these minerals ($D^{cr/aq} > 1$) under hydrothermal conditions (450 and 500 °C, 1 kbar, NH_4Cl-based solutions) even though only structural constituents are accounted for. The engagement of superficially bound modes makes these elements highly compatible although unsuitable to solving an inverse problem of metal content estimation in ore-forming fluids. This problem can be solved using only structurally bound constituents of element content. However, the comparison of experimental and natural data for pyrite from hydrothermal gold-ore deposits shows that the superficially bound mode can mask the structurally bound one producing the illusion that metal contents estimated from structural and superficial modes of trace element are close to one another. The evolution of NAP containing PM in chemically bound forms is accompanied by the exsolution of nano- and micro-sized phases including native PM forms. Therefore, NAPs can be considered as intermediate members in the chain of the PM concentration process from disseminated state in rocks to nuggets in high-grade deposits. The sensitivity of NAP to growth conditions allows principally predicting crystal surface composition and evaluation of trace element contents in different-size fractions. This is practically important for the analyses of ore samples and recovery processing of ore mineral resources. It should be taken into account that the behavior of indicative trace elements of variable valence like REE are dependent not only on physicochemical varieties (T, P, fO_2) but on the presence and composition of NAP as well.

Author Contributions: V.L.T. formulated the problem, organized the research team, and guided the study and interpretation of results. S.V.L. processed and attributed the XPS spectra, organized REE ICP-MS analysis and data handling. N.V.S. participated in experimental program and processed the PM analytical data; R.G.K. provided natural samples and contributed to interpretations. All authors participated in writing the manuscript.

Funding: The research was performed within the frame of the state order project IX.124.3, No. 0350-2016-0025 and partly was funded by the Russian Foundation for Basic Research grants Nos. 18-05-00077, 16-05-00104 and 17-05-00095.

Acknowledgments: We thank Yury L. Mikhlin and Yury V. Shchegol'kov for operational assistance with XPS analyses. Chemical analyses were undertaken at the Common Use Center (CUC) of Isotope and Geochemical Studies in Vinogradov Institute of Geochemistry with assistance of I.Yu.Voronova. ICP-MS analyses were done at CUC of Ultramicroanalysis in Limnological Institute SB RAS with assistance of A.P.Chebykin. We thank the authorities of both centers for providing the possibility of this study. The contributions of D.N.Babkin and T.M.Pastushkova to experimental part of this study are greatly appreciated. The authors would like to thank Galina A. Palyanova and two anonymous reviewers for valuable comments.

Conflicts of Interest: The authors declare no conflicts of interest.

References

1. Navrotsky, A. Thermodynamics of element partitioning: (1) Systematics of transition metals in crystalline and molten silicates and (2) Defect chemistry and the Henry's law problem. *Geochim. Cosmochim. Acta* **1978**, *42*, 887–902. [CrossRef]

2. Abramovich, M.G.; Shmakin, B.M.; Tauson, V.L.; Akimov, V.V. Mineral typochemistry: Anomalous trace-element concentrations in solid solutions with defect structures. *Int. Geol. Rev.* **1990**, *32*, 608–615. [CrossRef]

3. Urusov, V.S.; Dudnikova, V.B. The trace-component trapping effect: Experimental evidence, theoretical interpretation, and geochemical applications. *Geochim. Cosmochim. Acta* **1998**, *62*, 1233–1240. [CrossRef]

4. Tauson, V.L.; Akimov, V.V. Introduction to the theory of forced equilibria: General principles, basic concepts, and definitions. *Geochim. Cosmochim. Acta* **1997**, *61*, 4935–4943. [CrossRef]

5. Jones, J.H. Thoughts and reminiscences on experimental trace element partitioning. *Geochem. Perspect.* **2016**, *5*, 147–251. [CrossRef]

6. Hiraga, T.; Anderson, I.M.; Kohlstedt, D.L. Chemistry of grain boundaries in mantle rocks. *Am. Mineral.* **2003**, *88*, 1015–1019. [CrossRef]

7. Toulhoat, H.; Digne, M.; Arrouvel, C.; Raybaud, P. DFT studies of fluid-mineral interactions at the molecular level: Examples and perspectives. *Oil Gas Sci. Technol. Rev. IFP* **2005**, *60*, 417–433. [CrossRef]

8. Tauson, V.L.; Babkin, D.N.; Parkhomenko, I.Y.; Men'shikov, V.I.; Lipko, S.V.; Pastushkova, T.M. Distribution of heavy-metal (Hg, Cd, and Pb) chemical species between pyrite and hydrothermal solution. *Geochem. Int.* **2010**, *48*, 611–616. [CrossRef]

9. Tauson, V.L.; Parkhomenko, I.Y.; Babkin, D.N.; Men'shikov, V.I.; Lustenberg, E.E. Cadmium and mercury uptake by galena crystals under hydrothermal growth: A spectroscopic and element thermo-release atomic absorption study. *Eur. J. Mineral.* **2005**, *17*, 599–610. [CrossRef]

10. Tauson, V.L.; Babkin, D.N.; Lustenberg, E.E.; Lipko, S.V.; Parkhomenko, I.Y. Surface typochemistry of hydrothermal pyrite: Electron spectroscopic and scanning probe microscopic data. I. Synthetic pyrite. *Geochem. Int.* **2008**, *46*, 565–577. [CrossRef]

11. Tauson, V.L.; Kravtsova, R.G.; Grebenshchikova, V.I.; Lustenberg, E.E.; Lipko, S.V. Surface typochemistry of hydrothermal pyrite: Electron spectroscopic and scanning probe microscopic data. II. Natural pyrite. *Geochem. Int.* **2009**, *47*, 231–243. [CrossRef]

12. Tauson, V.L.; Kravtsova, R.G.; Smagunov, N.V.; Spiridonov, A.M.; Grebenshchikova, V.I.; Budyak, A.E. Structurally and superficially bound gold in pyrite from deposits of different genetic types. *Russ. Geol. Geophys.* **2014**, *55*, 273–289. [CrossRef]

13. Bancroft, G.M.; Jean, G. Gold deposition at low temperature on sulfide minerals. *Nature* **1982**, *298*, 730–731. [CrossRef]

14. Hyland, M.M.; Bancroft, G.M. An XPS study of gold deposition at low temperatures on sulphide minerals: Reducing agents. *Geochim. Cosmochim. Acta* **1989**, *53*, 367–372. [CrossRef]

15. Scaini, M.J.; Bancroft, G.M.; Knipe, S.W. An XPS, AES, and SEM study of the interactions of gold and silver chloride species with PbS and FeS_2: Comparison to natural samples. *Geochim. Cosmochim. Acta* **1997**, *61*, 1223–1231. [CrossRef]

16. Tauson, V.L. Isomorphism and endocrypty: New approaches to study the behavior of trace elements in mineral systems. *Russ. Geol. Geophys.* **1999**, *40*, 1468–1474.

17. Guevremont, J.M.; Strongin, D.R.; Schoonen, M.A.A. Thermal chemistry of H_2S and H_2O on the (100) plane of pyrite: Unique reactivity of defect sites. *Am. Mineral.* **1998**, *83*, 1246–1255. [CrossRef]

18. Tauson, V.L.; Ovchinnikova, O.V.; Bessarabova, O.I.; Smagunov, N.V.; Pastushkova, T.M. Distribution of gold deposited under reducing adsorption from $HAuCl_4$ solution on magnetite, sphalerite, and galena crystals. *Russ. Geol. Geophys.* **2000**, *41*, 1427–1430.

19. Koneev, R.I.; Knalmatov, R.A.; Mun, Y.S. Nanomineralogy and nanogeochemistry of ores of Uzbekistan gold deposits. *Zapiski Russ. Mineral. Obshchestva* **2010**, *139*, 1–14. (In Russian) [CrossRef]

20. Rusanov, A.I. *Phase Equilibria and Surface Phenomena*; Khimia: Leningrad, Russia, 1967. (In Russian)

21. Tauson, V.L.; Babkin, D.N.; Pastushkova, T.M.; Krasnoshchekova, T.S.; Lustenberg, E.E.; Belozerova, O.Y. Dualistic distribution coefficients of elements in the system Mineral-Hydrothermal solution. I. Gold accumulation in pyrite. *Geochem. Int.* **2011**, *49*, 568–577. [CrossRef]

22. Tauson, V.L.; Lipko, S.V.; Smagunov, N.V.; Arsent'ev, K.Y.; Loginov, B.A. Influence of surface nanophases on the processes of crystal formation in multiphase mineral systems. *Dokl. Earth Sci.* **2014**, *455*, 317–322. [CrossRef]

23. Tauson, V.L.; Babkin, D.N.; Pastushkova, T.M.; Akimov, V.V.; Krasnoshchekova, T.S.; Lipko, S.V.; Belozerova, O.Y. Dualistic distribution coefficients of elements in the system Mineral-Hydrothermal solution. II. Gold in magnetite. *Geochem. Int.* **2012**, *50*, 227–245. [CrossRef]

24. Tauson, V.L.; Babkin, D.N.; Pastushkova, T.M.; Smagunov, N.V.; Lipko, S.V.; Voronova, I.Y.; Men'shikov, V.I.; Bryanskii, N.V.; Arsent'ev, K.Y. Dualistic distribution coefficients of trace elements in the system Mineral-Hydrothermal solution. III. Precious metals (Au and Pd) in magnetite and manganmagnetite. *Geochem. Int.* **2016**, *54*, 149–166. [CrossRef]

25. Tauson, V.L.; Lipko, S.V.; Arsent'ev, K.Y.; Mikhlin, Y.L.; Babkin, D.N.; Smagunov, N.V.; Pastushkova, T.M.; Voronova, I.Y.; Belozerova, O.Y. Dualistic distribution coefficients of trace elements in the system Mineral-Hydrothermal solution. IV. Platinum and silver in pyrite. *Geochem. Int.* **2017**, *55*, 753–774. [CrossRef]

26. Tauson, V.L. Gold solubility in the common gold-bearing minerals: Experimental evaluation and application to pyrite. *Eur. J. Mineral.* **1999**, *11*, 937–947. [CrossRef]

27. Goldfarb, R.J.; Groves, D.I. Orogenic gold: Common or evolving fluid and metal sources through time. *Lithos* **2015**, *233*, 2–26. [CrossRef]

28. Gaboury, D. Does gold in orogenic deposits come from pyrite in deeply buried carbon-rich sediments? Insight from volatiles in fluid inclusions. *Geology* **2013**, *41*, 1207–1210. [CrossRef]

29. Tauson, V.L.; Akimov, V.V.; Lipko, S.V.; Spiridonov, A.M.; Budyak, A.E.; Belozerova, O.Y.; Smagunov, N.V. Typomorphism of pyrite of the Sukhoi Log deposit (East Siberia). *Russ. Geol. Geophys.* **2015**, *56*, 1394–1413. [CrossRef]

30. Arifulov, C.H.; Arsentieva, I.V.; Shchegolkov, Y.V. About the long-prism pyrite in gold-bearing black shale deposits of the Kirovsko-Kvarkenskiy ore district, East Orenburg region. *Rudy I Met.* **2011**, *5*, 78–84. (In Russian)

31. Bloomstein, E.L.; Kydd, R.A.; Levinson, A.A. Development of ammonium geochemistry as a new technique in precious and base metals exploration. *J. Geochem. Explor.* **1987**, *29*, 386. [CrossRef]

32. Men'shikov, V.I.; Voronova, I.Y.; Proidakova, O.A.; Malysheva, S.F.; Ivanova, N.I.; Belogorlova, N.A.; Gusarova, N.K.; Trofimov, B.A. Preconcentration of gold, silver, palladium, platinum, and ruthenium with organophosphorus extractants. *Russ. J. Appl. Chem.* **2009**, *82*, 183–189. [CrossRef]

33. Tauson, V.L.; Lipko, S.V. Pyrite as a concentrator of gold in laboratory and natural systems: A surface-related effect. In *Pyrite: Synthesis, Characterization and Uses*; Whitley, N., Vinsen, P.T., Eds.; Nova Science Publishers: Hauppauge, NY, USA, 2013; Chapter 1; pp. 1–40.

34. Palenik, C.S.; Utsunomiya, S.; Reich, M.; Kesler, S.E.; Wang, L.; Ewing, R.C. Invisible gold revealed: Direct imaging of gold nanoparticles in a Carlin-type deposit. *Am. Mineral.* **2004**, *89*, 1359–1366. [CrossRef]

35. Tauson, V.L.; Lustenberg, E.E. Quantitative determination of modes of gold occurrence in minerals by the statistical analysis of analytical data samplings. *Geochem. Int.* **2008**, *46*, 423–428. [CrossRef]

36. Tauson, V.L.; Babkin, D.N.; Akimov, V.V.; Lipko, S.V.; Smagunov, N.V.; Parkhomenko, I.Y. Trace elements as indicators of the physicochemical conditions of mineral formation in hydrothermal sulfide systems. *Russ. Geol. Geophys.* **2013**, *54*, 526–543. [CrossRef]

37. Bolanz, R.M.; Kiefer, S.; Göttlicher, J.; Steininger, R. Hematite (α-Fe$_2$O$_3$)—A potential Ce^{4+} carrier in red mud. *Sci. Total Envirn.* **2018**, *622–623*, 849–860. [CrossRef] [PubMed]

38. Naumkin, A.V.; Kraut-Vass, A.; Gaarenstroom, S.W.; Powell, C.J. NIST X-ray Photoelectron Spectroscopy Database. 2012. Available online: https://srdata.nist.gov/xps/EnergyTypeValSrch.aspx (accessed on 25 June 2018).

39. Moulder, J.E.; Sticle, W.F.; Sobol, P.E.; Bomben, K.D. *Handbook of X-ray Photoelectron Spectroscopy*; Perkin-Elmer Corp.: Eden Prairie, MN, USA, 1992.

40. Grosvenor, A.P.; Kobe, B.A.; Biesinger, M.C.; McIntyre, N.S. Investigation of multiplet splitting of Fe 2p XPS spectra and bonding in iron compounds. *Surf. Interface Anal.* **2004**, *36*, 1564–1574. [CrossRef]

41. Groves, D.I.; Goldfarb, R.J.; Gebre-Mariam, M.; Hagemann, S.G.; Robert, F. Orogenic gold deposits: A proposed classification in the context of their crustul distributoin and relationship to other gold deposit types. *Ore Geol. Rev.* **1998**, *13*, 7–27. [CrossRef]

42. Goncharov, V.I.; Voroshin, S.V.; Sidorov, V.A. *Natalkinskoe Gold-Ore Deposit*; North-East Complex Research Institute, Far-East Branch of RAS: Magadan, Russia, 2002. (In Russian)

43. Goryachev, N.A.; Vikent'eva, O.V.; Bortnikov, N.S.; Prokof'ev, V.Y.; Alpatov, V.A.; Golub, V.V. The world-class Natalka gold deposit, Northeast Russia: REE patterns, fluid inclusions, stable oxygen isotopes, and formation conditions of ore. *Geol. Ore Depos.* **2008**, *50*, 362–390. [CrossRef]

44. Mikhailov, B.K.; Struzhkov, S.F.; Natalenko, M.V.; Cimbalyuk, N.V. Multifactor model of large-volume gold-ore deposit Degdekan (Magadan region). *Otechestvennaya Geol.* **2010**, *2*, 20–31.

45. Khanchuk, A.I.; Plyusnina, L.P.; Nikitenko, E.M.; Kuzmina, T.V.; Barinov, N.N. The noble metal distribution in the black shales of the Degdekan gold deposit in Northeast Russia. *Russ. J. Pacif. Geol.* **2011**, *5*, 89–96. [CrossRef]

46. Palenova, E.E.; Belogub, E.V.; Novoselov, K.A.; Zabotina, M.V. Mineralogical and geochemical characteristics of gold-ore hosting carbonaceous series of Artemovskyi ore cluster (Bodaibo region). *Izv. SO RAEN. Geol. Prosp. Explor. Ore Depos.* **2013**, *43*, 29–36. (In Russian)

47. Palenova, E.E.; Kotliarov, V.A.; Belogub, E.V.; Yuminov, A.M.; Zateeva, P.A. Formation conditions of gold-ore objects of Artemovskyi ore cluster (Bodaibo region). *Vopr. Estestvoznaniya* **2015**, *3*, 53–56. (In Russian)

48. Smythe, D.J.; Brenan, J.M. Magmatic oxygen fugacity estimated using zircon-melt partitioning of cerium. *Earth Planet. Sci. Lett.* **2016**, *453*, 260–266. [CrossRef]

49. Smythe, D.J.; Brenan, J.M. Cerium oxidation state in silicate melts: Combined fO_2, temperature and compositional effect. *Geochim. Cosmochim. Acta* **2015**, *170*, 173–187. [CrossRef]

50. Nishimura, K. A trace-element geochemistry model for imperfect fractional crystallization associated with the development of crystal zoning. *Geochim. Cosmochim. Acta* **2009**, *73*, 2142–2149. [CrossRef]

51. Tauson, V.L. The principle of continuity of phase formation at mineral surfaces. *Dokl. Earth Sci.* **2009**, *425*, 471–475. [CrossRef]

52. Bochkarev, A.V.; Trefilova, A.N.; Bobrov, M.F.; Tsurkov, N.A. Isotope separation during ionic crystal growth from solution. *Russ. J. Phys. Chem.* **2003**, *77*, 1868–1872.

53. Burton, J.A.; Prim, R.C.; Slichter, W.P. The distribution of solute in crystals grown from the melt. Part I. Theoretical. *J. Chem. Phys.* **1953**, *21*, 1987–1991. [CrossRef]

54. Watson, E.B. Surface enrichment and trace-element uptake during crystal growth. *Geochim. Cosmochim. Acta* **1996**, *60*, 5013–5020. [CrossRef]

55. Graham, U.M.; Ohmoto, H. Experimental study of formation mechanisms of hydrothermal pyrite. *Geochim. Cosmochim. Acta* **1994**, *58*, 2187–2202. [CrossRef]

56. Chen, J.H.; Harvey, W.W. Cation self-diffusion in chalcopyrite and pyrite. *Met. Trans. B* **1975**, *6*, 331–339. [CrossRef]

57. Watson, E.B.; Cherniak, D.J.; Frank, E.A. Retention of biosignatures and mass-independent fractionations in pyrite: Self-diffusion of sulfur. *Geochim. Cosmochim. Acta* **2009**, *73*, 4792–4802. [CrossRef]

58. Zhang, Y.N.; Law, M.; Wu, R.Q. Atomic modeling of sulfur vacancy diffusion near iron pyrite surfaces. *J. Phys. Chem. C* **2015**, *119*, 24859–24864. [CrossRef]

59. Hallström, S.; Höglund, L.; Ågren, J. Modeling of iron diffusion in the iron oxides magnetite and hematite with variable stoichiometry. *Acta Mater.* **2011**, *59*, 53–60. [CrossRef]

60. Alibert, C. Rare earth elements in Hamersley BIF minerals. *Geochim. Cosmochim. Acta* **2016**, *184*, 311–328. [CrossRef]

Article

Sulfide Formation as a Result of Sulfate Subduction into Silicate Mantle (Experimental Modeling under High P,T-Parameters)

Yuliya Bataleva [1,2,*], **Yuri Palyanov** [1,2] and **Yuri Borzdov** [1,2]

[1] Sobolev Institute of Geology and Mineralogy, Siberian Branch of Russian Academy of Sciences, Koptyug ave 3, 630090 Novosibirsk, Russia; palyanov@igm.nsc.ru (Y.P.); borzdov60@mail.ru (Y.B.)

[2] Department of Geology and Geophysics, Novosibirsk State University, Pirogova str 2, 630090 Novosibirsk, Russia

* Correspondence: bataleva@igm.nsc.ru; Tel.: +7-3832-330-80-43

Received: 25 June 2018; Accepted: 23 August 2018; Published: 29 August 2018

Abstract: Ca,Mg-sulfates are subduction-related sources of oxidized S-rich fluid under lithospheric mantle P,T-parameters. Experimental study, aimed at the modeling of scenarios of S-rich fluid generation as a result of desulfation and subsequent sulfide formation, was performed using a multi-anvil high-pressure apparatus. Experiments were carried out in the Fe,Ni-olivine–anhydrite–C and Fe,Ni-olivine–Mg-sulfate–C systems (P = 6.3 GPa, T of 1050 and 1450 °C, t = 23–60 h). At 1050 °C, the interaction in the olivine–anhydrite–C system leads to the formation of olivine + diopside + pyrrhotite assemblage and at 1450 °C leads to the generation of immiscible silicate-oxide and sulfide melts. Desulfation of this system results in the formation of S-rich reduced fluid via the reaction olivine + anhydrite + C → diopside + S + CO_2. This fluid is found to be a medium for the recrystallization of olivine, extraction of Fe and Ni, and subsequent crystallization of Fe,Ni-sulfides (i.e., olivine sulfidation). At 1450 °C in the Ca-free system, the generation of carbonate-silicate and Fe,Ni-sulfide melts occurs. Formation of the carbonate component of the melt occurs via the reaction Mg-sulfate + C → magnesite + S. It is experimentally shown that the olivine-sulfate interaction can result in mantle sulfide formation and generation of potential mantle metasomatic agents—S- and CO_2-dominated fluids, silicate-oxide melt, or carbonate-silicate melt.

Keywords: sulfate; S-rich fluid; sulfide; sulfidation; desulfation; high-pressure experiment; subduction; mantle metasomatism; lithospheric mantle

1. Introduction

Sulfur, hydrogen, oxygen, and carbon are the major magmatic volatiles in the Earth's mantle and crust. Being a redox-sensitive element, sulfur can exist in reduced form (S^{2-} or S) as sulfides or native sulfur, sulfide melts or reduced fluid, intermediate form (S^{4+}) as an SO_2-component of erupted melts and in oxidized form (S^{6+}) as sulfates, oxidized fluids, or dissolved in silicate melts [1–5]. According to modern concepts, subduction processes play a key role in the transport of S in deep zones of the Earth and are intimately linked to the global geochemical cycle of sulfur, the genesis of arc-related sulfide ore deposits, and the long-term mantle redox evolution [3,6–9]. Existing data demonstrate that only 15–30% of sulfur, transported into the mantle with a subducted slab is released to the atmosphere via magmatic degassing at arcs [7]. Thus, most of the sulfur, which is retained in the down-going slab, either interacts with upper mantle rocks or is recycled to the deep mantle. The heterogeneity of the Earth's mantle as well as redox- and compositional contrast of slab and lithospheric mantle rocks results in numerous scenarios of sulfur behavior in the course of the subduction processes.

Calcium and magnesium sulfates ($CaSO_4$ and $MgSO_4$), being the most abundant sulfur-bearing minerals in the Earth's crust, are expected to play a principal role in recycling of oxidized form of sulfur into the deep mantle under subduction settings. There are two lines of direct evidence, demonstrating the possibility of sulfate subduction to the upper mantle depths, particularly (1) findings of anhydrite ($CaSO_4$) as inclusions in diamonds, in assemblages with coesite [10] and $CaCO_3$ [10,11], and (2) discovery of similarity of the $\delta^{34}S$ values of pyrite in eclogite with those of seawater-derived sulfates [12]. Experimental studies of anhydrite structural transitions under high pressure demonstrate that this mineral is stable under high-pressure, high-temperature (HPHT) conditions and does not decompose or melt up to 85 GPa [13–17].

It is known that sulfate stability under mantle conditions can be affected by the sulfur and oxygen fugacities and compositions of the host rock, as well as interactions with the interstitial fluid or melt. The recent experimental studies at variable oxygen fugacity [8,18] showed that anhydrite in the dehydrated slab below the region of formation of arc magmas may efficiently be recycled into the deep mantle. Thermodynamic calculations [3,9] suggest that anhydrite under high pressures should dominantly dissolve into fluids released across the transition from blueschist to eclogite facies, and enrich those fluids with SO_3, SO_2, S, HSO_4^-, or H_2S. The main process that results in the decomposition of sulfate with S-rich fluid formation under crustal conditions is desulfation which can occur at the interaction of anhydrite or other sulfates with carbon-bearing brines, resulting in the formation of carbonates and elemental sulfur. A recent investigation of S-bearing species in kimberlites demonstrates that the infiltration of brines into kimberlite can dissolve magmatic sulfates and initiate serpentinization accompanied by sulfidation of olivine [5]. Possible participation of sulfur (sulfide or S form) in the processes of diamond genesis was proposed in [19]. Diamond synthesis in the sulfur-carbon system was demonstrated experimentally [20–22].

Despite the growing interest in this area of research and a number of theoretical and experimental studies, the fate of subducted S-bearing species under high pressures and high temperatures is still poorly known. The most probable reactions involving S-bearing slab minerals, down-going into the silicate mantle, are the interactions of sulfate with silicates as well as carbon-bearing brines. Pioneering experimental work [23] demonstrated that sulfur readily reacts with many of the rock-forming silicates by sulfidation reactions. Recently, we have experimentally investigated these reactions involving olivine and subduction-related reduced S-species, such as S-rich fluid/melt, pyrite, and sulfide melt under high pressures and high temperatures [24]. Here, we report the results of the experimental modeling of interactions of natural Mg,Fe,Ni-olivine with anhydrite and Mg-sulfate (subduction-related oxidized S-species), which are sources of oxidized S-rich fluid under lithospheric mantle P,T-parameters. The main goal of the present study is to propose scenarios of desulfation, oxidized S-rich fluid generation, olivine-S-rich fluid interaction, and evaluate the possibility of sulfide formation via this interaction.

2. Materials and Methods

Experimental studies of the olivine-sulfate interactions were performed in the olivine–anhydrite–C ($Mg_{1.84}Fe_{0.14}Ni_{0.02}SiO_4$–$CaSO_4$–C) and olivine–Mg-sulfate–C ($Mg_{1.84}Fe_{0.14}Ni_{0.02}SiO_4$–$MgSO_4$–C) systems at 6.3 GPa, and temperatures of 1050 and 1450 °C, for 23 to 60 h using a multi-anvil high-pressure apparatus of a "split-sphere" type. Details on the pressure and temperature calibration, as well as high-pressure cell schemes, were published previously [25,26]. The starting materials were natural specimens of Fe,Ni-bearing forsterite (spinel lherzolite xenolith from kimberlite of the Udachnaya pipe, Yakutia, Russia), and powders of chemically pure anhydrite and Mg-sulfate (99.99%). Sulfates were preliminary heated at 250 °C for 4 h, to prevent contamination of the experimental charge with water, absorbed from the air. Compositions and proportions of starting materials are shown in Table 1. For every temperature, two or three different concentrations of sulfur (xS) were used. The procedure for the ampoule assembly, in which starting reagents are finely crushed (grain size ~20 μm) and homogenized, was used. The reaction charges were placed in graphite capsules (diameter of 5 mm, height of 2 mm), which are suitable for HPHT-experiments in sulfur- and iron-bearing systems [27–29]. These graphite capsules provide carbon-saturated conditions, which initiate the decomposition of sulfate ("desulfation") and S-rich fluid generation. Moreover, graphite acts as an oxygen buffer, providing fO_2 values below CCO (C–CO) buffer equilibrium during the experiments. To examine the effect of carbon-saturated conditions on anhydrite and Mg-sulfate stability under mantle P,T-parameters, two specific experiments were performed (CaSO$_4$–C and MgSO$_4$–C systems, P = 6.3 GPa, T = 1450 °C, t = 2 h); it was established, that the sulfate-carbon interaction resulted in the formation of carbonate (solid or melt) and elemental sulfur (melt), see Table 2.

Table 1. Composition of initial olivine, sulfates, and bulk compositions of the systems (wt %).

Phase or System	xS, mol %	Mass Concentrations, wt %						
		SiO$_2$	FeO	MgO	CaO	NiO	SO$_3$	Total
Olivine	-	41.05	9.25	49.25	-	0.45	-	100.0
Anhydrite	-	-	-	-	46.7	-	53.3	100.0
Magnesium sulfate	-	-	-	38.5	-	-	61.5	100.0
Olivine-anhydrite-C system	2	40.5	9.1	48.1	0.9	0.4	1.0	100.0
	5	39.4	8.8	46.7	2.2	0.4	2.5	100.0
Olivine-magnesium sulfate-C system	5	39.6	8.9	48.6	-	0.4	2.5	100.0
	10	37.9	8.5	48.2	-	0.4	5.0	100.0

Table 2. Experimental conditions and results.

Run N	T, °C	t, h	System	xS, mol %	Final Phases
933-2	1050	60	Ol–CaSO$_4$–C	2	Ol$_R$, Cpx, Po
933-5	1050	60	Ol–CaSO$_4$–C	5	Ol$_R$, Cpx, Po
934-2	1450	23	Ol–CaSO$_4$–C	2	Ol$_R$, L$_{sulf}$, L$_{sil-ox}$
934-5	1450	23	Ol–CaSO$_4$–C	5	Ol$_R$, L$_{sulf}$, L$_{sil-ox}$
1809-5	1450	40	Ol–MgSO$_4$–C	5	Ol$_R$, L$_{sulf}$, L$_{carb-sil}$
1809-10	1450	40	Ol–MgSO$_4$–C	10	Ol$_R$, L$_{sulf}$, L$_{carb-sil}$
571-M	1450	2	MgSO$_4$–C	-	Ms, L$_S$
571-C	1450	2	CaSO$_4$–C	-	L$_{carb}$, L$_S$

Ol—olivine, Ol$_R$—recrystallized olivine, Cpx—clinopyroxene, Po—pyrrhotite, L$_{sulf}$—sulfide melt, L$_{carb-sil}$—carbonate-silicate melt; L$_{carb}$—carbonate melt, L$_S$—sulfur melt, Ms—magnesite, xS—molar concentration of sulfur in the system.

The lower temperature of 1050 °C at 6.3 GPa was chosen according to P-T estimates of the formation of anhydrite-bearing diamond [10,11,30], see Figure 1. The higher temperature of 1450 °C at 6.3 GPa corresponds to the Ni-rich pyrrhotite melting under carbon-saturated conditions, see Figure 1. Following the experiments, phase identification of samples, analysis of phase relations, and measurements of energy dispersive spectra (EDS) of various phases were performed using a TESCAN MIRA 3 LMU scanning electron microscope (TESCAN, Brno, Czech Republic).

The chemical composition of mineral phases was analyzed using a Cameca Camebax-Micro microprobe (CAMECA, Gennevilliers, France). For the microprobe and EDS analysis, samples in the form of polished sections were prepared. An accelerating voltage of 20 kV and a probe current of 20 nA were used for all the analyses of crystalline phases. For the quenched melt, the accelerating voltage was lowered to 15 kV and the probe current to 10 nA. For the bulk of the analyses, a focused beam (d = 2–4 μm) for silicate and sulfide crystalline phases and a beam diameter of 20–100 μm for the quenched melt were used. Peak counting time was 60 s. Standards, used for the analyses of quenched sulfide melt were natural pyrite, pyrrhotite, and pentlandite. Analytical studies were performed at the Center for Collective Use of Multi-element and Isotopic Analysis of the Siberian Branch of the Russian Academy of Sciences.

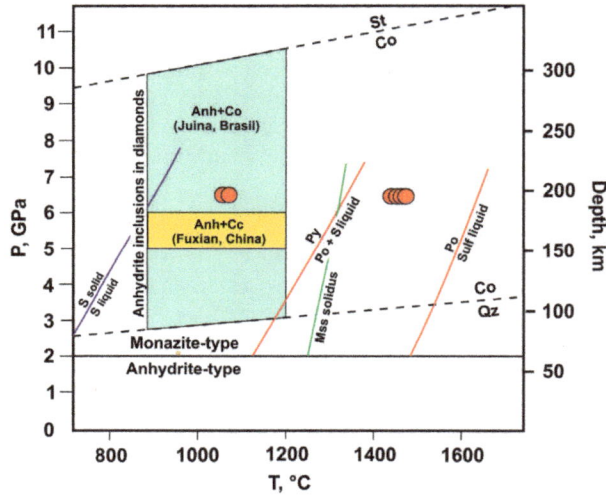

Figure 1. P-T diagram with experimental parameters (circles), line of structural transition of anhydrite [17], estimated field of the formation of diamonds with anhydrite inclusions [10,11,30], and experimentally determined melting curves of sulfur, pyrrhotite, and Mss and decomposition of pyrite [31–33]. **Sulf**—sulfide, **Po**—pyrrhotite, **Py**—pyrite, **Mss**—monosulfide solid solution ($Fe_{0.69}Ni_{0.23}Cu_{0.01}S_{1.00}$), **Anh**—anhydrite, **Co**—coesite, **St**—stishovite, **Qz**—quartz, **Cc**—calcite.

3. Results

3.1. Interaction in the Olivine–Anhydrite–C System

Experimental parameters and results are shown in Table 2. At a temperature of 1050 °C, 60 h, and xS = 2–5 mol %, a polycrystalline olivine aggregate with newly formed clinopyroxene and pyrrhotite in the interstices is formed, see Figure 2a,b and Figure 3a,b. The amount of these newly formed phases increases as the initial sulfur concentration in the system increases, as shown in Figure 2. Clinopyroxene and pyrrhotite occur as inclusions in olivine, see Figure 3b. The initial sulfate is absent in the samples after the experiments.

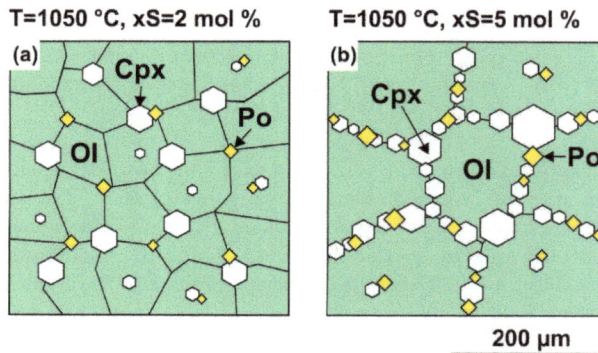

Figure 2. Schematic images of experimental products in olivine–anhydrite–C system, at 1050 °C, xS = 2 mol % (**a**) and xS = 5 mol % (**b**); **Ol**—olivine, **Cpx**—clinopyroxene, **Po**—pyrrhotite.

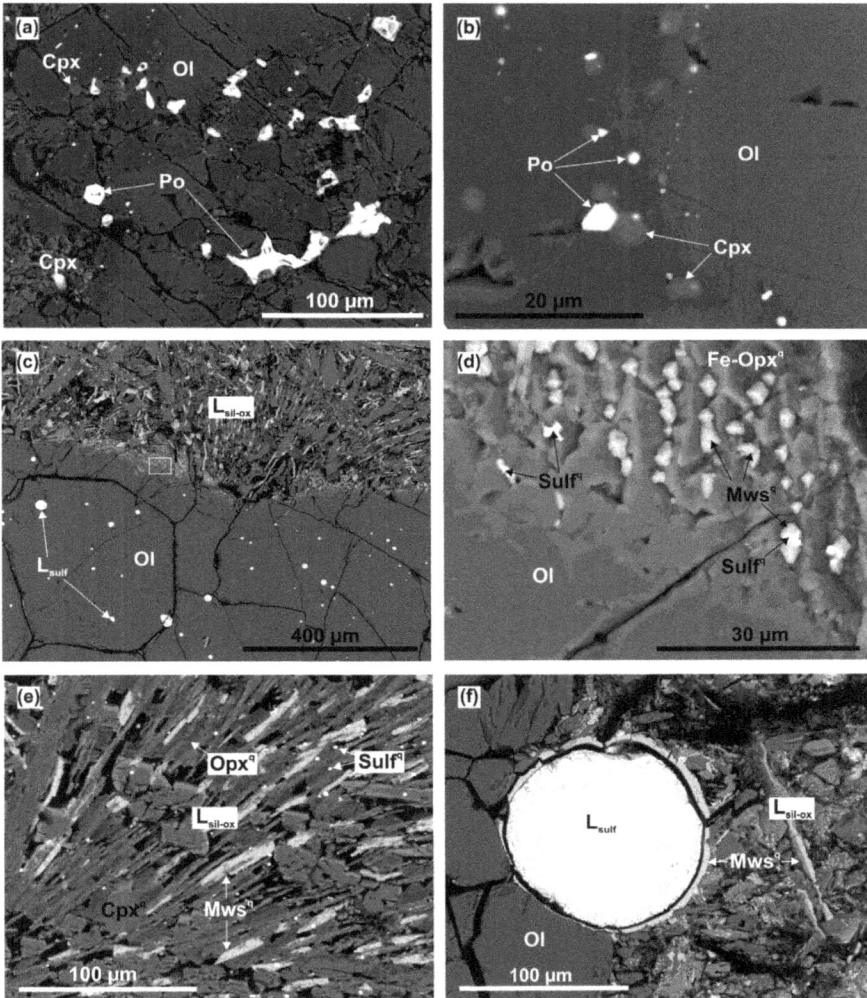

Figure 3. Scanning electron microscopy (SEM) images of polished fragments of the samples after experiments in olivine–anhydrite–C system, (**a–d**) 1050 °C, xS = 2 mol %, (**e–f**) 1450 °C, xS = 2 mol %: (**a**) Polycrystalline aggregate of recrystallized olivine with clinopyroxene and pyrrhotite in interstices; (**b**) inclusions of clinopyroxene and pyrrhotite in recrystallized olivine; (**c**) recrystallized olivine with sulfide melt inclusions, at the contact with the quenched silicate-oxide melt; (**d**) enlarged fragment of Figure 3c, quenched magnesiowüstite and ferrous orthopyroxene at the contact with olivine; (**e**) structure of the quenched silicate-oxide melt, consisted of pyroxenes and magnesiowüstite; (**f**) a contact of the quenched sulfide melt spherule with the quenched silicate-oxide melt, and a rim of magnesiowüstite between them; Ol—olivine, Cpx—clinopyroxene, Po—pyrrhotite, Sulf—sulfide, L_{sil-ox}—silicate-oxide melt, L_{sulf}—sulfide melt, Opx—orthopyroxene, Mws—magnesiowustite, q—quenched phase.

Olivine demonstrates recrystallization features, involving; (1) increase in the crystal size from the initial 20 μm to 150 μm (xS = 2 mol %) and 350 μm (xS = 5 mol %); (2) formation of inclusions;

and (3) zonal structure of crystals, with relicts of the initial olivine in cores and lower Fe and Ni rims. A similar decrease in FeO and NiO and an increase in MgO towards the rims of olivine are considered as the main signs of olivine sulfidation in nature [34]. The formation of zonal olivine crystals occurs only at xS = 2 mol %. The composition of the core and rim zones of this olivine is shown in Figure 4a. On average, the composition of the recrystallized olivine is characterized by a decrease in the FeO concentration (relative to the initial 9.3 wt %) to 7.5 wt % (xS = 2 wt %) and 8.0 wt % (xS = 5 wt %), as well as a negligible CaO content, see Table 3. Newly formed clinopyroxene compositions correspond to $Mg_{1.13}Ca_{0.85}Fe_{0.06}Si_2O_6$ (xS = 2 wt %) and $Mg_{1.12}Ca_{0.75}Fe_{0.09}Si_2O_6$ (xS = 5 wt %), which are identical to included (size 1–7 μm) and interstitial (crystal size 3–15 μm) pyroxenes. Pyrrhotite has an admixture of Ni of 6.3 wt % (xS = 2 mol %) and 11 wt % (xS = 5 wt %), see Figure 5a and Table 4.

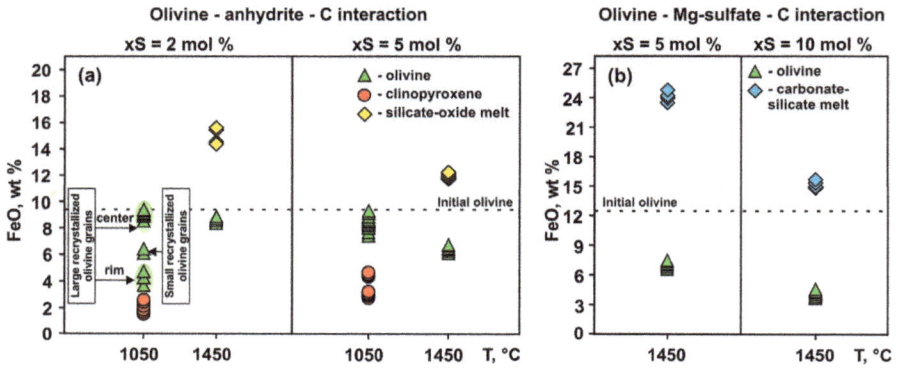

Figure 4. Temperature and sulfur content (xS) dependencies of FeO concentrations in silicates and silicate-oxide melts.

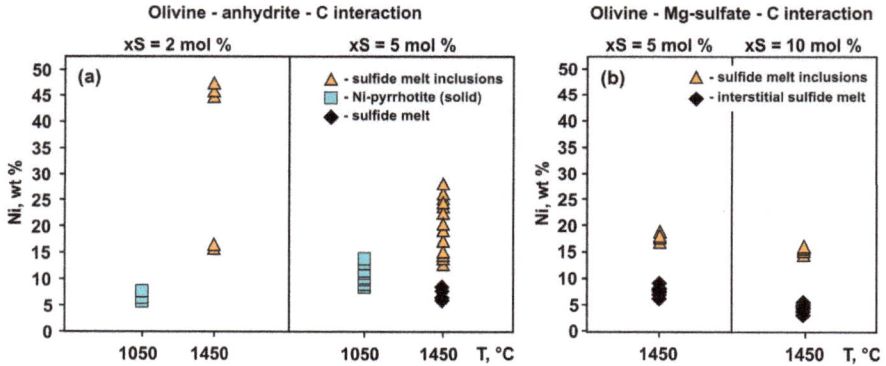

Figure 5. Temperature and sulfur content (xS) dependencies of Ni concentrations in sulfides and sulfide melts.

Table 3. Compositions of silicates, silicate-oxide and carbonate-silicate melts.

Run N	T, °C	xS, mol %	Phase	N_A	Composition, wt %							n(O)	Cations per Formula Unit					
					SiO_2	FeO	MgO	CaO	NiO	CO_2	Total		Si	Fe	Mg	Ca	Ni	Sum
								Ol–CaSO$_4$–C system										
933-2	1050	2	Ol	14	41.2(6)	7.5(28)	50.0(54)	1.0(7)	0.2(2)	–	99.9	4	1.00(1)	0.15(6)	1.82(17)	0.03(2)	–	2.99(1)
933-5	1050	5	Cpx	8	54.8(14)	2.1(5)	20.9(39)	22.0(28)	0.1(1)	–	(4) 99.9(3)	6	1.98(5)	0.06(2)	1.13(20)	0.85(11)	–	4.02(2)
			Ol	27	40.7(4)	8.6(6)	49.7(7)	0.2(2)	0.3(1)	–	99.6(3)	4	1.00(19)	0.17(2)	1.82(3)	–	0.01(0)	3.00(4)
934-2	1450	2	Cpx	10	56.5(5)	3.1(6)	23.6(66)	16.2(78)	0.1(1)	–	99.5(3)	6	2.02(1)	0.09(1)	1.12(4)	0.75(3)	–	3.98(1)
			Ol	9	41.1(1)	8.4(1)	50.6(3)	0.2(0)	–	–	100.3(3)	4	0.99(0)	0.17(0)	1.84(1)	0.01(0)	–	3.01(0)
			Liq1	19	41.8(8)	14.6(9)	29.6(4)	13.6(11)	–	–	99.6(5)	–	–	–	–	–	–	–
934-5	1450	5	Ol	8	41.7(3)	6.1(1)	51.6(2)	0.2(1)	–	–	99.5(3)	4	1.00(0)	0.12(0)	1.86(1)	0.01(0)	–	3.00(0)
			Liq1	15	38.0(1)	12.8(1)	22.2(18)	26.4(20)	–	–	99.4(3)	–	–	–	–	–	–	–
								Ol–MgSO$_4$–C system										
1809-5	1450	5	Ol	16	41.2(2)	6.7(1)	51.6(3)	–	0.1(1)	–	99.6(3)	4	1.00(0)	0.14(0)	1.87(1)	–	–	3.00(0)
			Fms^q	4	–	5.2(2)	42.3(2)	0.9(1)	–	51.6(4)	100	3	–	0.06(0)	0.91(0)	0.01(0)	–	0.98(0)
			Opx^q	10	57.6(3)	6.9(2)	35.4(3)	–	–	–	99.9	6	1.99(0)	0.20(0)	1.83(0)	–	–	4.01(0)
			Liq2	20	32.0(20)	24.1(48)	36.9(31)	–	–	6.5(5)	99.5(2)	–	–	–	–	–	–	–
1809-10	1450	10	Ol	12	41.8(2)	4.4(8)	53.4(6)	–	–	–	99.8(3)	4	1.00(0)	0.09(2)	1.91(2)	–	–	3.00(0)
			Liq2	21	31.7(36)	15.2(77)	44.3(78)	–	–	8.1(1)	99.3(3)	–	–	–	–	–	–	–

Ol—olivine, Cpx—clinopyroxene, Opx^q—quenched orthopyroxene, Fms^q—quenched ferromagnesite, Liq1—silicate-oxide melt, Liq2—carbonate-silicate melt, N_A—number of analyses, n(O)—number of oxygen atoms; The values in parentheses are one sigma errors of the means based on replicate electron microprobe analyses reported as least units cited.

Table 4. Compositions of sulfides.

Run N	T, °C	xS, mol %	Phase	N_A	Composition, wt %				n(S)	Cations per Formula Unit	
					Fe	Ni	S	Total		Fe	Ni
Ol–CaSO$_4$–C system											
933-2	1050	2	Ni-Po	9	56$_{(3)}$	6.3$_{(27)}$	37.0$_{(3)}$	99.6	1	0.87$_{(5)}$	0.09$_{(4)}$
933-5	1050	5	Ni-Po	8	50$_{(2)}$	11$_{(2)}$	38.4$_{(10)}$	99.3	1	0.74$_{(6)}$	0.16$_{(3)}$
934-2	1450	2	Liq^A	9	41$_{(11)}$	24$_{(14)}$	35$_{(3)}$	99.5	1	0.66$_{(13)}$	0.4$_{(3)}$
934-5	1450	5	Liq^A	16	45$_{(9)}$	21$_{(10)}$	33$_{(1)}$	99.6	1	0.78$_{(14)}$	0.34$_{(19)}$
			Liq^B	14	56$_{(1)}$	7$_{(1)}$	36.0$_{(7)}$	99.5	1	0.89$_{(4)}$	0.12$_{(1)}$
Ol–MgSO$_4$–C system											
1809-5	1450	5	Liq^A	11	46.1$_{(8)}$	17$_{(1)}$	36.4$_{(1)}$	99.7	1	0.72$_{(1)}$	0.26$_{(2)}$
			Liq^B	10	53.1$_{(5)}$	8.5$_{(4)}$	37.9$_{(2)}$	99.5	1	0.80$_{(1)}$	0.12$_{(1)}$
1809-10	1450	10	Liq^A	14	47$_{(1)}$	16$_{(2)}$	36.5$_{(2)}$	99.6	1	0.74$_{(2)}$	0.24$_{(3)}$
			Liq^B	10	56.6$_{(4)}$	5.5$_{(4)}$	37.5$_{(4)}$	99.5	1	0.86$_{(1)}$	0.08$_{(1)}$

Ni-Po—Ni-bearing pyrrhotite, Liq^A—sulfide melt from interstices or inclusions, Liq^B—sulfide melt drops from quenched silicate-oxide or carbonate-silicate melt; the values in parentheses are one sigma errors of the means based on replicate electron microprobe analyses reported as least units cited.

At 1450 °C, t = 23 h, and xS = 2 mol %, formation of a polycrystalline aggregate of olivine (crystal size of 100–400 μm) with inclusions of the quenched sulfide melt, in addition to a large pool of segregated quenched silicate-oxide melt, shown in Figures 3c–e and 6a, occur. At higher sulfur contents (5 mol %), the grain size of the recrystallized olivine increases to 400–700 μm, and the sulfide melt is present not only in inclusions but also in interstices. Its drops are also located as quenched silicate-oxide melt, as shown in Figures 3f and 6b.

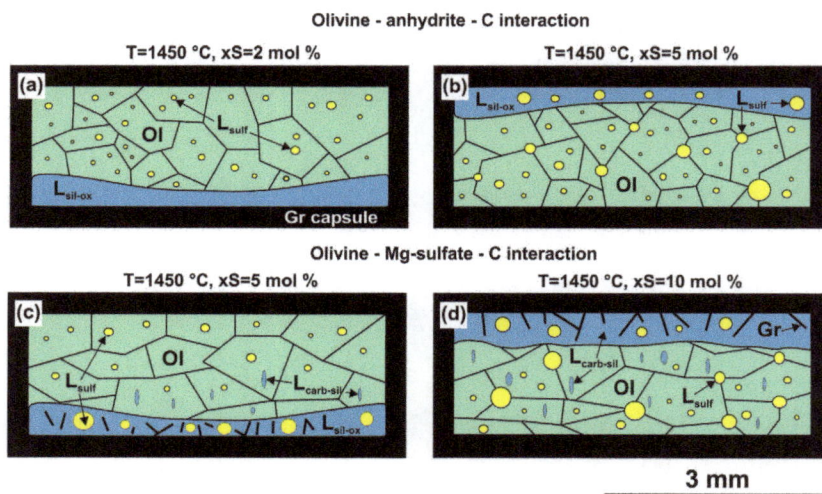

Figure 6. Schematic images of experimental results at 1450 °C in olivine-anhydrite-C system at xS = 2 mol % (**a**) and xS = 5 mol % (**b**) and in Mg-sulfate–olivine system at xS = 5 mol % (**c**) and xS = 10 mol % (**d**); Ol—olivine, Gr—graphite, L$_{sulf}$—sulfide melt, L$_{sil-ox}$—silicate-oxide melt.

Magnesiowustite rims form between immiscible quenched melts and are shown in Figure 3f. Olivine is homogeneous in composition, it shows a decrease in the FeO concentration (relative to the initial) to 8.4 wt % (xS = 2 mol %) and 6.1 wt % (xS = 5 mol %), as shown in Figure 4a. The composition of sulfide melt in inclusions at xS = 2 mol % corresponds to Fe$_{0.53-0.79}$Ni$_{0.21-0.47}$S, and is characterized by wide variations in Ni content (10–38 wt %). At xS = 5 mol %, the composition of the sulfide melt from the interstices of the polycrystalline olivine aggregate and the inclusions, corresponds to $Fe_{0.64-0.92}Ni_{0.15-0.53}S$; the sulfide melt drops originating from the silicate-oxide melt have lower Ni content, with a composition of $Fe_{0.85-0.93}Ni_{0.11-0.13}S$, see Figure 5a. The quenched aggregate of the silicate-oxide melt is represented by micro dendrites of clinopyroxene, orthopyroxene, and magnesiowüstite, as well as submicron sulfide species, as shown in Figure 3e. The bulk composition of this melt at xS = 2 mol % is characterized by 14.6 wt % FeO and 13.6 wt % CaO concentrations, and at the higher sulfur content in the system—12.8 wt % FeO and 26.4 wt % CaO. At xS = 2 mol %, specific quenched aggregates were established at the olivine/silicate-oxide melt contact, see Figure 3c,d.

The analysis of experimental results reveals the main regularities of phase formation in the olivine–anhydrite–C system depending on temperature and sulfur content, controlled by the amount of initial sulfate. It was experimentally established that with an increase in the sulfur content from 2 to 5 mol % in the system (irrespective of temperature), the average size of the olivine crystals increases, FeO and NiO concentrations in olivine decrease, and the number and grain size of the newly formed sulfide and clinopyroxene increase. Depending on the sulfur content in the system, formation of Fe-enriched reaction structures at the contact of olivine with the silicate-oxide melt (xS = 2 mol %) or segregation of sulfide melt droplets directly in the silicate-oxide melt (xS = 5 mol %) occur. An increase of the temperature of experiments (xS = const) also leads to an increase in the size of the olivine crystals, a decrease in the concentration of iron and nickel therein, as well as to the change of association of the newly formed phases from a clinopyroxene + pyrrhotite to a silicate-oxide melt + sulfide melt.

3.2. Interaction in the Olivine—Mg-sulfate—C System

At 1450 °C and xS = 5 mol %, the formation of a polycrystalline olivine aggregate, as well as the carbonate-silicate melt occur, as shown in Figure 7a–d. This melt is presented in the experimental samples as micro-dendritic aggregates of clinopyroxene, orthopyroxene, and magnesite, as shown in Figure 7d. In the quenched aggregate of a carbonate-silicate melt, segregated drops of sulfide melt and graphite crystals are situated as shown in Figure 7a–d,f. Additionally, quenched ferromagnesite and Ni-pyrrhotite are present in the interstices of the polycrystalline olivine aggregate, see Figure 7e. Recrystallized olivine (grain size of ~200–500 µm) contains inclusions of quenched melts—sulfide and carbonate-silicate, see Figure 7a–c.

The composition of the olivine corresponds to $Mg_{1.86}Fe_{0.14}SiO_4$; it demonstrates the complete absence of a nickel impurity, as well as a decrease in the iron concentration to 6.7 wt % relative to the initial concentration. The carbonate-silicate melt that co-exists with olivine is characterized by an FeO content of ~24 wt % and a CO_2 concentration of 6–7 wt %, as shown in Table 3 and Figure 4b. The composition of sulfide melt in the inclusions and interstices of the polycrystalline olivine aggregate corresponds to $Fe_{0.70-0.74}Ni_{0.24-0.28}S$, see Table 4 and Figure 5b, and the sulfide droplets in the carbonate-silicate melt are characterized by a composition of $Fe_{0.79-0.81}Ni_{0.11-0.13}S$. At higher sulfur contents (10 mol %), the structure of the experimental sample is generally similar to the above. The FeO content in the olivine is decreased to 4.4 wt %. The carbonate-silicate melt is characterized by FeO content of 15 wt % and CO_2 concentration of 8.5 wt %. The composition of the sulfide melt in inclusions in olivine and in interstices corresponds to $Fe_{0.72-0.76}Ni_{0.21-0.27}S$, see Figure 5b, while drops of sulfide in the carbonate-silicate melt are of the composition, $Fe_{0.85-0.87}Ni_{0.07-0.09}S$. Thus, it has been experimentally demonstrated that the increase in the sulfur concentration in the olivine—Mg-sulfate—C system at a relatively high temperature leads to an increase in the average size of the olivine crystals, a decrease in the concentration of iron and nickel in olivine and in the carbonate-silicate melt, and an increase in the amount of the newly formed sulfide melt.

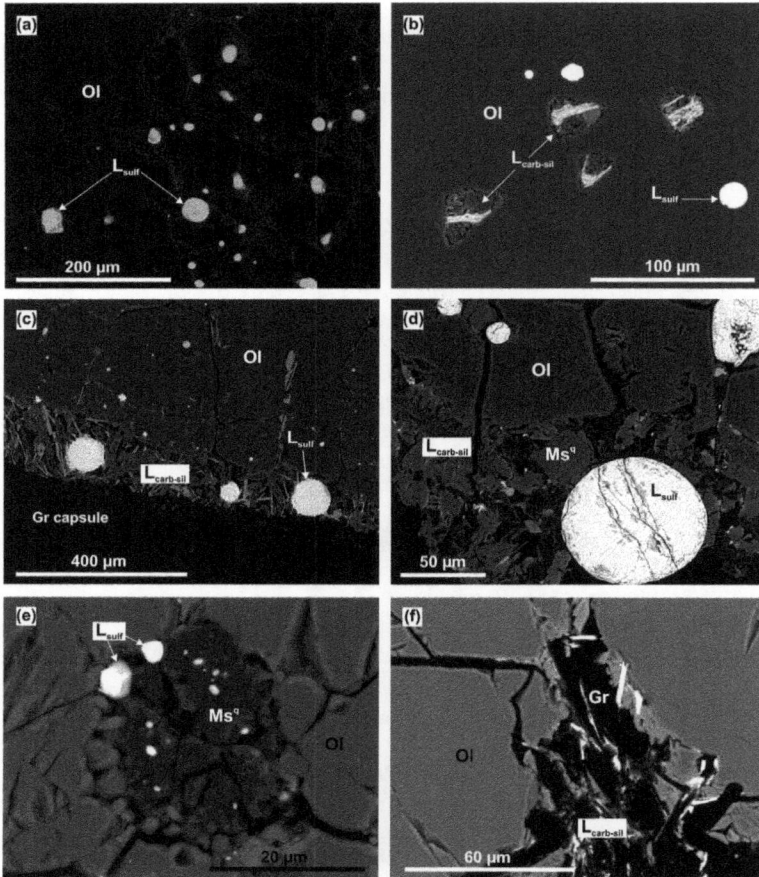

Figure 7. Scanning electron microscopy (SEM) images of polished fragments of the samples after experiments on the Mg-sulfate-olivine system, 1450 °C, xS = 5 mol %: (**a**) Polycrystalline aggregate of recrystallized olivine with sulfide and silicate-oxide melt in interstitions and inclusions; (**b**) inclusions of quenched sulfide and silicate-oxide melt in recrystallized olivine; (**c**,**d**) recrystallized olivine with inclusions, at the contact with the quenched silicate-oxide and sulfide immiscible melts; (**e**) aggregate of quenched ferromagnesite and sulfide melt; (**f**) recrystallized graphite in a quenched carbonate-silicate melt; **Ol**—olivine, **L$_{carb-sil}$**—carbonate-silicate melt, **L$_{sulf}$**—sulfide melt, **Ms**—magnesite, **Gr**—graphite, ^{q}—quenched phase.

4. Discussion

4.1. Reconstruction of Sulfate-Olivine Interaction Processes

The interaction processes in the olivine–anhydrite–C and olivine–Mg-sulfate–C systems are found to proceed via closely interrelated reactions and involving the generation of fluid or melt, recrystallization and crystallization of mineral phases, and the formation of inclusions therein. Given the complexity of the processes, they will be considered separately.

4.1.1. S-Rich Fluid Formation Processes in Olivine–Anhydrite–C System

The most informative experiments in terms of the reconstruction of fluid generation processes are those in the olivine–anhydrite–C system at sub-solidus temperatures, under averaged P,T-parameters estimated for the formation of sulfate inclusions in natural diamonds, see Figure 1. It has been established that the formation of a fluid, that has a direct influence on the phase formation in the system, takes place mainly through the desulfation reaction. This reaction under natural conditions (at high P and T) is thought to be realized by the interaction of Ca,Mg-sulfates with silicates or oxides, and leads to the formation of oxidized SO_3-fluid and enrichment of the associated phases with calcium or magnesium [3]. In our experiments on the olivine–anhydrite–C interaction, desulfation occurs via the scheme Ol + Anh → Di + SO_3 (1), and results in complete consumption of anhydrite. In this case, CaO, initially contained in the anhydrite, enters diopside, which forms in the interstices of the polycrystalline aggregate of olivine and in the inclusions therein. Under the carbon-saturated conditions, provided by the graphite ampoules, SO_3-fluid is reduced, according to the $2SO_3 + 3C \rightarrow 2S + 3CO_2$ (2) scheme. For a full-scale reconstruction of the fluid generation processes, it should be noted that there is a different understanding of the mechanisms of desulfation in nature. Desulfation is one of the characteristic processes that occur during the formation of infiltration-metasomatic deposits of sulfur in the Earth's crust. It can occur through the interaction of anhydrite (or oxidized SO_2, SO_3-fluid) with organic matter or carbon-bearing brines, and is accompanied by the formation of calcite and elemental sulfur. Moreover, a recent investigation of non-salty kimberlites, demonstrating an increased abundance of sulfides, showed that the infiltration of brines into kimberlites can dissolve magmatic sulfates and initiate serpentinization accompanied by the sulfidation of olivine [5]. The process of desulfation, realized by the interaction of sulfate + C, can also occur at great depths in the mantle, as evidenced by the presence of $CaCO_3$ in all described inclusions of anhydrite in diamonds [10,11]. Thus, as the main indicator of the realization of the desulfation reaction in carbon-saturated conditions, the formation of carbonate and a phase (mineral, fluid or melt) that bears sulfur in a reduced form can be considered. However, the experiments in the olivine–anhydrite–C system do not produce $CaCO_3$, but rather CaO in diopside. This result emphasizes the preference of the anhydrite + silicate interaction instead of the anhydrite + graphite under high pressure and temperatures. The formation of a diopside + Ni-pyrrhotite mineral association in inclusions in olivine and in interstices indicates that the reconstructed process of the fluid formation proceeds in stages, as a result of interactions (1) and (2), and not directly by the interaction of anhydrite + C.

4.1.2. Fe,Ni-Sulfide Formation via Olivine Interaction with S-rich Fluid in the Olivine–Anhydrite–C System

In the course of the interaction of Fe,Ni-olivine with S-rich fluid, generated by desulfation, even relatively low concentrations of the intergranular fluid provide conditions for the recrystallization of olivine–partial (1050 °C, xS = 2 mol %) and total (xS = 5 mol %). It was found that during recrystallization, extraction of iron, nickel, and other elements from olivine into intergranular S-rich fluid, results in a fluid that can be denoted as [S-Fe-Ni-O-Si-Mg]$_{fluid}$. When this fluid, enriched with metals and oxygen due to their extraction from olivine, reaches saturation conditions, formation of Ni-pyrrhotite as a separate phase, enrichment of interstitial clinopyroxene with FeO and MgO, and crystallization of Fe,Ni-poor olivine occur. The bulk of these processes, including olivine recrystallization, Fe and Ni extraction in S-rich fluid, sulfide formation, as well as pyroxene crystallization are realized in the course of olivine sulfidation [23,24,35]. Details on the reconstruction of the olivine sulfidation mechanism (xS = 10 mol %) under mantle P,T-parameters are presented in [24].

4.1.3. The Formation of Silicate-Oxide and Sulfide Melts in the Olivine–Anhydrite–C System

At 1450 °C, the generation of S-rich reduced fluid, recrystallization of olivine in this fluid, and the extraction of metals, accompanied by the formation of sulfide, occurs. In addition to these processes, partial melting leads to the formation of two immiscible melts—the sulfide and silicate-oxide ones.

In the olivine–anhydrite–C system at a sulfur content of 2 mol %, the formation of Ni-rich sulfide melt droplets occur only in the interstices, simultaneously with olivine recrystallization in the intergranular fluid and the subsequent capture of sulfide inclusions. The mechanism of the sulfide melt formation is similar to that of Ni-pyrrhotite crystallization at a lower temperature. However, the formation of submicron Ni-poor pyrrhotite crystals in the quenched silicate-oxide melt indicates that some sulfur is dissolved in this melt. As it is shown in the works on the solubility of sulfur in silicate melts under mantle P,T-parameters [4,36], the highest solubility is characteristic of the "sulfate" form of sulfur (S^{6+}). The "Sulfide" form of sulfur (S^{2-}) is known to be dissolved in the silicate melt as predominantly Fe-S. However, the solubility of "sulfide" sulfur is very low, and when the silicate melts are saturated with S^{2-}, the sulfide component of the melt is segregated [4,36,37]. Analysis of the Fe-rich structures, see Figure 3d, arising at the contact of olivine with the silicate-oxide melt (xS = 2 mol %) shows that the olivine recrystallization occurs not only in reduced S-rich fluid, but also in the silicate-oxide melt with dissolved sulfur. As a result of the extraction of iron, nickel, and other elements from olivine into the melt, and the subsequent interaction of Fe and Ni with the dissolved sulfur, Fe-S and Ni-S ligands are formed, that are quenched as submicron pyrrhotite species (xS = 2 mol %) or are segregated as large drops of sulfide directly in the silicate-oxide melt (xS = 5 mol %). The main indicators of sulfide in the silicate-oxide melt are: (1) The spatial confinement of sulfides to quenched magnesiowüstite, and (2) dramatic differences in the Ni concentration in the interstitial sulfide melt (high-nickel) and in the silicate-oxide melt (low-nickel), see Figure 5a.

4.1.4. The Formation of Carbonate-Silicate and Sulfide Melts in the Olivine–Mg-Sulfate–C System

The presence of quenched ferromagnesite and Ni-pyrrhotite in the interstices of a polycrystalline aggregate of olivine and the formation of a carbonate-bearing melt indicate that the desulfation scenario in the olivine–Mg-sulfate–C system differs entirely from that in the olivine–anhydrite–C system. In this case, the initiation of partial melting and the formation of the carbonate component of the melt occurs as a result of the desulfation reaction according to the scheme Mg-sulfate + C → magnesite + S_{fluid}. Reduced sulfur in the interstitial fluid causes olivine sulfidation–extraction of iron, nickel, and other elements into the fluid and formation of Ni-pyrrhotite, as well as enrichment of magnesite with FeO. The newly formed sulfide undergoes melting under the P,T-parameters of the experiments, and in the process of recrystallization of olivine, the inclusions of the sulfide melt are trapped. At the same time, sufficiently high sulfur concentrations in the system (5 and 10 mol %) lead to complete recrystallization of olivine and the formation of orthopyroxene—a product of olivine sulfidation.

As noted above, the capture of inclusions of the sulfide melt occurs directly from the interstitial spaces during the recrystallization of olivine. In this case, the newly formed ferromagnesite and Fe-orthopyroxene in the presence of a reduced sulfur fluid undergo partial melting and form droplets of carbonate-silicate melt with a small amount of dissolved sulfur. During the experiment, the droplets of the carbonate-silicate melt segregate and move from the interstices into a single melt pool. Inclusions of carbonate-silicate melt in the olivine indicate that its recrystallization is realized not only in the reduced sulfur fluid, but also in the carbonate-silicate melt containing dissolved sulfur. The direct result of this recrystallization and extraction of Fe, Ni, and other elements from olivine into the melt is the formation of droplets of sulfide in the carbonate-silicate melt. Most probably, recrystallization of graphite in carbonate-silicate melt proceeds via the redox reaction of a sulfide and a carbonate-bearing melt under carbon-saturated conditions [29,38,39].

4.2. Scenarios of Ca,Mg-Sulfates Interaction with Olivine Under High P,T-Conditions: Implications to Mantle Sulfide, Silicate and Carbonate Formation

During subduction, the main fluid-generating processes are supposed to be dehydration (water-releasing process with hydrous minerals decomposition), decarbonation (formation of CO_2-fluid with carbonates decomposition), and desulfation (formation of SO_3-fluid with sulfates decomposition). P,T-parameters of the realization of these processes vary widely and depend on many factors, such as fO_2, fS_2, and environment composition [9,18,40]. In the studies on the behavior of sulfur-bearing phases under subduction conditions, sulfate interactions with carbon-rich brines, reduced H-rich fluids, silicates, and silicate melts are proposed as specific processes for the formation of S-rich fluid. Results of the present study demonstrate that the interaction of Ca,Mg-sulfates with Fe,Ni-bearing forsterite (under carbon-saturated conditions) can be considered as a simplified model of these processes occurring during the subduction of oxidized crustal material into silicate mantle. However, revealed scenarios of the formation of mantle sulfides, silicates, carbonates and melts of various compositions are also applicable to more complex natural systems.

It is established that, in the process of the subduction of anhydrite into a silicate mantle, desulfation occurs as a result of its interaction with olivine, and leads to the generation of an oxidized SO_3-fluid and crystallization of a diopside, or the formation of a silicate-oxide melt enriched in calcium. In this case, despite carbon-saturated conditions, anhydrite does not participate in carbonate-producing reactions with C directly. In the presence of graphite, the SO_3-fluid is reduced to S (as fluid), and this process is accompanied by the formation of the CO_2-fluid. Subduction of Mg-sulfate into the silicate mantle under carbon-saturated conditions leads to the formation of carbonate (or carbonate-silicate melt) and reduced S-rich fluid. As a result of the interaction of this fluid with Fe,Ni-bearing forsterite, olivine is sulfidized, leading to the crystallization of Fe, Ni-sulfides, or the formation of a sulfide melt.

5. Conclusions

It is found that at a relatively low temperature (1050 °C), interaction in the Fe,Ni-olivine–anhydrite–C system leads to the formation of low-Fe, low-Ni olivine, diopside, and pyrrhotite. During this desulfation process a reduced sulfur fluid forms according to the reaction olivine + anhydrite + C → diopside + S + CO_2. The resulting fluid is a recrystallization medium for Fe,Ni-bearing olivine, extraction of Fe, Ni, and other elements and subsequent crystallization of sulfide ($Fe_{0.74-0.87}Ni_{0.09-0.16}S$) in association with pyroxene (i.e., olivine sulfidation).

As a result of the interaction in the Fe,Ni-olivine–anhydrite–C system at 1450 °C, olivine recrystallization occurs, accompanied by the formation of immiscible silicate-oxide (12.8–14.6 wt % FeO and 13.6–26.4 wt % CaO) and sulfide ($Fe_{0.66-0.89}Ni_{0.12-0.34}S$) melts. It was experimentally demonstrated that in the process of olivine recrystallization, a selective capture of inclusions of the sulfide melt takes place. It has been established that the recrystallization media for olivine are both a reduced S-rich fluid and a silicate-oxide melt with dissolved sulfur.

It is found that at 1450 °C in the Fe,Ni-olivine–Mg-sulfate–C system, olivine recrystallization occurs, accompanied by the capture of melt inclusions, generation of carbonate-silicate (15–24 wt % FeO and 6–8.5 wt % CO_2) and sulfide ($Fe_{0.70-0.87}Ni_{0.07-0.28}S$) melts, and the recrystallization of graphite in the carbonate-silicate melt. The formation of the carbonate component of the melt occurs as a result of the desulfation reaction, according to the scheme Mg-sulfate + C → magnesite + S.

It was experimentally demonstrated that as a result of anhydrite- and Mg-sulfate-bearing slab desulfation, sulfide mineralization of mantle silicate rocks, crystallization of the minerals of wehrlitic assemblage—olivine and diopside, and the generation of potential agents of mantle metasomatism—SO_3, S, CO_2 fluids, in addition to silicate-oxide melt or carbonate-silicate melt occur.

Author Contributions: Conceptualization, Y.P., Y.B. (Yuliya Bataleva) and Y.B. (Yuri Borzdov); methodology, Y.B. (Yuri Borzdov); formal analysis, Y.B. (Yuliya Bataleva); investigation, Y.B. (Yuliya Bataleva), Y.P.; writing-original draft preparation, Y.B. (Yuliya Bataleva); writing-review & editing, Y.P. and Y.B. (Yuliya Bataleva); visualization, Y.B. (Yuliya Bataleva); supervision, Y.P.

Funding: This work was supported by the Russian Science Foundation under Grant No. 14-27-00054.

Acknowledgments: The authors thank three anonymous reviewers for helpful and constructive reviews.

Conflicts of Interest: The authors declare no conflict of interest.

References

1. Alard, O.; Lorand, J.-P.; Reisberg, L.; Bodinier, J.-L.; Dautria, J.-M.; O'Reilly, S. Volatile-rich metasomatism in Montferrier xenoliths (Southern France): Implications for the abundances of chalcophile and highly siderophile elements in the subcontinental mantle. *J. Petrol.* **2011**, *52*, 2009–2045. [CrossRef]
2. Delpech, G.; Lorand, J.P.; Grégoire, M.; Cottin, J.-Y.; O'Reilly, S.Y. In-situ geochemistry of sulfides in highly metasomatized mantle xenoliths from Kerguelen, southern Indian Ocean. *Lithos* **2012**, *154*, 296–314. [CrossRef]
3. Evans, K.A.; Powell, R. The effect of subduction on the sulphur, carbon and redox budget of lithospheric mantle. *J. Metamorph. Geol.* **2015**, *33*, 649–670. [CrossRef]
4. Zajacz, Z. The effect of melt composition on the partitioning of oxidized sulfur between silicate melts and magmatic volatiles. *Geochim. Cosmochim. Acta* **2015**, *158*, 223–244. [CrossRef]
5. Kitayama, Y.; Thomassot, E.; Galy, A.; Golovin, A.; Korsakov, A.; d'Eyrames, E.; Assayag, N.; Bouden, N.; Ionov, D. Co-magmatic sulfides and sulfates in the Udachnaya-East pipe (Siberia): A record of the redox state and isotopic composition of sulfur in kimberlites and their mantle sources. *Chem. Geol.* **2017**, *455*, 315–330. [CrossRef]
6. Alt, J.C.; Shanks, W.C.; Jackson, M.C. Cycling of sulfur in subduction zones: The geochemistry of sulfur in the Mariana-Island Arc and back-arc trough. *Earth Planet. Sci. Lett.* **1993**, *119*, 477–494. [CrossRef]
7. Evans, K.A. The redox budget of subduction zones. *Earth Sci. Rev.* **2012**, *113*, 11–32. [CrossRef]
8. Jégo, S.; Dasgupta, R. The fate of sulfur during fluid-present melting of subducting basaltic crust at variable oxygen fugacity. *J. Petrol.* **2014**, *55*, 1019–1050. [CrossRef]
9. Tomkins, A.; Evans, K.A. Separate zones of sulfate and sulfide release from subducted mafic oceanic crust. *Earth Planet. Sci. Lett.* **2015**, *428*, 73–83. [CrossRef]
10. Wirth, R.; Kaminsky, F.; Matsyuk, S.; Schreiber, A. Unusual micro- and nano-inclusions in diamonds from the Juina Area, Brazil. *Earth Planet. Sci. Lett.* **2009**, *286*, 292–303. [CrossRef]
11. Leung, I.S. Silicon carbide cluster entrapped in a diamond from Fuxian, China. *Am. Mineral.* **1990**, *65*, 1110–1119.
12. Evans, K.A.; Tomkins, A.G.; Cliff, J.; Fiorentini, M.L. Insights into subduction zone sulfur recycling from isotopic analysis of eclogite-hosted sulfides. *Chem. Geol.* **2014**, *365*, 1–19. [CrossRef]
13. Stephens, D.R. The hydrostatic compression of eight rocks. *J. Geophys. Res.* **1968**, *69*, 2967–2978. [CrossRef]
14. Borg, I.Y.; Smith, D.K. A high pressure polymorph of $CaSO_4$. *Contrib. Mineral. Petrol.* **1975**, *50*, 127–133. [CrossRef]
15. Langenhorst, F.; Deutsch, A.; Homeman, U.; Ivano, B.A.; Lounejeva, E. On the shock ehaviour of anhydrite: Experimental results and natural observations. In Proceedings of the 34th Annual Lunar and Planetary Science Conference, League City, TX, USA, 17–21 March 2003.
16. Bradbury, S.E.; Williams, Q. X-ray diffraction and infrared spectroscopy of monazite-structured $CaSO_4$ at high pressures: Implications for shocked anhydrite. *J. Phys. Chem. Solids* **2009**, *70*, 134–141. [CrossRef]
17. Fujii, T.; Ohfuji, H.; Inoue, T. Phase relation of $CaSO_4$ at high pressure and temperature up to 90 GPa and 2300 K. *Phys. Chem. Miner.* **2016**, *43*, 353–361. [CrossRef]
18. Jégo, S.; Dasgupta, R. Fluid-present melting of sulfide-bearing ocean-crust: Experimental constraints on the transport of sulfur from subducting slab to mantle wedge. *Geochim. Cosmochim. Acta* **2013**, *110*, 106–134. [CrossRef]
19. Haggerty, S.E. Diamond genesis in a multiply-constrained model. *Nature* **1986**, *320*, 34–38. [CrossRef]
20. Sato, K.; Katsura, T. Sulfur: A new solvent-catalyst for diamond synthesis under high-pressure and high-temperature conditions. *J. Cryst. Growth* **2001**, *223*, 189–194. [CrossRef]
21. Pal'yanov, Y.N.; Borzdov, Y.M.; Kupriyanov, I.N.; Gusev, V.A.; Khokhryakov, A.F.; Sokol, A.G. High pressure synthesis and characterization of diamond from sulfur-carbon system. *Diam. Relat. Mater.* **2001**, *10*, 2145–2152. [CrossRef]
22. Palyanov, Y.N.; Kupriyanov, I.N.; Borzdov, Y.M.; Sokol, A.G.; Khokhryakov, A.F. Diamond Crystallization from a sulfur−carbon system at HPHT Conditions. *Cryst. Growth Des.* **2009**, *9*, 2922–2926. [CrossRef]
23. Kullerud, G.; Yoder, H.S., Jr. *Sulfide-Silicate Relations: Carnegie Institution of Washington Year Book, v. 62*; Carnegie Institution of Washington: Washington, DC, USA, 1963; pp. 215–218.

24. Bataleva, Y.V.; Palyanov, Y.N.; Borzdov, Y.M.; Sobolev, N.V. Sulfidation of silicate mantle by reduced S-bearing metasomatic fluids and melts. *Geology* **2016**, *44*, 271–274. [CrossRef]

25. Pal'yanov, Y.N.; Sokol, A.G.; Borzdov, Y.M.; Khokhryakov, A.F. Fluid-bearing alkaline–carbonate melts as the medium for the formation of diamonds in the Earth's mantle: An experimental study. *Lithos* **2002**, *60*, 145–159. [CrossRef]

26. Sokol, A.G.; Borzdov, Y.M.; Palyanov, Y.N.; Khokhryakov, A.F. High-temperature calibration of a multi-anvil high pressure apparatus. *High Press. Res.* **2015**, *35*, 139–147. [CrossRef]

27. Bataleva, Y.V.; Palyanov, Y.N.; Borzdov, Y.M.; Bayukov, O.A.; Zdrokov, E.V. Iron carbide as a source of carbon for graphite and diamond formation under lithospheric mantle P-T parameters. *Lithos* **2017**, *286–287*, 151–161. [CrossRef]

28. Dasgupta, R.; Buono, A.; Whelan, G.; Walker, D. High-pressure melting relations in Fe–C–S systems: Implications for formation, evolution, and structure of metallic cores in planetary bodies. *Geochim. Cosmochim. Acta* **2009**, *73*, 6678–6691. [CrossRef]

29. Palyanov, Y.N.; Borzdov, Y.M.; Bataleva, Y.V.; Sokol, A.G.; Palyanova, G.A.; Kupriyanov, I.N. Reducing role of sulfides and diamond formation in the Earth's mantle. *Earth Planet. Sci. Lett.* **2007**, *260*, 242–256. [CrossRef]

30. Cartigny, P.; Boyd, S.R.; Harris, J.W.; Javoy, M. Nitrogen isotopes in peridotitic diamonds from Fuxian, China: The mantle signature. *Terra Nova* **1997**, *9*, 175–179. [CrossRef]

31. Brazhkin, V.V.; Popova, S.V.; Voloshin, R.N. Pressure-temperature phase diagram of molten elements: Selenium, sulfur and iodine. *Phys. B Condens. Matter* **1999**, *265*, 64–71. [CrossRef]

32. Sharp, W.E. Melting curves of sphalerite, galena, and pyrrhotite and the decomposition curve of pyrite between 30 and 65 kilobars. *J. Geophys. Res.* **1969**, *74*, 1645–1652. [CrossRef]

33. Zhang, Z.; Lentsch, N.; Hirschmann, M.M. Carbon-saturated monosulfide melting in the shallow mantle: Solubility and effect on solidus. *Contrib. Mineral. Petrol.* **2015**, *170*, 47. [CrossRef]

34. Papike, J.J.; Spilde, M.N.; Fowler, G.W.; Layne, G.D.; Shearer, C.K. The Lodran primitive achondrite: Petrogenetic insights from electron and ion microprobe analysis of olivine and orthopyroxene. *Geochim. Cosmochim. Acta* **1995**, *59*, 3061–3070. [CrossRef]

35. Fleet, M.E.; MacRae, N.D. Sulfidation of Mg-rich olivine and the stability of niningerite in enstatite chondrites. *Geochim. Cosmochim. Acta* **1987**, *51*, 1511–1521. [CrossRef]

36. Zajacz, Z.; Candela, P.A.; Piccoli, P.M.; Sanchez-Valle, C.; Waelle, M. Solubility and partitioning behavior of Au, Cu, Ag and reduced S in magmas. *Geochim. Cosmochim. Acta* **2013**, *112*, 288–304. [CrossRef]

37. Kogarko, L.N.; Henderson, C.M.B.; Pacheco, H. Primary Ca-rich carbonatite magma and carbonate-silicate-sulphide liquid immiscibility in the upper mantle. *Contrib. Mineral. Petrol.* **1995**, *121*, 267–274. [CrossRef]

38. Gunn, S.C.; Luth, R.W. Carbonate reduction by Fe-S-O melts at high pressure and high temperature. *Am. Mineral.* **2006**, *91*, 1110–1116. [CrossRef]

39. Bataleva, Y.V.; Palyanov, Y.N.; Borzdov, Y.M.; Kupriyanov, I.N.; Sokol, A.G. Synthesis of diamonds with mineral, fluid and melt inclusions. *Lithos* **2016**, *265*, 292–303. [CrossRef]

40. Dasgupta, R.; Hirschmann, M.M. The deep carbon cycle and melting in Earth's interior. *Earth Planet. Sci. Lett.* **2010**, *298*, 1–13. [CrossRef]

minerals

MDPI

Article

Physicochemical Model of Formation of Gold-Bearing Magnetite-Chlorite-Carbonate Rocks at the Karabash Ultramafic Massif (Southern Urals, Russia)

Valery Murzin [1], Konstantin Chudnenko [2], Galina Palyanova [3,4,*], Aleksandr Kissin [1] and Dmitry Varlamov [5]

[1] A.N. Zavaritsky Institute of Geology and Geochemistry, Ural Branch of Russian Academy of Sciences, Vonsovskogo str., 15, Ekaterinburg 620016, Russia; murzin@igg.uran.ru (V.M.); kissin@igg.uran.ru (A.K.)
[2] A.P. Vinogradov Institute of Geochemistry, Siberian Branch of Russian Academy of Sciences, Favorskogo str., 1a, Irkutsk 664033, Russia; chud@igc.irk.ru
[3] V.S. Sobolev Institute of Geology and Mineralogy, Siberian Branch of Russian Academy of Sciences, Akademika Koptyuga pr., 3, Novosibirsk 630090, Russia
[4] Department of Geology and Geophysics, Novosibirsk State University, Pirogova str., 2, Novosibirsk 630090, Russia
[5] Institute of Experimental Mineralogy, Russian Academy of Sciences, Chernogolovka, Moscow Region 142432, Russia; dima@iem.ac.ru
*   Correspondence: palyan@igm.nsc.ru

Received: 8 June 2018; Accepted: 17 July 2018; Published: 20 July 2018

Abstract: We present a physicochemical model for the formation of magnetite-chlorite-carbonate rocks with copper gold in the Karabash ultramafic massif in the Southern Urals, Russia. The model was constructed based on the formation geotectonics of the Karabash massif, features of spatial distribution of metasomatically altered rocks in their central part, geochemical characteristics and mineral composition of altered ultramafic rocks, data on the pressure and temperature conditions of formation, and composition of the ore-forming fluids. Magnetite-chlorite-carbonate rocks were formed by the hydrothermal filling of the free space, whereas chloritolites were formed by the metasomatism of the serpentinites. As the source of the petrogenic and ore components, we considered rocks (serpentinites, gabbro, and limestones), deep magmatogenic fluids, probably mixed with metamorphogenic fluids released during dehydration and deserpentinization of rocks in the lower crust, and meteoric waters. The model supports the involvement of sodium chloride-carbon dioxide fluids extracting ore components (Au, Ag, and Cu) from deep-seated rocks and characterized by the ratio of ore elements corresponding to Clarke values in ultramafic rocks. The model calculations show that copper gold can also be deposited during serpentinization of deep-seated olivine-rich rocks and ore fluids raised by the tectonic flow to a higher hypsometric level. The results of our research allow predicting copper gold-rich ore occurrences in ultramafic massifs.

Keywords: Karabash ultramafic massif; magnetite-chlorite-carbonate rocks; chloritolites; Au-Cu mineralization; copper gold; thermodynamic modeling

1. Introduction

The Karabash ophiolite massif is located within a belt of ultramafic massifs stretching along the Main Ural fault zone in the Southern Urals (Figure 1) [1,2]. The massif became widely known after the discovery of the gold-copper (Au-Cu) mineralization at the Zolotaya Gora gold deposit. The main ore bodies at this deposit are represented by rodingites ("chlogropites"), which consist of chlorite, garnet, pyroxene and a minor quantity of calcite. These rocks contain a specific Au-Cu mineralization: tetra-auricupride (AuCu), auricupride (AuCu$_3$), Au$_3$Cu, and solid solutions of the gold-silver-mercury (Au-Ag-Hg) system [3–6]. The other gold occurrence at the Karabash massif was recognized in the magnetite-chlorite-carbonate rocks that are accompanied by strongly chloritized serpentinites (chloritolites) [7,8], where native gold is characterized by high fineness and Cu content of 1.3–2.6 wt %. These types of rocks are also known for their increased Y, Zr, REE, U and Th contents.

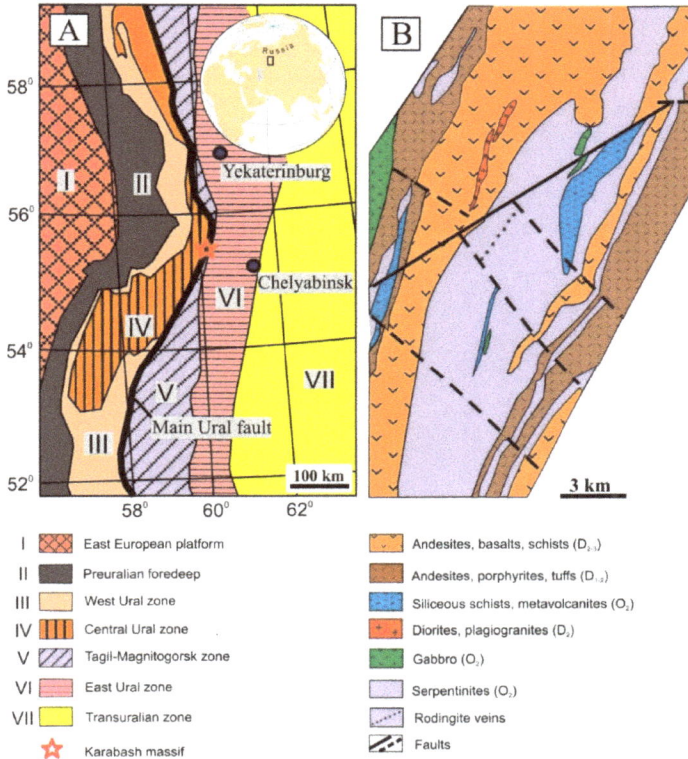

Figure 1. Karabash ultramafic massif (Southern Urals, Russia). (**A**) Position of the Karabash massif on the tectonic scheme of Urals per Puchkov [1]. (**B**) Schematic geological map of the Karabash massif (modified from Snachyov et al. [2] with authors' additions).

Earlier mineralogical-geochemical and isotope studies of magnetite-chlorite-carbonate and strongly chloritized rocks (chloritolites) [8,9] provide an opportunity for the reconstruction of a physicochemical model of the formation of these rocks and the deposition of copper gold within. Construction of this model was the main goal of our study. The model was calculated considering that: (1) the formation of magnetite-chlorite-carbonate rocks involves the participation of hydrothermal solution by the mechanism of free space filling, and chloritolites are formed by the mechanism of replacement of the serpentinite rocks; and (2) the most likely sources for petrogenic and ore components are: (a) rocks (serpentinite, gabbro, and limestone), (b) deep magmatogenic fluid, probably mixed with the metamorphogenic fluid released during dehydration and deserpentinization of the lower crustal rocks, and meteoric water.

2. Geological Background of the Studied Area

2.1. Geological Setting of the Karabash Massif and Hydrothermal-Metasomatic Rocks

The Karabash ultramafic massif is composed of antigorite and, to a lesser extent, chrysotile and lizardite serpentinites. Various types of hydrothermal-metasomatic rocks in the massif are spatially separated but have similar geologic settings, being localized in zones of tectonic mélange or at the contacts of the massif (Figure 2a) [10]. The belt of rodingite rocks extends continuously for about 2.5 km along the central part of the massif. The magnetite-chlorite-carbonate and riebeckite rocks occur locally only in its marginal parts. Listvenites are scarce; with a band up to 15 m thick with a diorite-porphyrite dike occurs in the western contact of the host volcano-sedimentary strata.

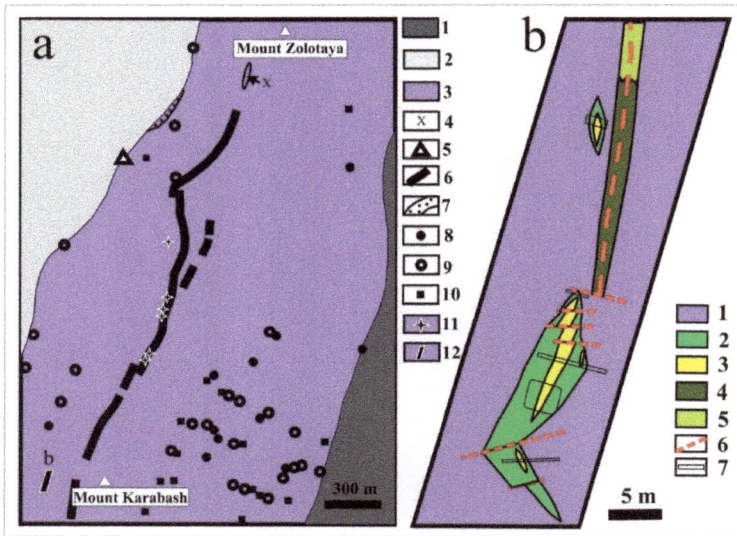

Figure 2. Geological position and types of hydrothermal-metasomatic rocks in the Karabash massif: (a) Spatial location of metasomatized rocks in the central part of the Karabash massif based on the geological map on a scale of 1:10,000 compiled by Lozhechkin [10] from the results of geological studies in 1933–1935. 1, Ordovician rocks of the Polyakovka Formation; 2, Devonian rocks of the Karamalytash and Ulutau Formations; 3, serpentinites; 4, diorite-porphyrites; 5, quartz diorite-porphyrites; 6–10, altered rocks: 6, rodingites: 7, listvenites; 8, epidote-chlorite-garnet; 9, magnetite-chlorite-carbonate; and 10, quartz-riebeckite; 11, mine workings (mines and adits) at the Zolotaya Gora deposit; and 12, location of the studied area. (b) Occurrence of lenses of magnetite-chlorite-carbonate rocks in the zone of foliated serpentinites in the west of the Karabash massif. 1, antigorite serpentinite; 2, chloritolite; 3, lenses of magnetite-chlorite-carbonate rocks; 4, intensely carbonatized serpentinite; 5, serpentinite with dispersed carbonate mineralization; 6, faults; 7, mine workings. Figure 2 is the reproduction of Figures 2 and 3 published in Murzin et al. [8].

Magnetite-chlorite-carbonate rocks and chloritolites occur nearby the zones of foliated serpentinites (Figure 2a, symbol 9). These zones are several hundred meters long and up to tens of meters thick. The lenticular bodies of magnetite-chlorite-carbonate rocks vary in size from few centimeters to several meters. The largest body that became the object for our study is about 20 m long and 2 m thick and occurs on the western slopes of Mt. Karabash (Figure 2a,b). The body is broken up by a series of transverse faults, which reveal the displacement of its contacts up to 1 m.

Zonation is visible between the ore body of the host antigorite serpentinites and the body of magnetite-chlorite-carbonate rocks, with intermediate zones of carbonatized and chloritized antigorite serpentinites and chloritolites (Figure 3). Zones of chloritolites bordering the magnetite-chlorite-carbonate body are nonsymmetrical in thickness. In the eastern contact, the thickness of the body is about 1.5 m, whereas in the western part it is only a few tens of centimeters.

Figure 3. Zonation around (**a**) a large body and (**b**) a small lens of magnetite-chlorite-carbonate rocks. Zones: 1—magnetite-chlorite-dolomite, 2—chloritolite, 3—serpentinite with carbonatization and chloritization, 4—antigorite serpentinite. (a) The central zone (2) is composed of nearly equal amounts of carbonate and magnetite and a minor amount of chlorite, the content of which increases in the marginal parts of the zone. (b) The central zone (1) is composed of dolomite and a small amount of chlorite.

2.2. Mineral Composition of Altered Ultramafic Rocks

A detailed description of the mineral composition of magnetite-chlorite-carbonate rocks as well as hosting chloritolites and antigorite serpentinites was reported in our previous works [7,8]. Their description is outlined below.

Antigorite serpentinite consists of stellar and tabular aggregates of antigorite with rare grains of magnetite scattered within. The rock is rich in tiny magnetite, which is often concentrated into veins and chains confined to both antigorite and magnesite (Figure 4a). Scattered pentlandite and chromspinel are the accessory minerals of serpentinites. Here, newly formed dolomite and chlorite (clinochlore) are present, the content of which is as high as 30 vol %. Chromspinel is intensely replaced by chromium (Cr)-bearing clinochlore, chromian magnetite and magnetite.

Chloritolites, consisting of fine-grained colorless chlorite, contain scattered carbonate and thin veinlets of coarse-grained chlorite (Figure 4b). They also contain impregnations of magnetite and chromspinel, which definitively indicates the aposerpentinite nature of chloritolites. The composition of relict chromspinel was established to be similar to the accessory spinel from ophiolite harzburgites [7].

The magnetite-chlorite-carbonate rocks have various contents of minerals. Typically, the central parts of these rock bodies consist of carbonate mass (dolomite and, in smaller quantity, calcite), and the marginal parts are enriched in chlorite, magnetite, ilmenite, and apatite, the total content of which is 40–45 vol % (Figure 4c). As accessory minerals, the magnetite-chlorite-carbonate rocks also contain native gold, zircon, monazite, allanite, thorianite, aeschynite-(Y), and very scarce grains of chalcopyrite up to 2 mm in size. Relict chromspinel was absent in these rocks.

Native gold is represented by the particles of various interstitial shapes, up to 3 mm in size. The gold particles contain inclusions of magnetite, ilmenite, dolomite, apatite, and chlorite (Figure 4d). Gold belongs to the copper-bearing variety of gold-silver solid solutions (fineness 833–865‰). Impurity components of native gold (wt %) are as follows: Ag (10.8–13.5), Cu (1.3–2.6), Pd (0.06–0.95) and Fe (to 0.07) (Tescan VEGA-II XMU electron scanning microscope (TESCAN, Brno, Czech Republic)) with attached INCA Energy 450 energy-dispersive spectrometer (Oxford Instruments, Oxford, UK), analyst D. Varlamov, Institute of Experimental Mineralogy, Russian Academy of Sciences) [8].

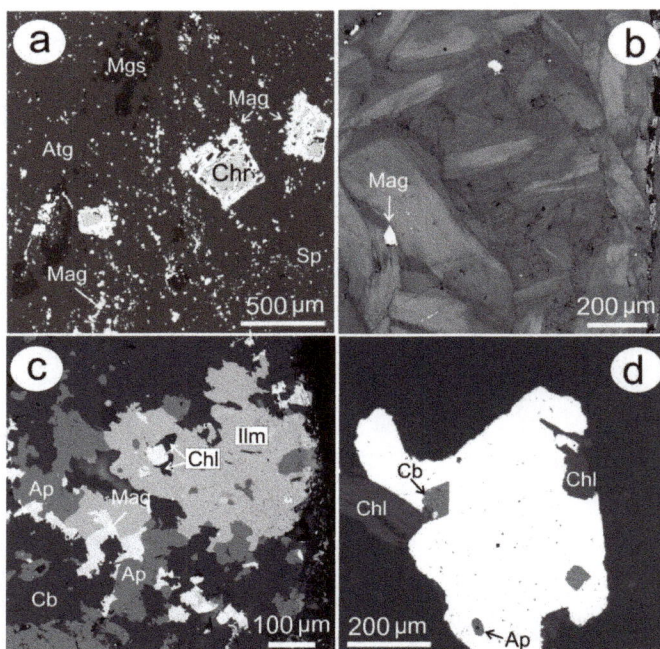

Figure 4. Relationships between minerals in altered ultramafic rocks from the Karabash massif. Back-scattered electron images of polished sections [8]: (**a**) Serpentinite including antigorite (Atg) with dispersed magnesite (Mgs), magnetite (Mag), and chromspinel (Chr). Chromspinel is replaced by magnetite. (**b**) Chloritolite composed of zoned chlorite grains with scarce impregnation of magnetite (Mag). Central parts of chlorite grains (light gray) contain 12–16 wt % FeO, marginal parts (dark gray), 5–7 wt % FeO. Magnetite-chlorite-carbonate rock: (**c**) intergrowths of apatite (Ap), ilmenite (Ilm), magnetite (Mag), chlorite (Chl), and carbonate (Cb); (**d**) copper gold with inclusions of chlorite (Chl), apatite (Ap), and carbonate (Cb).

No signs of multistage formation of magnetite-chlorite-carbonate rocks were clearly revealed. They only demonstrate post-ore tectonics. However, many minerals of these rocks, such as dolomite, chlorite, zircon, and aeschynite have two generations with distinct signs of the replacement of early generations by late ones.

2.3. Geochemical Characteristics of Rocks

The most complete data, including the distribution of rare earth elements (REE) in the rocks, reported in our previous work [8]. Here, we only outline the general regularities of trace elements in rock samples (Table S1). The sampling spots in the ditch are shown in Figure 3a. The trace element composition of rocks was determined by ICP MS on an ELAN 9000 Perkin Elmer mass spectrometer (PerkinElmer, Waltham, MA, USA, analyst D.V. Kiseleva, Institute of Geology and Geochemistry, Ural Branch of Russian Academy of Sciences).

Analysis of the trace element composition of rocks showed that antigorite serpentinite has increased contents of elements typical of ultramafic rocks: Cr, Ni, Co, as well as Mn and B. Some samples contained Cu, Zn, Hg, Cd, and Sb. In chloritolites and magnetite-chlorite-carbonate rocks, the following elements are concentrated: Mn, Ba, Sr, Ti, V, Cu, As, Zr, Hf, Nb, Y, Mo, Ga, Ge, Sc, Ta, Li, Rb, Cs, U, Th, and REE. The total content of REE in serpentinites was less than 2 g/t, whereas in chloritolites and magnetite-chlorite-carbonate rocks it reached 304 g/t.

2.4. P,T-Conditions of Formation and Composition of Ore-Forming Fluid

The temperature regime of the formation of magnetite-chlorite-carbonate rocks estimated using oxygen isotope and dolomite-calcite geothermometers [8] showed that these rocks formed at 480–280 °C. The homogenization temperature of gas-liquid (fluid) inclusions in apatite (THMSG-600 "Linkam" heating-freezing stage, Linkam Scientific, Tadworth, UK, analyst A.A Garaeva, Institute of Geology and Geochemistry, Ural Branch of Russian Academy of Sciences), reflecting the minimal crystallization temperature of this mineral, was 142–221 °C [9]. The eutectic temperature varies from −19.0 to −23.0 °C, which corresponds to the H_2O-NaCl system [11]. The melting temperature of ice ranges from −2.2 to −5.7 wt %, which corresponds to NaCl concentration in solution equal to 3.7–8.8 wt % [12].

The mineral formation derived temperature from the oxygen isotope and dolomite-calcite geothermometers exceeded the homogenization temperatures of fluid inclusions by 140–260 °C. Considering the pressure correction, which is 80–90 °C/kbar [13], the pressure during the formation of magnetite-chlorite-carbonate rocks could have reached 2–3 kbar.

The gas component content in fluid inclusions in the minerals of chloritolites and magnetite-chlorite-carbonate rocks [9], extracted by pyrolysis to 450 °C (below the decomposition temperature of carbonate), can be described using the C-H-O system with minor nitrogen. Composition of gases was analyzed on the Tsvet-800 gas chromatograph (SPE ACADEMPRYLAD Ltd., Sumy, Russia, analyst S.N. Shanina, Institute of Geology, Komi Science Centre, Ural Branch of Russian Academy of Sciences). The fluid was dominated by water and carbon dioxide and contained small amounts of reduced gases: CO, H_2, CH_4, and heavy hydrocarbons including C_2H_4, C_2H_6, C_3H_6, C_3H_8, etc. The degree of oxidation of the gas components ($CO_2/(CO_2 + CO + H_2 + CH_4)$) in the fluid for magnetite-chlorite-carbonate rock was higher than that of chloritolites, at 0.92 and 0.73, respectively.

2.5. Sources of Metals and Ore-Bearing Fluid

High content of gold, Y, Zr, REE, U and Th, the absence of any essential input of K and Na, and rather low scales of mineralization allow us to regard the same rock reservoir as the source of these elements. Data on the isotope composition of C, O, and Sr in carbonates [8] show that in the generation of ore-forming fluids the marine carbonate reservoir was also involved. The restoration of the isotope composition of the water of these fluids suggests that they have been of metamorphogenic origin and have been composed of a mixture of at least two water sources—rock (dehydration of water derived from serpentinite) and deep-seated reservoirs. Taking into consideration that the isotope composition of serpentinite hydrogen, which typically varies from −65 to −90‰, but may even decrease to −128‰, the involvement of meteoric water in serpentinization, is not ruled out.

2.6. Geodynamic Model

Our model of the formation of the Karabash massif (Figure 5A) is based on the concept of the "blocky folding of the Earth's crust" [14]. According to this model, the upper elastic crust is divided into huge blocks by the thrusts dipping toward each other as a result of applied horizontal force. The thrusts result in bends and deformation of blocks, thickening in the vertical plane, and shortening along the compression axis. The thickening and shortening of the upper crust generated a tectonic flow in the lower elastic crust and caused its thermodynamic destabilization.

The Moho boundary is the base of the tectonic flow of the lower crustal rocks. In areas of decompressed mantle, this tectonic flow descends downward. The lower crust rocks undergo high-temperature/high-pressure—metamorphism and dehydration and mix with the rocks of the decompressed upper mantle. The flow of the crust-mantle mixture, directed upward, penetrated the lower and then the upper crust. The upper crust rocks ascended and stretched to form mountain relief.

During the ascending process, the crust-mantle mixture trapped the portions of the non-metamorphosed lower crust and the sialic upper crust experienced decompressing, fracturing, and autometamorphism.

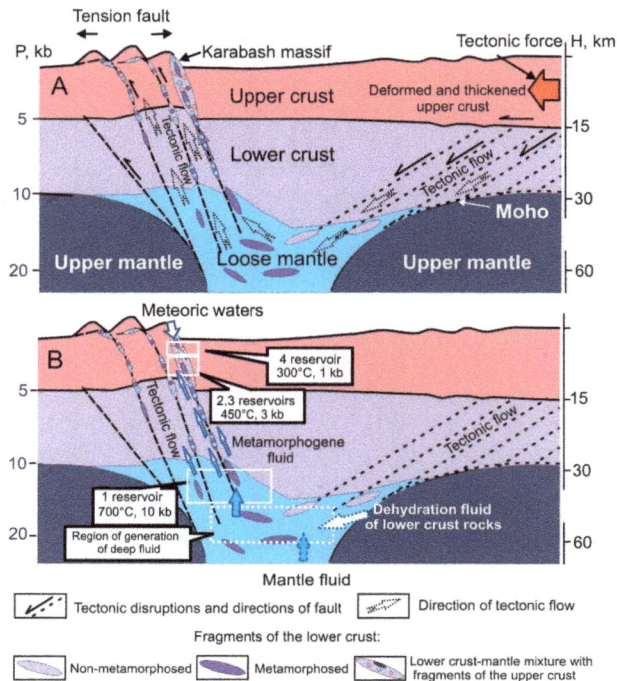

Figure 5. Model of the formation of the Karabash ultramafic massif and sources of ore-forming fluids. (**A**) Geodynamic model and (**B**) fluid model.

The application of the model of blocky folding of the Earth's crust to explain the tectonic setting of the Karabash massif is supported by the results of our research into minor structural forms, bodies of rotation, authors' interpretation of the seismic profile URSEIS-95, and deciphering space photographs [15]. Figure 5B shows the main sources of the fluid and the fluid components involved in the formation of the Karabash massif and the magnetite-chlorite-carbonate rocks therein. The main source of fluid was the juvenile water formed during the oxidation of the mantle hydrogen, as well the water separated during the dehydration and deserpentinization of the lower crust rocks. The fluid reacted with the rocks of the crust-mantle mixture, extracting various components. On the way to the surface, the tectonic flow was enriched in volcano-sedimentary rocks, including marine limestones, which also interacted with the fluid. Rapid decompression on the upper horizons was accompanied by the formation of large steeply dipping en-échelon extension cracks in which the deep fluid was percolated, filling the free space. Simultaneously, meteoric waters that arrived from the top along the extension cracks mixed with the deep fluid and reacted with serpentinites.

3. Methods

3.1. Software and Thermodynamic Dataset for Modeling

Thermodynamic modeling was performed using the Selektor-C software that employs the Gibbs energy minimization method, including minerals, aqueous solution components, and gasses in given T, P-conditions [16,17]. We applied a similar approach for studying the gold-bearing rodingites from the Zolotaya Gora deposit in the Karabash massif [6] and three stages of epithermal gold-silver mineralization of primary and placer gold deposits [18,19]. The main calculation procedures implementing the determination of thermodynamic parameters of compounds of a system at high T, P values were based on the following methods. The calculation of the thermodynamics for aqueous solution components was performed using the revised Helgeson-Kirkham-Flowers (HKF) equation [20]. The dependence of the thermodynamic characteristics of gases on pressure was calculated using the modified Benedict-Webb-Rubin equation of state [21]. In the high-pressure zone, a deviation from ideal gas mixture for real gases was calculated using the van der Waals equation [22] and data from Breedveld and Prausnitz [23]. Fugacity coefficients and molar gas volumes were calculated using the two- [23] and three-parameter equation of state [21].

Modeling was completed for the Na-K-Mg-Ca-Al-Si-Ti-Mn-Fe-Cu-Ag-Au-Hg-S-P-Cl-C-H-O system. The thermodynamic properties of various compounds were calculated using the Selektor-C database. The list of minerals, aqueous, and gaseous species considered in the model is provided in Tables S2 and S3 of the Supplementary Materials.

Thermodynamic constants for chlorites, ilmenites, pyroxenes, carbonates, olivines, and plagioclases, which are natural binary and ternary solid solutions, as well as solid solutions of quaternary system Ag-Au-Cu-Hg, were calculated considering the activity coefficients of end members for the accepted models of solid solutions [17,18]. Serpentinite was introduced as an ideal solid solution consisting of antigorite and lizardite.

3.2. Initial Data for Thermodynamic Modeling

Our calculations were based on the reactive transport model involved in the Selektor-C software. Modeling was conducted at initial temperature (T) and pressure (P) characterizing the lower crust conditions: 700 °C and 10 kbar, respectively. The initial hydrothermal liquid consisted of (mole/kg H_2O): NaCl = 1.5, CO_2 = 0.5, H_2 = 0.01, NaOH = 0.3, and H_2S = 0.0001. The concentrations of NaCl were specified following the results of fluid inclusion studies [9] in magnetite-chlorite-carbonate rock. As the ore-forming system has minor sodium and sulfur contents, we added NaOH and H_2S into the model solution. The following parameters were applied to the model solution: $\log f_{O_2}$ = −14.3, $\log f_{S_2}$ = −5.44, pH = 6.9. The amount of ore components in the model solution was used depending on the content of Cu, Ag, Au, and Hg in the ultramafic rocks from Southern Urals (Table 1).

Table 1. Average contents of gold (Au), silver (Ag), copper (Cu), and mercury (Hg) (ppm) in the massif of ultramafic rocks (Southern Urals, Russia).

	Ultramafic Rocks	Mafic Rocks
Au	0.01135 [1]	0.003925 [1]
Ag	0.14 [2]	0.11 [2]
Cu	6.15 [1]	31.7 [1]
Hg	0.02 [2]	0.07 [2]

[1] harzburgites from the Nuralinsky massif [24]; [2] upper part of continental crust [25].

The hydrothermal model evolved given its gradual flowing through the system of reservoirs up to final discharge at 300 °C and 1 kbar. The composition of meteoric waters diluting the deep fluid was specified as a nonmineralized rainwater: 1 kg H_2O + 0.015 m CO_2 + 0.00032 m O_2 (25 °C, 1 bar), pH = 5.6, Eh = 0.89 [26].

The model included a 4-reservoir calculation scheme depicted in (Figure 6). In the first reservoir at 700 °C and 10 kbar, 1 kg of initial solution reacts with 100 g of rock consisting of 80% harzburgite + 20% gabbro, enriched in the components of these rocks. In the next reservoirs, a discharge zone was simulated as a result of the movement of fluid to upper horizons (T = 450 °C and P = 3 kbar) under different scenarios. In the second reservoir, we assumed interaction of the solution with serpentinite (100 g), and in the third reservoir, with limestone (5 g) and serpentinite (0.5 g) in the flow regime. In the fourth reservoir, mineral formation takes place in a closed system represented initially by serpentinite

(100 g), which was affected by the solution flowing from the second reservoir and meteoric waters that diluted the deep fluid and led to a decrease in T and P to 300 °C and 1 kbar, respectively.

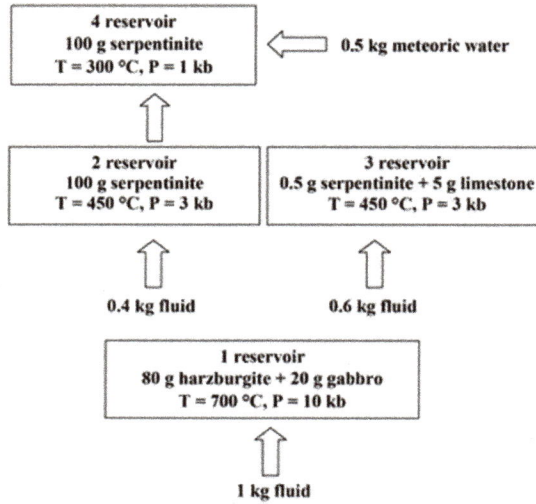

Figure 6. Scheme of the four-reservoir model.

We additionally constructed a version of a particular model in which we calculated the alterations in the composition of deep fluid from the first reservoir during ascent to the surface into the region of decreased temperatures and pressures (450–300 °C and 1–3 kbar).

The chemical compositions of rocks used in the model are provided in Table 2.

Table 2. Chemical compositions of rocks involved in the thermodynamic model.

	Gabbro (n = 7)	Harzburgite (n = 9)	Limestone	Serpentinite
SiO_2	48.81	40.43	0.19	40.47
TiO_2	0.37	0.04	-	0.04
Al_2O_3	14.65	1.52	0.18	1.44
Fe_2O_3	7.97	4.26	-	5.31
FeO	7.20	3.79	0.23	1.61
MnO	0.18	0.09	0.17	0.09
MgO	8.57	38.41	0.39	38.21
CaO	12.03	1.62	55.13	0.48
Na_2O	2.75	0.11	-	0.12
K_2O	0.44	0.03	-	0.03
P	0.09 *	0.035 *	-	0.013 **
S	-	-	-	0.75
CO_2	-	-	43.75	0.46
LOI	1.96	6.77	-	10.59

Data for gabbro and harzburgite from Southern Urals (Talovsky massif) were borrowed from Gritsuk [27], for limestones at the contact of the Karabash massif, from Spiridonov and Pletniov [4], for serpentinites from the Karabash massif, from Berzon [28]. Content of phosphorus: * average content in mafic and ultramafic rocks of the upper mantle [25]; ** in serpentinites from the Karabash massif, our data (ICP-MS, n = 8).

4. Results of Thermodynamic Modeling

4.1. Fluid Composition and Forms of Transfer and Deposition of Au, Ag, Cu, and Hg

Sodium chloride-carbon dioxide ore-bearing fluid (log f_{O_2} = −14.3, log f_{S_2} = −5.44, pH = 6.9) increased total dissolved solids (TDS) to 100 g/L during interaction with the rocks of the second and third reservoirs, and thereafter, being diluted with meteoric water, stabilized at TDS = 50 g/L and pH = 7.4. Oxygen and sulfur fugacities depend mainly on temperature and gradually decreased in the range of log f_{O_2} = −14/(−34) and log f_{S_2} = −5/(−11) (Figure 7).

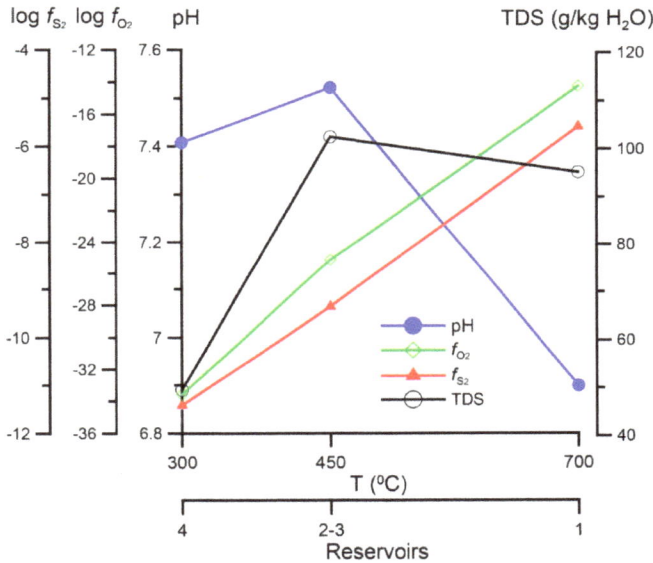

Figure 7. Variations in the values of pH, total dissolved solids (TDS), sulfur fugacity (log f_{S_2}), and oxygen fugacity (log f_{O_2}) in different reservoirs.

In the suggested model, the fluid extracted ore components from deeply-seated rocks, and the contents of metals in the fluid were taken to be close to those in ultramafic and mafic rocks of the upper crust and Nuralinsky massif in the Southern Urals (Table 1). To estimate the contents of deposited gold and to determine the composition of Au-Ag-Cu solid solutions, we performed calculations for four different metal contents in the fluid. The first variant included the contents of metals in the fluid equivalent to that in ultramafic rocks (U), the intermediate variants, in ultramafic and mafic rocks (U + M), and the last variant, in mafic rocks (M) (Figure 8a–h). These calculations show that the gold content in the rocks of the third reservoir increased with the supply of every new portion of solution from 0.7 to 20 g/t in the first variant and from 0.5 to 10 g/t in the last. Gold content in the fourth reservoir was high (about 10 g/t) only in the first case; in all other cases, the content of the deposited gold was less than 0.3 g/t.

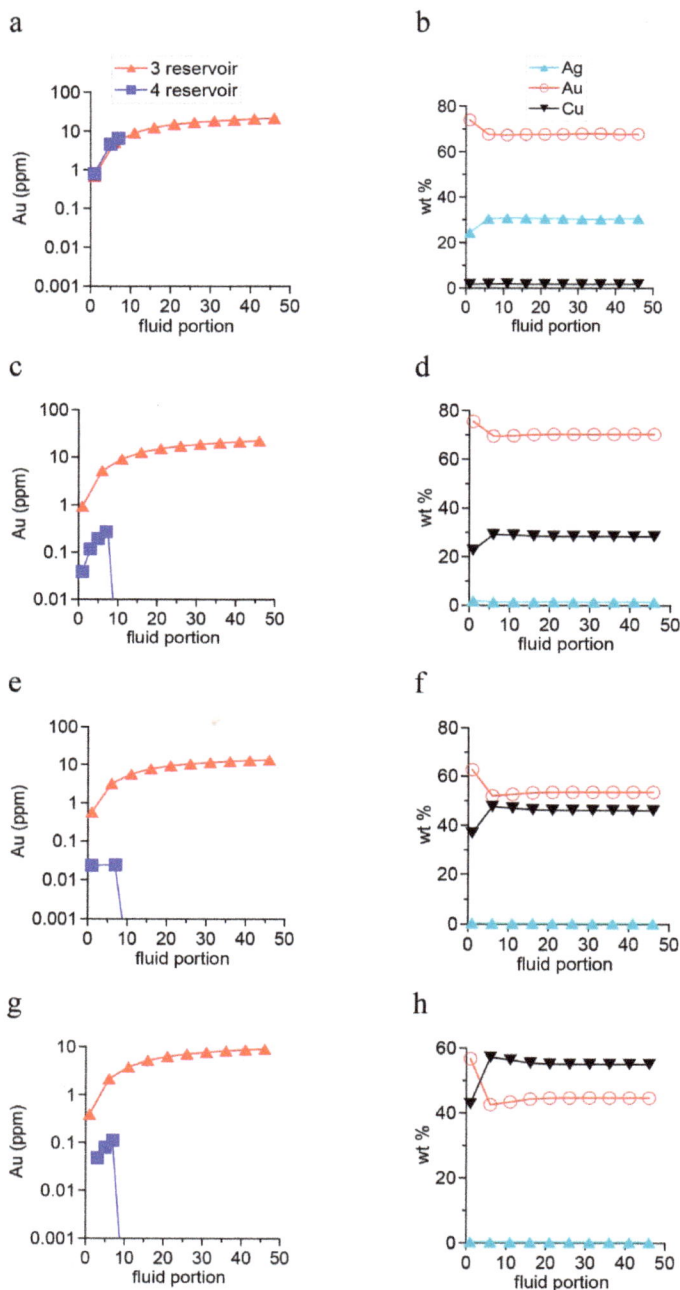

Figure 8. Distribution of gold in the model at the ratios of Au, Ag, Cu, and Hg in the fluid according to their average concentrations in ultramafic (U) and mafic (M) rocks (Table 1), specified in different proportions: (**a,b**) U, (**c,d**) 0.75 U and 0.25 M, (**e,f**) 0.25 U and 0.75 M, and (**g,h**) M. (**a,c,e,g**) Au in rock (third and fourth reservoirs); (**b,d,f,h**) components in Ag-Au-Cu solid solution (third reservoir).

The Au-Ag-Cu solid solutions in the third reservoir at T = 450 °C were formed with the first portions of aqueous solution and their composition remained stable with further incoming solution. When the fluid source was in ultramafic rocks, the Cu-bearing Au-Ag solid solution with ratios (wt %) Au:Ag:Cu = 68:30:2 occurred. When the ratio of mafic rocks in the reservoir and the amount of ore components in the fluid increased, the solid phase contained silver-bearing Au-Cu solid solution with the amount of copper increasing from 28 to 55 wt % (Figure 8b,d,f,h). The sulfur fugacity (log fS_2) was approx. −9.4. The comparison of different variants of the Au, Ag, and Cu ratios in the model solution at the preliminary stage became the basis for selecting ultramafic rocks with a minor fraction of mafic rocks as a source of ore components for further calculations.

Notably, the absence of degassing during the formation of magnetite-chlorite-carbonate rocks contributed to the fact that Au, Ag, Cu, and Hg do not preferably enter into a gas phase but accumulate in the hydrothermal solution, and when saturation is achieved, Au-Ag-Cu-Hg solid solution precipitates. Depending on the composition of hydrothermal solution and the evolution of its physicochemical parameters, the forms of metal complexes change during transfer and deposition of ore elements (Figure 9).

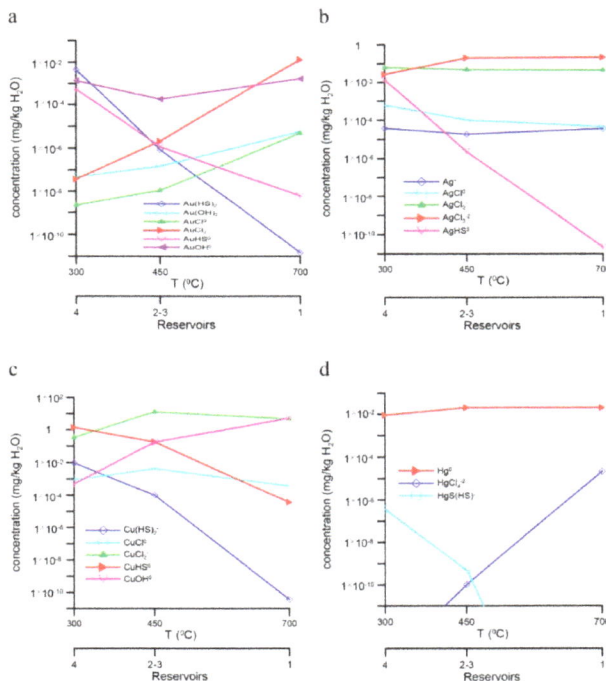

Figure 9. Content of different forms of metal complexes in ore-bearing solution containing average contents of ore components in ultramafic rocks: (**a**) Au, (**b**) Ag, (**c**) Cu, and (**d**) Hg.

In the suggested model, gold in the fluid (Figure 9a) in the first reservoir (700 °C) is mainly represented by chloride complex $AuCl_2^-$ [29]. Thereafter, when the fluid was transported to the next reservoirs and the temperature decreased, its amount decreased significantly: at 450 °C, the main gold-bearing complex is hydroxocomplex $AuOH^0$ in the presence of small amounts of $Au(HS)_2^-$ and $AuHS^0$ [30], and at 300 °C, the role of gold sulfide complexes increased drastically [29].

NaCl-rich fluid determines the predominance of chloride complexes of Ag and Cu in the 1.–3. reservoirs (Figure 9b,c). In contrast to silver, the copper hydrocomplex $CuOH^0$ is also present in the fluid. Mercury, opposed to Au, Ag, and Cu, is primarily present in the solution as Hg^0, whereas chloride and sulfide complexes are minor (Figure 9d).

4.2. Specific Features of Mineral Composition of Rocks in Different Reservoirs of the Model

Mineral compositions of rocks in different reservoirs of the model are presented in Figure 10. The first reservoir corresponds to the lower crust rocks and has a stable set of minerals, dominated by olivine (to 70 wt %) accompanied by smaller amounts of garnet, magnetite, and chlorite (Figure 10a). In the second reservoir, imitating the deep fluid infiltration through serpentinites, magnetite (less than 10%) and carbonate (less than 3%) appear as minerals in equilibrium with serpentine. The other stages of the process are marked by the occurrence of pyroxene, talc, and tremolite (Figure 10b). Through the third reservoir, the fluid passed in a flow mode, interacting with a minor amount of host rocks (serpentinites), which can be interpreted as precipitation of minerals from the fluid in an open crack system. Within the third reservoir, a mixture of carbonate (to 70 wt %), magnetite (to 30 wt %) and a small amount of chlorite (to 5 wt %) was within open cavities and cracks. Pyroxene formed from the first portions of solution became unstable (Figure 10c). The composition of carbonate solid solution was dominated by calcite. In the fourth reservoir, mineral formation occurred in a closed system in which serpentinite was replaced by chloritolite, a rock composed of chlorite (to 90%) and small amounts of talc and carbonate (Figure 10d). Ilmenite and hydroxyapatite were accessory minerals in all reservoirs. In addition, chalcopyrite up to 0.2 wt % was formed in the fourth reservoir. Gold was deposited at all stages of the rock-water interaction during the formation of magnetite-chlorite-carbonate rocks (third reservoir) and at the initial stages in the fourth reservoir, when the transformation of serpentinites initialized (Figure 10c,d).

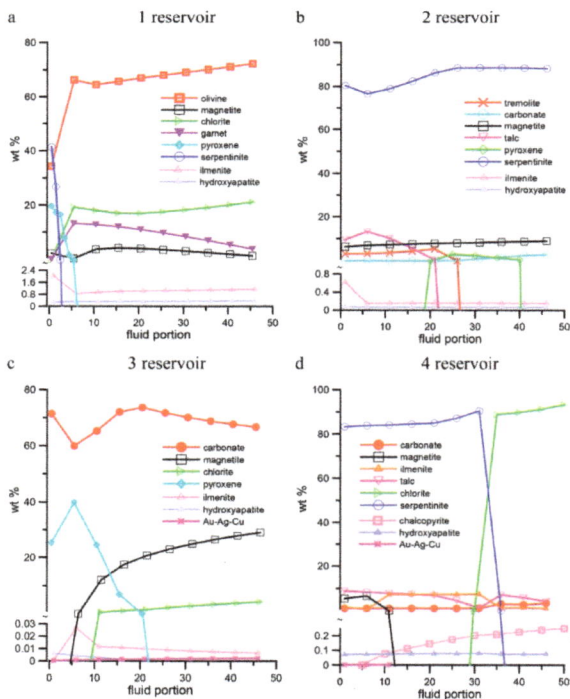

Figure 10. (**a–d**) Changes in the mineral composition of model reservoirs 1–4, respectively, as the amount of incoming fluid increases.

4.3. Uplift of Deep-Seated Rocks under Low T,P-Conditions

To analyze the changes in the mineral composition during the ascent of deep-seated rocks into the region of lower temperatures and pressures, we considered a particular model in which mineral associations were calculated in the range of T = 300–700 °C and P = 1–10 kbar based on the chemical composition of chlorite-olivine rocks (Figure 11).

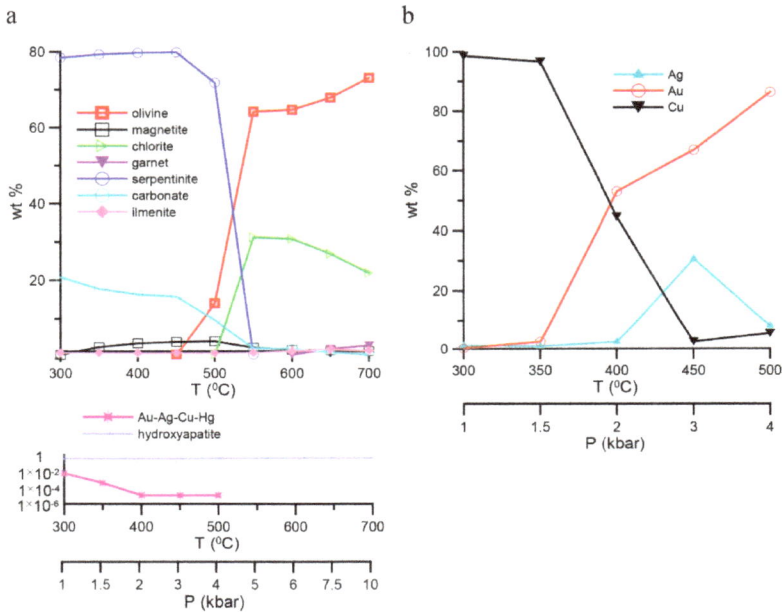

Figure 11. Mineral composition of (**a**) rocks and (**b**) Au-Ag-Cu solid solutions in changing pressure (P) and temperature (T) conditions.

When T = 450–300 °C and P = 3–1 kbar, the olivine-rich rock was replaced by the carbonate-serpentine rocks with a mineral composition similar to serpentinite from the second reservoir of the basic model. It contained up to 15–20 wt % carbonate (mainly of magnesian composition), a small amount (up to 5 wt %) magnetite and less than 1 wt % ilmenite, apatite, and Au-Ag-Cu in solid solution (Figure 11a).

The Au-Ag-Cu composition of solid solution varied from copper gold (8% Ag + 87% Au + 5% Cu) at 500 °C to copper-bearing electrum (30.5% Ag + 67% Au + 2.5% Cu) at 450 °C, which, after a further decrease in temperature to 300 °C, was replaced by native copper with minor admixture of silver and gold (98.8% Cu + 1% Ag + 0.2% Au) (Figure 11b).

5. Discussion

5.1. Similarity of Natural and Model Composition of Rocks and Gold Mineralization

Our calculations provide good similarity of the modeled and natural minerals compositions of altered rocks from the Karabash massif. Although garnet-chlorite-olivine rocks corresponding to the mineral composition of the first reservoir have not been found near the present-day erosion level, they are suggested to occur at a considerably deeper level or, most likely, their mineral composition was altered during the movement of rocks and fluids to the region of lower temperatures and pressures.

The calculated particular model (Figure 11) shows the possibility of transformation of essentially olivine-rich rocks to serpentinite with a small amount of carbonate and magnetite. Similar serpentinites are a significant part of the Karabash massif. They commonly contain native copper, whose origin is confirmed by the thermodynamic calculations at 300 °C. Model calculations at 450 °C corroborate the formation of copper-bearing Au-Ag solid solution and copper sulfides in serpentinites.

The gold-bearing rocks composed of magnetite, chlorite, and carbonate formed in the third reservoir modeling the decompression in open cracks. The P,T-parameters accepted for this reservoir were similar to the formation conditions of magnetite-chlorite-carbonate rocks (480–280 °C, 2–3 kbar). The quantitative ratios of the main minerals (magnetite, chlorite, and carbonate) in the model reservoir correspond to their variations in natural parageneses. Good correlation of model and natural paranegeses was observed in the set of accessory minerals of rocks. In both cases, they are represented by ilmenite, apatite, chalcopyrite, and Au-Ag-Cu solid solutions.

The calculated composition of Au-Ag-Cu solid solutions in the third reservoir is similar to the composition of native gold in magnetite-chlorite-carbonate rocks. The model solid solution differs from copper gold of the Karabash ultramafic massif only in terms of a slightly higher silver content. Moreover, the calculation results of various metal ratios in solution showed that the deep-seated rocks of ultramafic composition with a minor fraction of mafic rocks (0.75 U and 0.25 M) (Figure 8c,d) were the most probable source of these metals. An increase in the fraction of mafic rocks in the model led to the formation of Au-Cu solid solutions that were not observed in magnetite-chlorite rocks.

Association of chloritolite and magnetite-chlorite-carbonate rock is reflected in one of the variants of the model calculations—on cooling of serpentinites of the second reservoir to 300 °C and dilution of the equilibrium fluid by meteoric waters approximately in equal ratios. As this association of rocks in the Karabash massif created submeridionally stretching schist zones, we suggest that meteoric waters infiltrated from the top along tectonic faults into serpentinites toward the deep-seated fluid ascending through the same faults.

The presence of gold and REE in the deep-seated rock source was the main factor influencing the high concentrations of these elements in the magnetite-chlorite-carbonate rocks. We analyzed the model devoid of Y, Zr, REE, U, and Th mineralization due to the complex mineral composition of altered ultramafic rocks from the Karabash massif. This is one of the prospective tasks that will be solved to reveal the simultaneous formation conditions of gold and Y, Zr, REE, U, Th mineralization.

5.2. Scales of Gold Mineralization in the Karabash Massif and the Variations in the Compositions of Gold Minerals

Gold mineralization, including various intermetallides and solid solutions of the Au-Ag-Cu-(Hg) system, occurred in virtually all types of altered ultramafic rocks in the Karabash massif. Our previous physicochemical models of formation of gold-bearing rodingites [6] and magnetite-chlorite-carbonate rocks, serpentinites, and chloritolites, presented in this work, support the variety of minerals forms and different concentrations of Au, Ag, and Cu (Hg) in the rocks of the massif. The maximum gold concentration was determined to be in the rodingites. The gold content in the Zolotaya Gora deposit reaches several hundred ppm, averaging 5 ppm. These rocks are also characterized by the greatest variety of mineral forms of gold—Au-Cu intermetallides (tetra-auricupride AuCu, auricupride $AuCu_3$, and Au_3Cu) and solid solutions of the Au-Ag-Hg system.

Gold concentrations in the third and fourth reservoirs in the model calculated in this work (0.3–30 ppm) correspond to the concentration levels in the magnetite-chlorite-carbonate rocks and chloritolites. In magnetite-chlorite-carbonate rocks, the amount of copper-bearing Au-Ag solid solution may be high, but the small sizes of these rocks do not allow them to be considered objects for mining.

In addition to rodingites and magnetite-chlorite-carbonate rocks, other altered rocks, mainly listvenites and quartz-riebeckite metasomatites, have also been found in the Karabash massif. These rocks either do not contain or are poor in gold. However, in the Southern Urals within the zone of the Main Ural Fault, many auriferous ore deposits have been found in listvenites (Nailinskoe,

Tyelginskoe, Altyn-Tashskoe, and Mechnikovskoe, etc.) [31]. The chemical composition of the native gold in these deposits is similar to the most widespread Au-Ag solid solution containing no less than 1 wt % copper.

Our calculations show that the copper-bearing Au-Ag solid solution could have formed via the participation of the ore fluid during the serpentinization of deep-seated olivine rocks pumped by the tectonic flow to a higher hypsometric level. Data on gold content in serpentinites from the Karabash massif are unknown. However, in the other Ural massifs, antigorite serpentinites confined to the fault zones bear gold mineralization and are the source of gold in the placers [32].

6. Conclusions and Perspectives

We developed a physicochemical model of the formation of ophiolite serpentinites of specific rocks that host gold mineralization with a very low sulfide content. Using thermodynamic calculations, we obtained mineral associations of gold-bearing (magnetite-chlorite-carbonate) and weakly gold-bearing rocks (serpentinites and chloritolites) similar to those occurring in the Karabash massif in the Southern Urals. These rocks formed with the participation of meteoric waters and the deep-seated fluid that changed its composition during the tectonic movement of the crustal-mantle mixture to the surface. The copper gold can precipitate during serpentinization of deep-seated olivine-rich rocks uplifted by the tectonic flow to a higher hypsometric level.

The model supports the involvement of sodium chloride-carbon dioxide fluids extracting ore components (Au, Ag, and Cu) from deep-seated ultramafic rocks with a minor admixture of mafic rocks. The developed model of tectonic squeezing of the deep-seated fluid, crust and mantle rocks allows the prediction of the copper gold ore occurrences in ultramafic ore massifs. At the same time, a more complex model should be developed that also involves Y, Zr, REE, U, and Th mineralization.

Supplementary Materials: The following are available online at: http://www.mdpi.com/2075-163X/8/7/306/s1, Table S1: Trace-element composition of rocks (ppm), Table S2: Components of aqueous solution and gases considered in thermodynamic models, Table S3: Solid phases components used in thermodynamic models.

Author Contributions: V.M., K.C. and C.P. made substantial contributions to the conception of the work and interpretation of data and wrote the paper. K.C. performed the thermodynamic calculations. A.K. revised the geological section. D.V. studied the mineral associations in the rocks and analyzed the compositions of minerals.

Funding: This work was supported by the Russian Foundation for Basic Research (grant 16-05-00407a) and state assignment projects (Nos. 0330-2016-0001, 0350-2016-0033 and 0393-2016-0024).

Acknowledgments: We thank the reviewers for important comments and constructive suggestions which helped us to significantly improve the quality of the manuscript. Special thanks go to the reviewers for correcting the English language of the manuscript. We thank Kiselev, V.D., Garaeva, A.A., Velivetskaya, T.A. and Shanina, S.N. for the analytical data.

Conflicts of Interest: The authors declare no conflict of interest.

References

1. Puchkov, V.N. General features relating to the occurrence of mineral deposits in the Urals: What, where, when and why. *Ore Geol. Rev.* **2017**, *85*, 4–29. [CrossRef]
2. Snachyov, A.V.; Kuznetsov, N.S.; Snachyov, V.I. The Chernoe ozero gold occurrence in carbonaceous deposits of the ophiolite association: The first object of such a type in the Soutern Urals. *Dokl. Earth Sci.* **2011**, *439*, 906–908. [CrossRef]
3. Novgorodova, M.I.; Tsepin, A.I.; Kudrevich, I.M.; Vyal'sov, L.N. New data on crystal chemistry and properties of natural intermetallic compounds in the copper-gold system. *Zap. Vses. Miner. Obshch.* **1977**, *106*, 540–552. (In Russian)
4. Spiridonov, E.M.; Pletniov, P.A. *Zolotaya Gora Deposit of Cupriferous Gold (About "Golden-Rodingite" Formation)*; Scientific World: Moskow, Russia, 2002; p. 220. ISBN 5-89176-169-6. (In Russian)
5. Murzin, V.V.; Varlamov, D.A.; Ronkin, Y.L.; Shanina, S.N. Origin of Au-bearing Rodingite in the Karabash Massif of Alpine-Type Ultramafic Rocks in the Southern Urals. *Geol. Ore Depos.* **2013**, *55*, 278–297. [CrossRef]

6. Murzin, V.V.; Chudnenko, K.V.; Palyanova, G.A.; Varlamov, D.A.; Naumov, E.A.; Pirajno, F. Physicochemical model of formation of Cu-Ag-Au-Hg solid solutions and intermetallic alloys in the rodingites of the Zolotaya Gora gold deposit (Urals, Russia). *Ore Geol. Rev.* **2018**, *93*, 81–97. [CrossRef]
7. Murzin, V.V.; Popov, V.A.; Erokhin, Y.V.; Rakhov, E.V. Mineralogic and Geochemical Features of Gold-Rare-Metal-Rare-Earth Mineralization of Chlorite-Carbonate Rocks from the KARABASH Massif of Ultrabasic Rocks (Southern Urals). In *Ural'skii Mineralogicheskii Sbornik*; IMin UrO, RAS: Miass, Russia, 2005; Volume 13, pp. 123–145. (In Russian)
8. Murzin, V.V.; Varlamov, D.A.; Palyanova, G.A. Conditions of formation of gold-bearing magnetite–chlorite–carbonate rocks of the Karabash ultrabasic massif (South Urals). *Russ. Geol. Geophys.* **2017**, *58*, 803–814. [CrossRef]
9. Murzin, V.V.; Shanina, S.N. Physic-chemical conditions of gold-bearing magnetite-chlorite-carbonate rocks' formation of the Karabash ultramafic massif (the Southern Urals). *Litosphera* **2017**, *17*, 110–117. [CrossRef]
10. Lozhechkin, M.P. *The Karabash Deposit of Cupriferous Gold*; Trudy UFAN SSSR: Sverdlovsk, Russia, 1935; pp. 35–44. (In Russian)
11. Borisenko, A.S. Study of the salt composition of solutions in gasliquid inclusions in minerals by the cryometric method. *Sov. Geol. Geophys.* **1977**, *18*, 11–19.
12. Bodnar, R.J.; Vityk, M.O. Interpretation of microthermometric data for H_2O-NaCl fluid inclusions. In *Fluid Inclusions in Minerals: Methods and Applications*, 2nd ed.; De Vivo, B., Frezzotti, M.L., Eds.; Virginia Tech: Blacksburg, VA, USA, 1994; pp. 117–130.
13. Roedder, E. Fluid inclusions. *Rev. Mineral.* **1984**, *12*, 1–644.
14. Kisin, A.Y.; Koroteev, V.A. *Block Folding and Oreogenesis*; IGG UrB RAS: Ekaterinburg, Russia, 2017; p. 349. (In Russian)
15. Kissin, Y.; Murzin, V.V.; Pritchin, M.E. Tectonic position of the gold mineralization of the Karabash Mountain (Southern Urals): Examination of small structural forms. *Litosphera* **2016**, *16*, 79–91. (In Russian)
16. Karpov, I.K.; Chudnenko, K.V.; Kulik, D.A. Modeling chemical mass-transfer in geochemical processes: Thermodynamic relations, conditions of equilibria and numerical algorithms. *Am. J. Sci.* **1997**, *297*, 767–806. [CrossRef]
17. Chudnenko, K.V. *Thermodynamic Modeling in Geochemistry: Theory, Algorithms, Software, Applications*; Academic Publishing House Geo: Novosibirsk, Russia, 2010; 287p, ISBN 978-5-904682-18-7. (In Russian)
18. Chudnenko, K.V.; Palyanova, G.A. Thermodynamic modeling of native formation Cu-Ag-Au-Hg solid solutions. *Appl. Geochem.* **2016**, *66*, 88–100. [CrossRef]
19. Zhuravkova, T.V.; Palyanova, G.A.; Chudnenko, K.V.; Kravtsova, R.G.; Prokopyev, I.R.; Makshakov, A.S.; Borisenko, A.S. Physicochemical models of formation of gold–silver ore mineralization at the Rogovik deposit (Northeastern Russia). *Ore Geol. Rev.* **2017**, *91*, 1–20. [CrossRef]
20. Tanger, J.C.; Helgeson, H.C. Calculation of the thermodynamic and transport properties of aqueous species at high pressures and temperatures: Revised equations of state for standard partial molal properties of ions and electrolytes. *Am. J. Sci.* **1988**, *288*, 19–98. [CrossRef]
21. Lee, B.I.; Kesler, M.G. Generalized thermodynamic correlation based on three parameter corresponding states. *AIChE J.* **1975**, *21*, 510–527. [CrossRef]
22. Walas, S.M. *Phase Equilibria in Chemical Engineering*; Butterworth Publishers: Boston, MA, USA, 1985; 671p, ISBN 978-0-409-95162-2.
23. Breedveld, G.J.F.; Prausnitz, J.M. Thermodynamic properties of supercritical fluids and their mixtures at very high pressure. *AIChE J.* **1973**, *19*, 783–796. [CrossRef]
24. Garuti, G.; Fershtater, G.B.; Bea, F.; Montero, P.; Pushkarev, E.V.; Zaccarini, F. Platinum Group Elements As a Pertrological Indicators in Mafic-Ultramafic Complexes of the Central and Southern Urals: Preliminary Results. *Tectonophysics* **1997**, *276*, 181–194. [CrossRef]
25. Grigoriev, N.A. *Chemical Element Distribution in the Upper Continental Crust*; UB RAS: Ekaterinburg, Russia, 2009; p. 382. (In Russian)
26. Karpov, I.K.; Chudnenko, K.V.; Kravtsova, R.G.; Bychinskiy, V.A. Simulation modeling of physical and chemical processes of dissolution, transport and deposition of gold in epithermal gold-silver deposits of the North-East Russia. *Russ. Geol. Geophys.* **2001**, *3*, 393–408.
27. Gritsuk, N.A. Petrogeochemical Features and Pre Potential of the Talov Massif of Gabbro-Hyperbasic Rocks. Ph.D. Thesis, Moscow State University, Russia, 2003; p. 148. (In Russian)

28. Berzon, R.O. *Gold Resource Potential of Ultramafics*; VIEMS: Moscow, Russia, 1983; p. 47. (In Russian)
29. Zotov, A.; Kuzmin, N.; Reukov, V.; Tagirov, B. Stability of $AuCl_2^-$ from 25 to 1000 °C at pressures to 5000 bar, and consequences for hydrothermal gold mobilization. *Minerals* **2018**, *8*, 286. [CrossRef]
30. Seward, T.M.; Williams-Jones, A.E.; Migdisov, A.A. The chemistry of metal transport and deposition by ore-forming hydrothermal fluids. In *Treatise on Geochemistry*, 2nd ed.; Turekian, K., Holland, H., Eds.; Elsevier: Amsterdam, The Netherlands, 2014; Volume 13, pp. 29–57.
31. Belogub, E.V.; Melekestseva, I.Y.; Novoselov, K.A.; Zabotina, M.V.; Tret'yakov, G.A.; Zaykov, V.V.; Yuminov, A.M. Listvenite-related gold deposits of the South Urals (Russia): A review. *Ore Geol. Rev.* **2017**, *85*, 247–270. [CrossRef]
32. Murzin, V.V.; Varlamov, D.A.; Shanina, S.N. New Data on the Gold–Antigorite Association of the Urals. *Dokl. Earth Sci.* **2007**, *417A*, 1436–1439. [CrossRef]

![minerals logo] *minerals*

MDPI

Article

Numerical Modeling of REE Fractionation Patterns in Fluorapatite from the Olympic Dam Deposit (South Australia)

Sasha Krneta [1,†], **Cristiana L. Ciobanu** [2,*], **Nigel J. Cook** [2] and **Kathy J. Ehrig** [3]

[1]   School of Physical Sciences, The University of Adelaide, Adelaide, SA 5005, Australia;
     sasha.krneta@adelaide.edu.au
[2]   School of Chemical Engineering, The University of Adelaide, Adelaide, SA 5005, Australia;
     nigel.cook@adelaide.edu.au
[3]   BHP Olympic Dam, Adelaide, SA 5000, Australia; Kathy.J.Ehrig@bhpbilliton.com
*   Correspondence: cristiana.ciobanu@adelaide.edu.au; Tel.: +61-405-826-057
†   Present address: Boart Longyear Asia-Pacific, 6 Butler Boulevard, Adelaide Airport, SA 5950, Australia.

Received: 11 July 2018; Accepted: 6 August 2018; Published: 8 August 2018

Abstract: Trace element signatures in apatite are used to study hydrothermal processes due to the ability of this mineral to chemically record and preserve the impact of individual hydrothermal events. Interpretation of rare earth element (REE)-signatures in hydrothermal apatite can be complex due to not only evolving fO_2, fS_2 and fluid composition, but also to variety of different REE-complexes (Cl-, F-, P-, SO_4, CO_3, oxide, OH^- etc.) in hydrothermal fluid, and the significant differences in solubility and stability that these complexes exhibit. This contribution applies numerical modeling to evolving REE-signatures in apatite within the Olympic Dam iron-oxide-copper-gold deposit, South Australia with the aim of constraining fluid evolution. The REE-signatures of three unique types of apatite from hydrothermal assemblages that crystallized under partially constrained conditions have been numerically modeled, and the partitioning coefficients between apatite and fluid calculated in each case. Results of these calculations replicate the measured data well and show a transition from early light rare earth element (LREE)- to later middle rare earth element (MREE)-enriched apatite, which can be achieved by an evolution in the proportions of different REE-complexes. Modeling also efficiently explains the switch from REE-signatures with negative to positive Eu-anomalies. REE transport in hydrothermal fluids at Olympic Dam is attributed to REE–chloride complexes, thus explaining both the LREE-enriched character of the deposit and the relatively LREE-depleted nature of later generations of apatite. REE deposition may, however, have been induced by a weakening of REE–Cl activity and subsequent REE complexation with fluoride species. The conspicuous positive Eu-anomalies displayed by later apatite with are attributed to crystallization from high pH fluids characterized by the presence of Eu^{3+} species.

Keywords: apatite; numerical modeling; Olympic Dam; rare earth elements; ore genesis

1. Introduction

The concentrations of trace elements (<<1 wt %) and their variation within hydrothermal minerals can provide valuable information on fluid parameters and conditions of ore deposition for assemblages that are well constrained with respect to paragenetic position. Studies have demonstrated the interdependency between hydrothermal conditions and the compositions of specific minerals (e.g., [1–3]). Many such studies have focused on the rare earth elements (REE), which display a coherent behavior to one another due to similar electronic configurations, common trivalent oxidation state, and systematic decrease in atomic radius with increased atomic number. This typically leads to smooth fractionation across the group (e.g., [4]). However, since Eu and Ce may also occur as Eu^{2+} and Ce^{4+} ions, redox-sensitive anomalies may result.

The behavior of the REE in hydrothermal fluids is affected by parameters such as pH, temperature, salinity, redox conditions and fluid composition (e.g., [4–7]), thus allowing REE to be used as geochemical tracers in hydrothermal systems. Emphasis has been placed on determining the thermodynamic properties of various REE complexes in hydrothermal fluids at temperatures typical of ore deposit formation [6,7], which can support numerical modeling of REE behavior.

Modeling of REE patterns for zoned calcic garnet [1] and scheelite [2,8] have shown the sensitivity of these minerals to changes in fluid parameters. In both cases, modeling addressed compositionally-zoned minerals, in which core-to-rim compositional variation was modeled in terms of partitioning between mineral and fluid, involving evolving fluid parameters, successfully reproducing the patterns measured. The work of Brugger et al. [2], which focused on high-grade orogenic gold ores, emphasized the sensitivity of REE patterns in scheelite to pH variation, and suggested that apatite should behave similarly and display analogous signatures enriched in middle rare earths (MREE).

Chondrite-normalized REE fractionation patterns of apatite-group minerals are recognized as valuable tools for understanding hydrothermal processes [9]. Despite being widespread in a wide range of magmatic to metamorphic rocks [10–14], apatite chemistry has not previously been modeled in the same way as scheelite or calc-silicate minerals. Chondrite-normalized REE fractionation trends for apatite are widely reported from a variety of rocks but granite related apatite is conspicuous by consistent, downward-sloping trends featuring relative enrichment in light rare earths (LREE) [12–14]. Variation in the size and sign of Eu anomalies across rock suites from across metallogenic provinces have been used to infer variability or change in redox conditions (e.g., [12]).

Apatite is an abundant component of Iron-Oxide-Copper-Gold (IOCG) systems, including those within the Olympic Cu-Au Province, South Australia [15–18]. In deposits and prospects from the Olympic Dam District, apatite is particularly abundant within early apatite-magnetite assemblages [19]. Recent study of apatite from Olympic Dam [15–17], by far the largest deposit in the Olympic Cu-Au Province [20], has demonstrated a systematic compositional variation that can be correlated with changes in the host intrusive or hydrothermal assemblage. Distinct geochemical signatures in terms of F, Cl, S, As, and REE are characteristic of apatite from certain hydrothermal assemblages. They vary from LREE-enriched types within early, high-temperature magnetite–chlorite $\pm$ pyrite $\pm$ chalcopyrite assemblages preserved on the margins of the deposit to MREE-enriched signatures towards the margins of the orebody in which hematite (+sericite) is the dominant alteration. Finally, apatite within late-stage massive bornite mineralization shows the greatest MREE-enrichment and a marked positive Eu-anomaly [17]. MREE-enrichment in association with hematite-sericite alteration has also been documented from the Wirrda Well and Acropolis IOCG prospects SSE and SSW of Olympic Dam, suggesting that such a feature may be a generic feature linked to ore genesis [16,17].

The transition from early high-temperature magnetite-dominant to later hematite-dominant assemblages in the Olympic Dam District coincided with significant changes in the mineralizing fluids as determined by fluid inclusion studies [21,22]. Given the well documented association between apatite with a specific chemistry and hydrothermal assemblages formed under known fluid conditions, numerical modeling of apatite/fluid partitioning can offer valuable insights into the behavior of REE during formation of the deposit. In this study, REE behavior and partitioning coefficients between apatite and fluid are modeled numerically employing empirical mineral compositional data and newly published thermodynamic values for REE complexes [7], along with assumptions for other fluid parameters grounded in the stabilities of the host assemblage. Using three examples, each with unique apatite REE-signatures, it can be demonstrated that the transition from LREE- to MREE-enriched and finally MREE-enriched signatures with positive Eu-anomalies, is a direct result of changes in the character of the hydrothermal fluids.

2. Description of the Deposit, Apatite-Group Minerals and Rare Earth Element Behavior

2.1. The Olympic Dam Deposit

Olympic Dam is by far the largest single IOCG deposit within the Olympic Cu-Au Province (10,400 million tonnes at 0.77% Cu, 0.25 kg/t U_3O_8, 0.32 g/t Au and 1 g/t Ag; [23]). The deposit is hosted within the Olympic Dam Breccia Complex (ODBC), comprising dominant granite breccias and subordinate intrusive rocks and other lithologies [20,24]. The ODBC is in turn hosted within the Roxby Downs Granite (RDG), a pink, hydrothermally altered, two-feldspar granite belonging to the ~1.59 Ga Hiltaba Intrusive Suite [25,26]. Geochronological U-Pb data for magmatic and hydrothermal zircon [27] and for hydrothermal hematite in the deposit [28,29] indicate that mineralization and associated alteration is associated with RDG.

2.2. Apatite Mineralogy, Geochemistry and Crystal Structure

The apatite supergroup [30] constitutes a large group of named minerals made possible by the extraordinary flexibility of the apatite structure (e.g., [31]), which allows for incorporation of approximately half the elements in the periodic table [32]. The general formula of apatite supergroup minerals is defined as $A_5(XO_4)_3Z$, where the A position is most commonly occupied by Ca^{2+} but can be substituted by a variety of other di-, tri- and tetravalent cations such as Na^+, Sr^{2+}, Pb^{2+}, Ba^{2+}, Mn^{2+}, Fe^{2+}, Mg^{2+}, Ni^{2+}, Co^{2+}, Cu^{2+}, Zn^{2+}, Sn^{2+}, Cd^{2+}, Eu^{2+}, REE^{3+}, Y^{3+}, Zr^{4+}, Ti^{4+}, Th^{4+}, U^{4+}, and S^{4+}, as well as U^{6+}. The X position is dominantly occupied by P, as PO_4^{3-} but can also accommodate SO_4^{2-}, AsO_4^{3-}, VO_4^{3-}, SiO_4^{4-}, CO_3^{2-}, CrO_4^{2-}, CrO_4^{3-}, GeO_4^{4-}, SeO_4^{4-} and WO_4^{3-}. Lastly, the Z position hosts F, Cl and OH^- defining the three end-members fluorapatite, chlorapatite and hydroxyapatite [33].

Three cation polyhedra make up the apatite structure (Figure 1), a single, rigid PO_4 tetrahedron and two Ca polyhedra, Ca1 and Ca2 [33]. Of these, the Ca2 position dominantly hosts LREE and the Ca1 position the heavy-REE (HREE), with Nd expressing no preference for either position [32]. The size of these positions exerts the dominant control on trace element substitution through the proximity principle [34], meaning that elements with atomic radii closest to that of the substituting position are most readily substituted. In instances where an elements valance state is different to that of the position it is substituting, such as in the case of REE^{3+} substituting for Ca^{2+}, overall charge balance is maintained through a variety of different coupled charge-compensated substitutions [35,36].

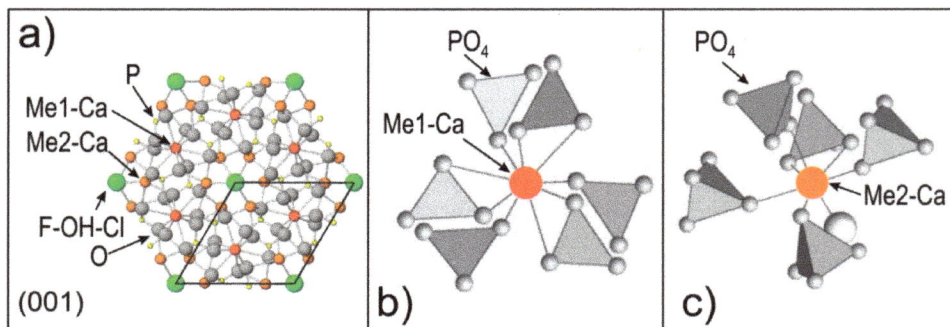

Figure 1. Crystal structure of apatite. (**a**) Atomic arrangement projected down to (001) (from Hughes [37]); (**b**,**c**) Environment of Me1 and Me2 sites, respectively, (from Luo et al. [38]). The different grey shadings of the PO_4 tetrahedra show their orientation relative to projection planes.

In the magmatic environment, once the effects of the proximity principle, whole rock composition and the co-partitioning of REE between apatite and other minerals are taken into account, apatite REE-signatures are largely predictable. In contrast, REE-trends in hydrothermal apatite could be more varied since the fluids capable of crystallizing apatite can carry variable concentrations of REE as a range of soluble complexes (e.g., [7]). These factors, combined with temperature, pH, fO_2 and others, dictate the dominance or absence of an individual REE at the conditions of apatite crystallization, in turn resulting in a range of REE-signatures among hydrothermal apatites.

For example, redox-sensitive Eu, can be present as Eu^{3+} and Eu^{2+} giving it the ability to partition away from the other trivalent REE. This behavior has been used to infer variability or change in redox conditions (e.g., [12]), as well as changes in fluid parameters due to the sensitivity of Eu^{3+}/Eu^{2+} complexing to pH [2]. No evidence is seen in the deposit for the presence Ce^{4+} (e.g., as cerianite), which might also influence partitioning trends. Similarly, the common tendency for the LREE to speciate as stable and soluble LREE-chloride complexes to a greater extent than for HREE can lead to spatial fractionation between the LREE and HREE [7].

2.3. REE Trends in Fluorapatite from the Olympic Dam Deposit

Mineralogical zoning is expressed across the Olympic Dam deposit by variation in the dominant Fe-oxide and Cu-(Fe)-sulfide species [20] whereby assemblages of magnetite ± pyrite ± chalcopyrite grade laterally from the peripheries of the deposit and upwards from depth to hematite-dominant chalcopyrite+bornite assemblages, and finally to chalcocite-dominant assemblages at shallow levels. Similarly, the dominant silicate alteration minerals vary from chlorite-dominant in association with magnetite assemblages (in which magmatic feldspars are also often preserved) to sericite-dominant in association with hematite [20].

Krneta et al. [15,17] defined the morphological and chemical characteristics of apatite across the Olympic Dam deposit and showed that apatite associated with a specific intrusive rock or hydrothermal assemblage displays chemical characteristics unique to that particular assemblage. Moreover, apatite was found to record subsequent hydrothermal overprinting events expressed within zoned grains. The three apatite REE trends previously described in the literature [17] and described below, are illustrative of changes from environments representing the magmatic-hydrothermal transition (altered granite from the deposit outer shell) to high-grade Cu-mineralization.

Hydrothermal apatite, abundant within the magnetite-bearing assemblages such as the '*deep mineralization*' [15] representing one of the deepest mineralized intervals (2.3–2.33 km) intersected by drillholes just outside the orebody (~1 km E) is illustrative of the earliest hydrothermal stage. Here, apatite increases markedly in abundance and grain size across the contact between RDG and a porphyritic felsic unit, both intensively altered. Apatite occurs as individual grains up to several mm in width, or as cm-scale aggregates consisting of multiple ~500 μm grains commonly interstitial to magnetite. Compositionally, such apatite is LREE-enriched with moderate negative Eu-anomalies (Figure 2, Table 1).

A second trend is recorded within apatite from granite displaying locally abundant sericite + hematite alteration (*hematite-sericite altered RDG*). Although hematite-sericite alteration is developed throughout the orebody [39], this also partially overprints the weakly-mineralized (magnetite + chalcopyrite + pyrite) granite breccias forming an outer shell around the orebody at Olympic Dam. Early generations of hydrothermal apatite from such locations immediately adjacent to the orebody, are altered along fractures and grain rims and depleted in LREE, S and Cl. This apatite is MREE-enriched with a weak negative Eu-anomaly (Figure 2, Table 1; [15,17]).

The third trend, with characteristic highest MREE-enrichment, is representative of apatite associated with massive, high-grade bornite mineralization from the orebody at Olympic Dam. Unlike the hematite-sericite associated mineralization in altered RDG, such apatite displays a positive Eu-anomaly (*high-grade bornite ore hosted apatite*, Figure 2, Table 1; [17]). Such a signature is very unusual for IOCG apatite in the Olympic Province (e.g., [15–18]), however it has also been observed from the Acropolis deposit [16], where it is shown to be the result of crystallization from late-stage neutral to alkaline fluids based on mineral stabilities and Eu-speciation.

Table 1. REE concentrations (in ppm) of three apatite types hosted within the Olympic Dam mineralized system [15,17] used as examples in the numerical modeling of partitioning between apatite and fluid.

	Apatite in Altered Granite		Ore-Hosted Apatite
	Deep Mineralization REE + Y-, S- and Cl-rich cores	**Hematite-Sericite Altered**	**High-Grade Bornite Ore**
La	1520	84	7
Ce	4192	258	49
Pr	504	47	17
Nd	1868	280	194
Sm	298	195	631
Eu	27	45	391
Gd	247	290	1217
Tb	32	32	97
Dy	187	125	365
Ho	37	16	49
Er	104	33	92
Tm	13	4	9
Yb	83	24	39
Lu	11	3	4

2.4. Fluid Evolution within IOCG Systems, REE Speciation and the Controls on Apatite/Fluid Partitioning

Study of fluid inclusions within IOCG mineralization in the Olympic Cu-Au Province [21,22,40,41], as well as from IOCG deposits globally [42–44] has provided insights into the formation of these deposits by defining the temperatures and salinities of fluids. In the case of Olympic Dam, early magnetite-dominant mineralization was found to have formed at temperatures exceeding 400 °C from fluids with salinities between 20 and 45 wt % NaCl equiv. [21]. Similar magnetite-dominant assemblages throughout the Olympic Cu-Au Province formed from analogous fluids [22]. Hematite-dominant mineralization at Olympic Dam may, however, have crystallized from cooler (150–300 °C), low-salinity (1–8 wt % NaCl) fluids [21]. Several authors have speculated that this transition was largely responsible for deposition of Cu-Au within many of the IOCG systems given that the early magnetite-stage ore fluids were copper-rich (>500 ppm Cu based on PIXE analysis; [22]) but lacked a suitable depositional mechanism. Similarly, Haynes et al. [45] modeled this transition and found that changes in fluid salinity, temperature and other parameters, notably pH, were sufficient to bring about deposition of Cu-Au-U-Ag mineralization.

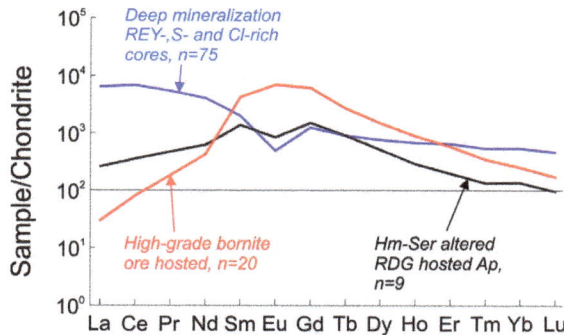

Figure 2. Chondrite-normalized REE-signatures of three apatite types hosted within the Olympic Dam mineralized system [17] used in the numerical modeling of partitioning between apatite and fluid. *n* refers to the number of laser-ablation inductively coupled plasma mass spectrometry analyses used to defining each mean trend. REE = Σ(REE + Y).

None of these authors considered, however, the behavior of the REE during this transition despite the presence of REE within the deposit at concentrations well above crustal values, particularly in the case of La and Ce. The work of Migdisov et al. [7] provides the most comprehensive account of REE behavior in hydrothermal fluids along with thermodynamic properties for a suite of REE complexes derived from experiments conducted at elevated temperatures. The new thermodynamic data are inconsistent with the ambient temperature extrapolated values of Haas et al. [4] and can constrain the roles of the various REE complexes within hydrothermal systems formed at higher temperatures and neutral to acidic conditions. Migdisov et al. [7] emphasize that the low solubility of REE–F, –P, –oxide, –OH$^-$ and –CO$_3$ complexes makes them unlikely to be involved in REE transport. Conversely, the high solubility and stability of REE–Cl and –SO$_4$ complexes, along with their likely high concentration in hydrothermal fluids (e.g., highly saline hydrothermal fluids), suggests that they play a dominant role in REE transport. Significant variability is, however, observed among the REE with regards to complexing behavior. LREE are much more stable as Cl-complexes than HREE, suggesting that they should also be more readily transported and concentrated upon precipitation. Modeling can suggest and illustrate that such a hypothesis is very likely correct [7] and is supported indirectly by the predominance of LREE-enriched hydrothermal systems as opposed to those enriched in HREE (e.g., [46,47], even though the unique geological settings of specific hydrothermal systems may facilitate further complexities.

3. Methodology

Numerical Modeling of Apatite/Fluid Partitioning Coefficients

Brugger et al. [2,8], Smith et al. [1] and Migdisov et al. [7] have outlined the methodology for calculation of fluid partitioning coefficients in minerals where Ca is a major component. Substitution mechanisms must be defined as a preliminary requisite step. These can either be direct, e.g., Eu^{2+} for Ca^{2+} substitution, or via charge-compensated coupled substitutions involving elements such as Na or Si in the case of REE^{3+}, as reported and discussed by Krneta et al. [17]. For calculation of partitioning coefficients between apatite (Ap) and fluid (aq), the two substitution mechanisms proposed are:

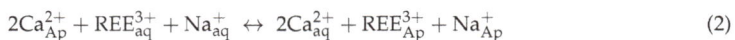

$$Ca^{2+}_{Ap} + Eu^{2+}_{aq} \leftrightarrow Ca^{2+}_{aq} + Eu^{2+}_{Ap} \tag{1}$$

$$2Ca^{2+}_{Ap} + REE^{3+}_{aq} + Na^+_{aq} \leftrightarrow 2Ca^{2+}_{aq} + REE^{3+}_{Ap} + Na^+_{Ap} \tag{2}$$

Although REE incorporation into the apatite types used here (Figure 1, Table 1) appears to be primarily accommodated by Si^{4+}- for -P^{5+} substitution based on EPMA analysis [15], the simplifying assumption that the dominant substitution involves charge compensation by Na$^+$ has been made to eliminate the complications involved in considering multiple elements in this role. This is considered in line with analogous approaches (e.g., [1]).

Using Equations (1) and (2) the equilibrium constants (K) expressed in Equations (3) and (4), respectively, can be defined, where 'a_i' is the activity of element 'i' in the fluid, or in apatite:

$$K_1(P, T) = \frac{a_{Eu^{2+}_{Ap}} \, a_{Ca^{2+}_{aq}}}{a_{Eu^{2+}_{aq}} \, a_{Ca^{2+}_{Ap}}} \tag{3}$$

$$K_2(P, T) = \frac{a^2_{Ca^{2+}_{aq}} \, a_{REE^{3+}_{Ap}} \, a_{Na^+_{Ap}}}{a^2_{Ca^{2+}_{Ap}} \, a_{REE^{3+}_{aq}} \, a_{Na^+_{aq}}} \tag{4}$$

The apatite/fluid partitioning coefficient (D_i) is related to the equilibrium constant K_i by Equations (5) and (6), where 'Y_i' denotes the activity coefficient of element 'i' and the square brackets in (6) define the concentration of the elements in the fluid in molal terms, and as mole fraction in the apatite:

$$D_{Eu^{2+}}^{\frac{Ap}{fluid}} = K_1(P, T) \frac{Y_{Eu^{2+}_{aq}}}{Y_{Ca^{2+}_{aq}}} \frac{Y_{Ca^{2+}_{Ap}}}{Y_{Eu^{2+}_{Ap}}} \tag{5}$$

$$D_{REE^{3+}}^{\frac{Ap}{fluid}} = K_2(P, T) \frac{Y_{REE^{3+}_{aq}}}{Y^2_{Ca^{2+}_{aq}}} \frac{Y_{Na^+_{aq}}}{Y_{REE^{3+}_{Ap}}} \frac{Y^2_{Ca^{2+}_{Ap}}}{Y_{Na^+_{Ap}}} \frac{\left[Na^+_{aq}\right]}{\left[Ca^{2+}_{aq}\right]} \frac{\left[Ca^{2+}_{Ap}\right]}{\left[Na^+_{Ap}\right]} \tag{6}$$

Assuming that the solid solution is dilute and ideal, the activity coefficients for Ca and each individual REE can be assumed to be at unity. Moreover, if the fluid is considered dilute and ideal, then Ca and each of the other REE are equally at unity allowing (6) to be reduced to (7):

$$D_{REE^{3+}}^{Ap/fluid} = K_2(P, T) \frac{\left[Na^+_{aq}\right]}{\left[Ca^{2+}_{aq}\right]} \frac{\left[Ca^{2+}_{Ap}\right]}{\left[Na^+_{Ap}\right]} \tag{7}$$

Using values obtained for individual REE from Equations (5) and (7) together with Equation (8) modified from Smith et al. [1], the crystallization of apatite can be modeled in an open system involving a constantly replenished fluid accounting for the common lack of intra-grain zoning observed in the apatite types used as the basis for the preceding models. This methodology also largely follows Brugger et al. [8]:

$$C_{Ap} = D_{REE^{3+}}^{\frac{Ap}{fluid}} \times C_{aq}^{(D-1)} \tag{8}$$

Activities of REE and other relevant components were calculated using Geochemist's Workbench® 11 [48] after updating the database for REE–Cl and –F aqueous species according to thermodynamic values given by Migdisov et al. [49]. In the case of LREE, the activity is most commonly represented by a single chloride complex whereas the activity of MREEs and HREEs are distributed among multiple species (chloride, fluoride, hydroxide and oxide). As such these were combined and a total value, $a_{REE_{total}}$ was used in the calculations. Although this method may be a simplification of the natural hydrothermal system, it is nevertheless in line with the approach followed elsewhere. Previous studies (e.g., [1]) have resolved the complication of an individual REE being present as multiple complexes by not defining the various species and treating the activity of an individual REE as a single parameter. In other published work, notably Brugger et al. [2], the scenario investigated involved the speciation of Eu under two sets of conditions, each of which contained Eu exclusively as a single species.

4. Results

4.1. Study Cases and Determination of Fluid Conditions

The three apatite sub-types with REE trends, as shown in Figure 2, and corresponding to measured REE mean compositions tabulated in Table 1 (see also [17]), are considered as study cases illustrative of fluid evolution from the magmatic-hydrothermal transition to high-grade mineralization stages. To perform the numerical modeling for the apatite/fluid REE partitioning, a set of fluid conditions for each case must be defined along with the concentrations of key components, such as the REE, complexing ligands, Ca and Na (Table 2).

Given that no empirical measurements of fluid REE, Ca or complexing ligand concentrations currently exist, these need to be assumed. Following Smith et al. [1], the various element concentrations are assumed to be equal to the whole rock concentrations of the interpreted fluid source (this is unaltered RDG in the case of the deep mineralization), or the whole rock concentrations in the rock host in the case of the hematite-sericite associated and the high-grade bornite ore hosted apatite. Whole rock values given by Ehrig et al. [20] for unaltered RDG, sericite-altered RDG and mineralized RDG (10–20 wt % Fe, Cu ≥ 3000 ppm) were used for the three cases as outlined in Table 2. In the absence of fluid inclusion data specific to the present assemblages (granite or high-grade bornite ore), salinity conditions showing a decrease from 20 to 10 and 5 NaCl wt % equiv. are chosen for cases 1 to 3. This is concordant with previous fluid inclusions studies of IOCG deposits in the Olympic Dam deposit/district showing such a trend from early to late hydrothermal evolution. For purposes of simplicity, a temperature (T) of 300 °C is considered in all cases. This is considered feasible considering the ranges of T obtained for assemblages-bearing both magnetite or hematite.

Table 2. Numerical model fluid parameters and fluid chemistry.

	Case 1 Mt + Py + Cp K-Feldspar Stable		Case 2 Hm + Py + Cp Sericite Stable		Case 3 Hm + Bn Silicates Absent	
Model	**1.1**	**1.2**	**2.1**	**2.2**	**3.1**	**3.2**
O_2 aq (log g)	−36.8	−36.8	−31.7	−31.7	−31.7	−32
H^+ (pH)	5.2	5.2	4.4	4.4	8.5	6.6
NaCl (wt %)	20	20	10	10	5	5
Ca (wt %)	0.61	3	0.37	4	1.06	3.5
HCO_3^- (wt %)	0.37	0.37	0.94	0.94	2.40	2.40
F- (wt %)	0.005	0.005	0.005	0.005	0.005	0.005
SO_4^- (wt %)	0.15	0.15	0.15	0.15	2.00	2.00
Temperature (°C)	300	300	300	300	300	300
La (ppm)	100	100	100	100	1200	1200
Ce (ppm)	200	200	200	200	1700	1700
Pr (ppm)	21	21	24	24	150	150
Nd (ppm)	70	70	80	80	393	393
Sm (ppm)	12.4	12.4	14	14	41.2	41.2
Eu (ppm)	1.7	1.7	1.9	1.9	12.4	12.4
Gd (ppm)	10.3	10.3	11.8	11.8	28	28
Tb (ppm)	1.6	1.6	1.7	1.7	3.4	3.4
Dy (ppm)	9.5	9.5	10.3	10.3	16.7	16.7
Ho (ppm)	1.9	1.9	2	2	3	3
Er (ppm)	5.7	5.7	6.1	6.1	9.1	9.1
Tm (ppm)	0.5	0.5	0.7	0.7	1.2	1.2
Yb (ppm)	5.6	5.6	6.3	6.3	8.6	8.6
Lu (ppm)	0.9	0.9	1.1	1.1	1.3	1.3

Estimation of wt % Ca, HCO_3^-, F and SO_4^- were made considering the whole rock data as well as the mineralogy of each assemblage. In each case, two subsets of conditions were considered by varying each of these parameters, except for F, which was kept constant, as shown in Table 2. Variation in Ca from ≤1 to 3–4 wt % was introduced for each case to test the effect this has on the obtained trends. A substantial increase in SO_4^- was considered for case 3 relative to the other two cases to reach precipitation of a sulfide-rich mineral assemblage. The presence of fractures in apatite containing micron-scale broken fragments of sulfide minerals [17] infers fluids with a high volatile component, which were approximated by increasing HCO_3^- in case 3. Numerical estimates for these parameters were varied to obtained a best fit between the calculated and measured REE trends. Variation in the activity of REE complexes is shown in Table 3 for the six sets of conditions.

4.2. Apatite/Fluid REE Partitioning and the Effects of Evolving Fluid Conditions

Using the methods and fluid parameters described and chemistry as outlined in Table 2, six models were generated, two for each apatite type (deep mineralization, hematite-sericite associated and massive bornite hosted apatite). Results of the modeling and calculations are shown in Appendix A with the model apatite and apatite/fluid partitioning coefficient D shown graphically as Figure 3.

Within the generated models, by far the best fits to the measured apatite compositions are provided by models 1.2 and 2.2. The remaining models significantly understate the observed absolute REE concentrations but nonetheless successfully replicate the shape of the measured chondrite-normalized REE fractionation trends. Many of these however are within the concentration ranges obtained for the measured apatite. To replicate the absolute mean REE concentrations, models 1.2 and 2.2 required an assumption of high Ca concentrations in the fluid (3 and 4 wt %), respectively which may be excessive for such fluids, even if it is noted that Haynes et al. [45] proposed fluids containing ~2.5 wt % Ca. As suggested, the primary effect of increasing Ca concentrations is an increase in the absolute REE levels within model apatite. However, this is not the case in models 3.1 and 3.2, suggesting that at the pH values at 300 °C proposed for these models (8.5 and 6.6, respectively) the effects of increasing Ca concentration are minor. Similarly, varying Na concentrations within all models, in isolation, primarily effects the absolute levels of REE but to a lesser extent than variation in Ca.

The various model groups are primarily distinguished with respect to changes in apatite/fluid partitioning coefficients, D, and the speciation of individual REE. For example, models 1.1 and 1.2 display relatively flat fractionation behavior, varying across a single order of magnitude, whereas in the case of the models attempting to replicate the MREE-enriched apatite varieties, they vary across at least 4 orders of magnitude with very low D values for LREE, increasing towards the MREE, and decreases slightly for HREE (Figure 3). These very low rates of LREE fractionation are, to a certain extent, artefacts of the highly LREE-enriched nature of the fluids chosen to represent models 2.1, 2.2, 3.1 and 3.2. However, in the case of models 2.1 and 2.2, an explanation for this fractionation can be observed in the increased proportion of fluid LREE activity being represented by Cl complexes (Table 3), a change that can be expected to increase the likelihood of these elements remaining mobile [7] and thus inhibit incorporation into apatite. Moreover, this would explain the commonly observed LREE depletion associated with hematite-sericite overprinting of pre-existing apatite [15,17]. An increase in the dominance of Cl complexes is primarily caused by a drop in pH from 5.2 to 4.4 and does not exhibit a sensitivity to changes in NaCl concentrations since sufficient Cl is available in all the models to facilitate formation of REE–Cl species.

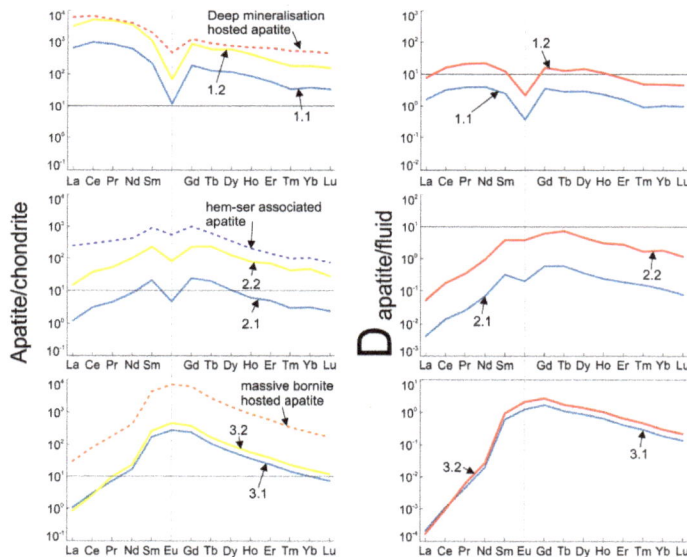

Figure 3. (a) Measured (dashed line) and model (full lines) chondrite-normalized concentrations of REE; and (b) the corresponding calculated apatite/fluid partitioning coefficients D for each model.

Within all models chosen, with the exceptions of 3.1 and 3.2, REE–F species account for a portion of each REE's activity. This is particularly true in the case of models 2.1 and 2.2 where a major proportion of HREE activity is represented by these complexes. In comparison to models 1.1 and 1.2 where these species are carried largely by oxide and hydroxide species, the transition to lower pH conditions would inhibit the crystallization of F as fluorite and increase its availability for REE complexing [7].

Table 3. Percentages of REE activity represented within the model fluids as various complexes. Chloride complexes dominate LREE speciation in models 1.1 and 1.2. This proportion increases within models 2.1 and 2.2, in which Cl complexes increasingly represent all REE. Models 3.1 and 3.2 are entirely dominated by oxide and hydroxide species. REE–SO$_4$ species are absent in most models and only minor constituents in models that assume very high SO$_4$ concentrations (~2 wt %).

	% of REE Activity as Cl Complexes						Proportion of REE Activity as Oxide and OH$^-$ Complexes					
Model	1.1	1.2	2.1	2.2	3.1	3.2	1.1	1.2	2.1	2.2	3.1	3.2
La	91.9	94.0	97.5	99.1	0	3.5×10^{-3}	6.21	4.57	0.20	0.11	100	100
Ce	63.7	70.8	93.8	97.1	1.2×10^{-10}	4.9×10^{-4}	31.7	27.0	1.89	1.09	100	100
Pr	67.0	73.6	96.1	98.2	0	1.6×10^{-4}	29.6	25.0	1.67	0.97	100	100
Nd	30.7	36.9	84.6	92.6	0	2.9×10^{-5}	65.3	61.9	6.01	4.39	100	100
Sm	10.2	13.1	64.5	79.7	0	1.7×10^{-6}	85.5	84.8	12.9	10.4	100	100
Eu	100	100	100	100	0	0	0	0	0	0	100	100
Gd	13.3	18.3	57.3	77.1	0	1.7×10^{-6}	72.4	73.9	7.40	5.32	100	100
Tb	5.04	7.65	34.4	58.8	0	7.4×10^{-7}	79.4	83.8	8.49	8.37	100	100
Dy	5.96	2.02	30.3	56.0	0	5.1×10^{-7}	69.3	83.4	5.36	5.05	100	100
Ho	7.82	6.79	31.2	56.5	0	4.7×10^{-7}	75.2	83.8	7.85	7.24	100	100
Er	1.06	1.47	16.4	34.7	0	0	89.1	93.3	12.3	13.6	100	100
Tm	1.30	1.80	3.72	36.0	0	8.4×10^{-8}	91.8	95.2	17.8	47.9	100	100
Yb	0.60	0.82	11.2	26.7	0	0	93.4	97.2	20.5	24.5	100	100
Lu	0.18	0.25	3.01	6.51	0	0	88.7	94.1	14.0	30.7	100	100

	% of REE Activity as F Complexes						Proportion of REE Activity as SO$_4$ Complexes					
Model	1.1	1.2	2.1	2.2	3.1	3.2	1.1	1.2	2.1	2.2	3.1	3.2
La	1.85	1.38	1.94	0.74	2.3×10^{-8}	0	0	0	0.38	0.10	3.3×10^{-9}	1.5×10^{-4}
Ce	4.54	2.21	4.28	1.76	3.1×10^{-9}	7.9×10^{-10}	0	0	0	0	0	0
Pr	3.44	1.42	2.19	0.83	4.2×10^{-10}	0	0	0	0	0	0	0
Nd	4.02	1.20	5.76	2.00	1.5×10^{-10}	1.2×10^{-10}	0	0	3.59	1.04	7.5×10^{-11}	1.3×10^{-5}
Sm	4.34	2.09	18.4	8.57	1.9×10^{-11}	0	0	0	4.17	1.34	6.5×10^{-12}	1.4×10^{-6}
Eu	0	0	0	0	0	0	0	0	0	0	0	0
Gd	14.3	7.77	35.3	17.5	7.5×10^{-11}	2.9×10^{-10}	0	0	0	0	0	0
Tb	15.5	8.60	53.8	32.8	8.5×10^{-11}	4.4×10^{-10}	0	0	3.38	0	2.9×10^{-12}	0
Dy	24.7	14.6	62.7	38.9	1.1×10^{-10}	3.5×10^{-10}	0	0	1.73	0	1.7×10^{-12}	0
Ho	17.0	9.41	59.2	36.3	8.6×10^{-11}	1.1×10^{-10}	0	0	1.76	0	1.5×10^{-12}	0
Er	9.81	5.19	69.7	51.7	2.5×10^{-11}	7.2×10^{-11}	0	0	1.59	0	3.5×10^{-13}	7.7×10^{-8}
Tm	6.89	2.96	74.2	10.7	2.7×10^{-11}	8.2×10^{-11}	0	0	4.32	5.34	4.0×10^{-13}	8.7×10^{-8}
Yb	5.96	1.96	66.6	48.8	1.7×10^{-11}	5.5×10^{-10}	0	0	1.69	0	3.5×10^{-13}	7.7×10^{-8}
Lu	11.1	5.68	82.3	62.8	3.1×10^{-11}	1.2×10^{-10}	0	0	0.68	0	0	2.8×10^{-8}

In all cases, the majority of HREE activity is represented by low-solubility F-, oxide- and hydroxide-complexes, all of which are considered highly immobile [7]. This may explain the much smaller variability that these elements exhibit compared to LREE with respect to both REE-signatures and absolute concentrations.

The significant variabilities in REE speciation present in models 1.1, 1.2, 2.1 and 2.2 are lacking in models 3.1 and 3.2, in which all REE species are dominated by oxide and hydroxide species. This suggests that the unusual MREE-enriched signature of this apatite cannot be explained in terms of solubility and stability discrepancies between the REE. Given the relative lack of co-crystallizing LREE-enriched species within these samples, except for very minor florencite, it is reasonable to suggest that the fluids from which this apatite formed were already slightly MREE-enriched and that the preferential partitioning of REE closest to the Sm-Gd range into apatite [35] led to further MREE-enrichment as was suggested for scheelite under similar conditions [8]. Significantly, within these models, Eu is present as the trivalent species EuO$_2{}^-$, a form much more easily incorporated into apatite compared to the EuCl$_4{}^{2-}$ species present within the other models. Given that the di- vs. trivalent speciation of Eu is primarily controlled by pH in hydrothermal systems [2], the presence of a positive Eu anomaly in the *massive bornite hosted apatite* can be readily attributed to crystallization under pH conditions significantly higher than those proposed for the other apatites that display negative Eu anomalies. Moreover, Brugger et al. [2] noted that transition of $EuCl_4{}^{2-}$ to $EuO_2{}^-$ is very sensitive to Cl activity, as shown in Figure 4. This suggests that lower salinity, in addition to higher pH, was responsible for formations of this apatite.

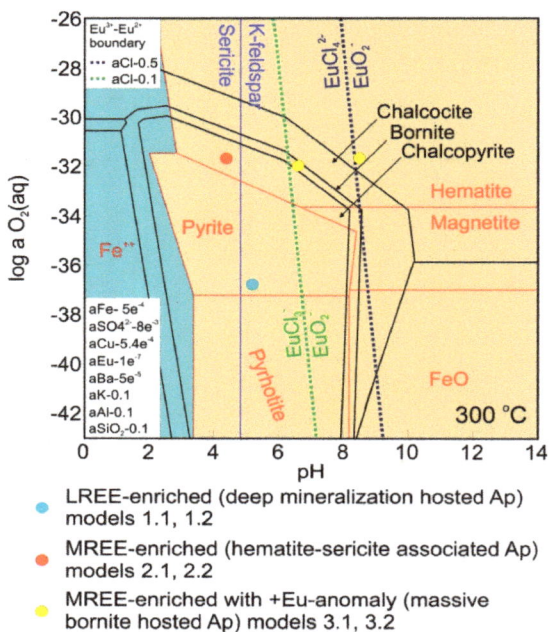

LREE-enriched (deep mineralization hosted Ap) models 1.1, 1.2

MREE-enriched (hematite-sericite associated Ap) models 2.1, 2.2

MREE-enriched with +Eu-anomaly (massive bornite hosted Ap) models 3.1, 3.2

Figure 4. Diagram showing the relative stabilities of Cu-Fe sulfides, Fe-oxides and K-feldspar-sericite. The various model scenarios are shown as colored circles demonstrating their agreement with the wider mineral assemblage [15,17].

5. Discussion

Modeling of REE partitioning between apatite and fluid along with the consideration of REE-speciation in fluid during formation of various hydrothermal assemblages within Olympic Dam offers valuable insights into the transport and deposition of REE within the deposit. It is apparent that REE transport was primarily facilitated by REE–Cl complexes due to their high solubility and dominance (at least for LREE) in both early magnetite-stable and later hematite-stable ore-forming conditions. Such a scenario is concordant with the findings of van Dongen et al. [50] in their studies of REE behavior within porphyry Cu-Au mineralization and with arguments put forward by Migdisov et al. [7]. The latter authors also emphasized the importance of SO_4 complexes for REE transport. This is, however, found to have been negligible at Olympic Dam due to the very low activity of REE–SO_4 complexes under all conditions considered, even at high SO_4 concentrations (e.g., models 3.1 and 3.2).

Despite the mobility of REE during formation of both magnetite- and hematite-dominant mineralization, the latter appears to have made a much more considerable contribution to the overall enrichment of the deposit in REE. Specifically, pervasive sericite alteration results in the replacement of feldspars and apatite [15,26], releasing significant volumes of REE (as these minerals account for most of the REE budget within the RDG and other intrusive rocks) under conditions exceptionally well suited for their transport. Such a coupled process of REE release and transport could account for a portion of the elevated REE concentrations within the deposit. Moreover, the significantly higher enrichment in LREE is easily explained through such a model given their dominance as REE–Cl complexes within fluids causing hematite-sericite alteration.

Within the deposit, REE are dominantly hosted in the REE-fluorocarbonate mineral bastnäsite [(REE)CO$_3$F], in florencite [(REE)Al$_3$(PO$_4$)$_2$(OH)$_6$], and in subordinate amounts within synchysite [Ca(REE)(CO$_3$)$_2$F], Sr- and Ca-dominant Al-phosphate-sulfates of the crandallite- and beudantite groups, as well as in xenotime and monazite [20,51]. Outside the deposit, a further REE–fluorocarbonate, parasite, is also observed [52,53], forming micron- to nanoscale intergrowths with bastnäsite and an unnamed species as part of a complex sericite + quartz + molybdenite assemblage replacing magmatic feldspar.

All these minerals contain P, F and/or CO$_3$, supporting the hypothesis that deposition of REE was achieved via a weakening of REE–Cl activity and subsequent REE complexation with the aforementioned species. In the presence of REE-fluorocarbonate-rich assemblages outside the deposit, hydrothermal apatite is S-bearing but shows little change in the REE fractionation trends compared to magmatic apatite [15,17]. On the other hand, florencite + sericite replace apatite, the latter displaying REE fractionation trends represented here by case 2 [17]. The modeling presented here does not reflect the many, diverse environments of apatite formation at Olympic Dam. It is, however, illustrative of the underlying reasons behind the magmatic to hydrothermal transition (cases 1 and 2) and the high-grade ore (case 3).

6. Conclusions and Recommendations

The modeling outlined here shows a relatively good fit with measured REE fractionation in apatite from Olympic Dam and satellite prospects in the region. This not only highlights that the observed trends can be efficiently and plausibly explained in terms of IOCG fluid evolution but also that these trends can be used to predict regional-scale variation within IOCG systems comparable to Olympic Dam.

Further modeling is required to precisely determine under which parameters REE deposition occurred. Models should also explain the dominance of fluorocarbonates as REE-carriers at Olympic Dam. Speciation of REE as F-complexes is easily achieved by increasing F concentrations within fluid at low pH and may also need to consider the potential role of REE-CO$_3$ complexes. These data, from one of the largest REE accumulations on Earth complement published empirical datasets and thermodynamic modeling in a broad range of different ore deposits [1–8,54–56].

Author Contributions: S.K. performed the work presented, guided by C.L.C. and N.J.C. K.J.E. supplied sample material and provided advice. S.K. wrote the manuscript assisted by C.L.C., N.J.C., and K.J.E.

Acknowledgments: This work is supported by BHP Olympic Dam. We gratefully acknowledge microanalytical assistance from the staff at Adelaide Microscopy, notably Ben Wade. N.J.C. and K.E. acknowledge support from the ARC Research Hub for Australian Copper–Uranium. CLC acknowledges support from the 'FOX' project (Trace elements in iron oxides), supported by BHP Olympic Dam and the South Australian Mining and Petroleum Services Centre of Excellence. A draft version of this work benefitted from helpful comments and encouragement to publish from David R. Lentz and Daniel Harlov. Last but not least, we thank the three anonymous *Minerals* reviewers and special issue editor Galina Palyanova, which have helped us improve expression and presentation of this work.

Conflicts of Interest: The authors declare no conflict of interest.

Appendix A. Results of Modeling (Models 1.1, 1.2, 2.1, 2.2, 3.1 and 3.2).

MODEL 1.1	a (aq.) Total	a (Apatite)	Concentration in Apatite (Mole Fraction)	Concentration Fluid (Modal)	K	$D_{apatite/fluid}$	Concentration REE Fluid (ppm)	Model Apatite (ppm)	Model Apatite/Chondrite
Ca	0.141	0.974	0.974	0.195					
Na	3.05	0.013	0.013	4.39					
La	1.35×10^{-4}	2.25×10^{-3}			1.51×10^{-3}	1.58	98.4	156	658
Ce	1.82×10^{-4}	6.15×10^{-3}			3.05×10^{-3}	3.20	194	622	1014
Pr	1.80×10^{-5}	7.35×10^{-4}			3.70×10^{-3}	3.89	21.1	81.9	881
Nd	6.35×10^{-5}	2.66×10^{-3}			3.79×10^{-3}	3.98	72.2	287	629
Sm	1.57×10^{-5}	4.07×10^{-4}			2.35×10^{-3}	2.47	13.0	32.0	216
Eu	1.03×10^{-7}	3.70×10^{-5}			0.032	0.37	1.76	0.65	11.6
Gd	8.63×10^{-6}	3.23×10^{-4}			3.39×10^{-3}	3.56	10.8	38.3	192
Tb	1.41×10^{-6}	4.20×10^{-5}			2.69×10^{-3}	2.83	1.65	4.68	130
Dy	7.59×10^{-6}	2.36×10^{-4}			2.81×10^{-3}	2.95	9.92	29.3	119
Ho	1.90×10^{-6}	4.63×10^{-5}			2.21×10^{-3}	2.32	2.08	4.82	87.5
Er	7.62×10^{-6}	1.28×10^{-4}			1.52×10^{-3}	1.59	5.81	9.26	57.9
Tm	1.70×10^{-6}	1.64×10^{-5}			8.72×10^{-4}	0.916	0.922	0.84	33.8
Yb	9.21×10^{-6}	9.90×10^{-5}			9.72×10^{-4}	1.02	6.04	6.17	38.3
Lu	1.30×10^{-6}	1.34×10^{-5}			9.38×10^{-4}	0.985	0.846	0.83	33.9

Y Eu^{2+} (aq.) 7.22×10^{-3}

Minerals **2018**, *8*, 342

MODEL 1.2	a (aq.) Total	a (Apatite)	Concentration in Apatite (Mole Fraction)	Concentration Fluid (Modal)	K	$D_{apatite/fluid}$	Concentration REE Fluid (ppm)	Model Apatite (ppm)	Model Apatite/Chondrite
Ca	0.789	0.974	0.974	1.05					
Na	3.39	0.013	0.013	4.80					
La	1.59×10^{-4}	2.25×10^{-3}			0.036	7.68	98.4	756	3189
Ce	2.08×10^{-4}	6.15×10^{-3}			0.075	16.1	194	3117	5084
Pr	1.89×10^{-5}	7.35×10^{-4}			0.099	21.1	21.1	445	4785
Nd	6.59×10^{-5}	2.66×10^{-3}			0.103	22.0	72.2	1587	3473
Sm	1.82×10^{-5}	4.07×10^{-4}			0.057	12.2	13.0	158	1067
Eu	9.78×10^{-8}	3.70×10^{-5}			0.965	2.19	1.76	3.85	68.7
Gd	1.10×10^{-5}	3.23×10^{-4}			0.075	16.0	10.8	172	864
Tb	1.78×10^{-6}	4.20×10^{-5}			0.060	12.8	1.65	21.2	588
Dy	8.80×10^{-6}	2.36×10^{-4}			0.068	14.6	9.92	145	589
Ho	2.31×10^{-6}	4.63×10^{-5}			0.051	10.9	2.08	22.6	412
Er	9.50×10^{-6}	1.28×10^{-4}			0.034	7.32	5.81	42.5	266
Tm	1.84×10^{-6}	1.64×10^{-5}			0.023	4.86	0.922	4.48	179
Yb	1.13×10^{-5}	9.90×10^{-5}			0.022	4.77	6.04	28.8	179
Lu	1.61×10^{-6}	1.34×10^{-5}			0.021	4.54	0.846	3.84	156

Y Eu^{2+} (aq.) 6.25×10^{-3}

MODEL 2.1	a (aq.) Total	a (Apatite)	Concentration in Apatite (Mole Fraction)	Concentration Fluid (Modal)	K	$D_{apatite/fluid}$	Concentration REE Fluid (ppm)	Model Apatite (ppm)	Model Apatite/Chondrite
Ca	0.068	0.996	0.996	0.105					
Na	1.28	1.96×10^{-3}	1.96×10^{-3}	1.94					
La	1.42×10^{-4}	9.31×10^{-5}			4.73×10^{-6}	4.10×10^{-3}	100	0.41	1.73
Ce	1.49×10^{-4}	3.27×10^{-4}			1.58×10^{-5}	0.014	200	2.74	4.47
Pr	1.62×10^{-5}	6.56×10^{-5}			2.93×10^{-5}	0.025	24.0	0.61	6.54
Nd	3.74×10^{-5}	4.15×10^{-4}			8.01×10^{-5}	0.069	80.0	5.55	12.1
Sm	5.63×10^{-6}	2.86×10^{-4}			3.67×10^{-4}	0.318	14.0	4.45	30.0
Eu	1.60×10^{-7}	6.52×10^{-5}			27.9	0.199	1.90	0.38	6.76
Gd	4.26×10^{-6}	4.01×10^{-4}			6.78×10^{-4}	0.587	11.8	6.93	34.8
Tb	4.50×10^{-7}	4.29×10^{-5}			6.88×10^{-4}	0.596	1.70	1.01	28.2
Dy	2.89×10^{-6}	1.66×10^{-4}			4.15×10^{-4}	0.359	10.3	3.70	15.0
Ho	5.46×10^{-7}	2.10×10^{-5}			2.78×10^{-4}	0.241	2.00	0.48	8.75
Er	1.37×10^{-6}	4.14×10^{-5}			2.17×10^{-4}	0.188	6.10	1.15	7.18
Tm	1.79×10^{-7}	4.40×10^{-6}			1.78×10^{-4}	0.154	0.70	0.11	4.31
Yb	1.48×10^{-6}	2.75×10^{-5}			1.34×10^{-4}	0.116	6.30	0.73	4.53
Lu	2.48×10^{-7}	3.10×10^{-6}			9.03×10^{-5}	0.078	1.10	0.086	3.50

$Y Eu^{2+}$ (aq.)　0.012

MODEL 2.2	a (aq.) Total	a (Apatite)	Concentration in Apatite (Mole Fraction)	Concentration Fluid (Modal)	K	$D_{apatite/fluid}$	Concentration REE Fluid (ppm)	Model Apatite (ppm)	Model Apatite/Chondrite
Ca	0.906	0.996	0.996	1.28					
Na	1.48	1.96×10^{-3}	1.96×10^{-3}	2.19					
La	1.58×10^{-4}	9.31×10^{-5}			6.46×10^{-4}	0.052	100	5.18	21.8
Ce	1.67×10^{-4}	3.27×10^{-4}			2.14×10^{-3}	0.172	200	34.4	56.0
Pr	1.70×10^{-5}	6.56×10^{-5}			4.22×10^{-3}	0.338	24.0	8.11	87.2
Nd	3.85×10^{-5}	4.15×10^{-4}			0.012	0.946	80.0	75.7	166
Sm	6.55×10^{-6}	2.86×10^{-4}			0.048	3.83	14.0	53.6	362
Eu	1.27×10^{-7}	6.52×10^{-5}			468	3.79	1.90	7.21	129
Gd	5.71×10^{-6}	4.01×10^{-4}			0.077	6.15	11.8	72.6	365
Tb	5.09×10^{-7}	4.29×10^{-5}			0.092	7.38	1.70	12.5	349
Dy	3.19×10^{-6}	1.66×10^{-4}			0.057	4.56	10.3	47.0	191
Ho	6.08×10^{-7}	2.10×10^{-5}			0.038	3.03	2.00	6.06	110
Er	1.31×10^{-6}	4.14×10^{-5}			0.035	2.78	6.10	16.9	106
Tm	2.30×10^{-7}	4.40×10^{-6}			0.021	1.67	0.70	1.17	46.9
Yb	1.33×10^{-6}	2.75×10^{-5}			0.023	1.81	6.30	11.4	70.9
Lu	2.32×10^{-7}	3.10×10^{-6}			0.015	1.17	1.10	1.29	52.5

$Y Eu^{2+}$ (aq.) 8.10×10^{-3}

MODEL 3.1	a (aq.) Total	a (Apatite)	Concentration in Apatite (Mole Fraction)	Concentration Fluid (Modal)	K	$D_{apatite/fluid}$	Concentration REE Fluid (ppm)	Model Apatite (ppm)	Model Apatite/Chondrite
Ca	0.224	0.992	0.992	0.292					
Na	0.439	4.10×10^{-3}	4.10×10^{-3}	0.944					
La	3.47×10^{-3}	1.01×10^{-5}			1.39×10^{-6}	2.10×10^{-4}	120	0.252	1.06
Ce	4.71×10^{-3}	6.99×10^{-5}			7.10×10^{-6}	1.07×10^{-3}	170	1.82	2.98
Pr	3.94×10^{-4}	2.42×10^{-5}			2.93×10^{-5}	4.44×10^{-3}	150	0.665	7.16
Nd	1.01×10^{-3}	2.69×10^{-4}			1.28×10^{-4}	0.019	393	7.58	16.6
Sm	1.01×10^{-4}	8.41×10^{-4}			3.99×10^{-3}	0.603	41.2	24.9	168
Eu	2.99×10^{-5}	5.15×10^{-4}			8.22×10^{-3}	1.24	12.4	15.4	275
Gd	6.54×10^{-5}	1.55×10^{-3}			0.011	1.71	28.0	47.9	241
Tb	7.87×10^{-6}	1.22×10^{-4}			7.40×10^{-3}	1.12	3.40	3.80	106
Dy	3.78×10^{-5}	4.58×10^{-4}			5.80×10^{-3}	0.877	16.7	14.6	59.5
Ho	6.69×10^{-6}	6.10×10^{-5}			4.36×10^{-3}	0.659	3.00	1.98	36.0
Er	2.00×10^{-5}	1.15×10^{-4}			2.74×10^{-3}	0.415	9.10	3.78	23.6
Tm	2.61×10^{-6}	1.08×10^{-5}			1.98×10^{-3}	0.299	1.20	0.359	14.4
Yb	1.83×10^{-5}	4.82×10^{-5}			1.26×10^{-3}	0.191	8.60	1.64	10.2
Lu	2.73×10^{-6}	5.25×10^{-6}			9.18×10^{-4}	0.139	1.30	0.180	7.34
Y Eu^{2+} (aq.)	NA								

MODEL 3.2	a (aq.) Total	a (Apatite)	Concentration in Apatite (Mole Fraction)	Concentration Fluid (Modal)	K	$D_{apatite/fluid}$	Concentration REE Fluid (ppm)	Model Apatite (ppm)	Model Apatite/Chondrite
Ca	0.645	0.992	0.992	1.05					
Na	0.652	4.10×10^{-3}	4.10×10^{-3}	1.03					
La	7.37×10^{-3}	1.01×10^{-5}			3.63×10^{-6}	1.66×10^{-4}	120	0.199	0.841
Ce	9.16×10^{-3}	6.99×10^{-5}			2.03×10^{-5}	9.29×10^{-4}	170	1.58	2.58
Pr	4.88×10^{-4}	2.42×10^{-5}			1.32×10^{-4}	6.03×10^{-3}	150	0.904	9.72
Nd	1.23×10^{-3}	2.69×10^{-4}			5.81×10^{-4}	0.027	393	10.4	22.9
Sm	1.10×10^{-4}	8.41×10^{-4}			0.020	0.933	41.2	38.4	260
Eu	2.99×10^{-5}	5.15×10^{-4}			0.046	2.10	12.4	26.0	464
Gd	7.01×10^{-5}	1.55×10^{-3}			0.059	2.69	28.0	75.3	378
Tb	8.49×10^{-6}	1.22×10^{-4}			0.038	1.75	3.40	5.93	165
Dy	4.02×10^{-5}	4.58×10^{-4}			0.030	1.39	16.7	23.2	94.3
Ho	7.17×10^{-6}	6.10×10^{-5}			0.023	1.04	3.00	3.11	56.5
Er	2.11×10^{-5}	1.15×10^{-4}			0.014	0.661	9.10	6.01	37.6
Tm	2.77×10^{-6}	1.08×10^{-5}			0.010	0.475	1.20	0.570	22.8
Yb	1.95×10^{-5}	4.82×10^{-5}			6.59×10^{-3}	0.301	8.60	2.59	16.1
Lu	2.89×10^{-6}	5.25×10^{-6}			4.83×10^{-3}	0.221	1.30	0.287	11.7
Y Eu^{2+} (aq.)	NA								

References

1. Smith, M.; Henderson, P.; Jeffries, T.; Long, J.; Williams, C. The rare earth elements and uranium in garnets from the Beinn an Dubhaich Aureole, Skye, Scotland, UK: Constraints on processes in a dynamic hydrothermal system. *J. Pet.* **2004**, *45*, 457–484. [CrossRef]
2. Brugger, J.; Etschmann, B.; Pownceby, M.; Liu, W.; Grundler, P.; Brewe, D. Oxidation state of europium in scheelite: Tracking fluid-rock interaction in gold deposits. *Chem. Geol.* **2008**, *257*, 26–33. [CrossRef]
3. van Hinsberg, V.J.; Migdisov, A.A.; Williams-Jones, A.E. Reading the mineral record of fluid composition from element partitioning. *Geology* **2010**, *38*, 847–850. [CrossRef]
4. Haas, J.R.; Shock, E.L.; Sassani, D.C. Rare earth elements in hydrothermal systems: Estimates of standard partial molal thermodynamic properties of aqueous complexes of the rare earth elements at high pressures and temperatures. *Geochim. Cosmochim. Acta* **1995**, *59*, 4329–4350. [CrossRef]
5. Bau, M. Rare-earth element mobility during hydrothermal and metamorphic fluid-rock interaction and the significance of the oxidation state of europium. *Chem. Geol.* **1991**, *93*, 219–230. [CrossRef]
6. Migdisov, A.A.; Williams-Jones, A.E. Hydrothermal transport and deposition of the rare earth elements by fluorine-bearing aqueous liquids. *Miner. Deposita* **2014**, *49*, 987–997. [CrossRef]
7. Migdisov, A.A.; Williams-Jones, A.E.; Brugger, J.; Caporuscio, F.A. Hydrothermal transport, deposition, and fractionation of the REE: Experimental data and thermodynamic calculations. *Chem. Geol.* **2016**, *439*, 13–42. [CrossRef]
8. Brugger, J.; Lahaye, Y.; Costa, S.; Lambert, D.; Bateman, R. Inhomogenous distribution of REE in scheelite and dynamics of Archean hydrothermal systems (Mt. Charlotte and Drysdale gold deposits, Western Australia). *Contrib. Miner. Pet.* **2000**, *139*, 251–264. [CrossRef]
9. Harlov, D.E. Apatite: A fingerprint for metasomatic processes. *Elements* **2015**, *11*, 171–176. [CrossRef]
10. Belousova, E.A.; Walters, S.; Griffin, W.L.; O'Reilly, S.Y. Trace-element signatures of apatites in granitoids from the Mt Isa Inlier, northwestern Queensland. *Aust. J. Earth Sci.* **2001**, *48*, 603–619. [CrossRef]
11. Belousova, E.A.; Griffin, W.L.; O'Reilly, S.Y.; Fisher, N.I. Apatite as an indicator mineral for mineral exploration: Trace-element compositions and their relationship to host rock type. *J. Geochem. Explor.* **2002**, *76*, 45–69. [CrossRef]
12. Cao, M.; Li, G.; Qin, K.; Seitmuratova, E.Y.; Liu, Y. Major and Trace element characteristics of apatites in granitoids from Central Kazakhstan: Implications for petrogenesis and mineralization. *Res. Geol.* **2012**, *62*, 63–83. [CrossRef]
13. Teiber, H.; Marks, M.A.W.; Arzamastsev, A.A.; Wenzel, T.; Markl, G. Compositional variation in apatite from various host rocks: Clues with regards to source composition and crystallization conditions. *J. Miner. Geochem.* **2015**, *192*, 151–167. [CrossRef]
14. Mao, M.; Rukhlov, A.S.; Rowins, S.M.; Spence, J.; Coogan, L.A. Apatite trace element compositions: A robust new tool for mineral exploration. *Econ. Geol.* **2016**, *111*, 1187–1222. [CrossRef]
15. Krneta, S.; Ciobanu, C.L.; Cook, N.J.; Ehrig, K.; Kontonikas-Charos, A. Apatite at Olympic Dam, South Australia: A petrogenetic tool. *Lithos* **2016**, *262*, 470–485. [CrossRef]
16. Krneta, S.; Ciobanu, C.L.; Cook, N.J.; Ehrig, K.; Kontonikas-Charos, A. The Wirrda Well and Acropolis prospects Gawler Craton, South Australia: Insights into evolving fluid conditions through apatite chemistry. *J. Geochem. Explor.* **2017**, *181*, 276–291. [CrossRef]
17. Krneta, S.; Ciobanu, C.L.; Cook, N.J.; Ehrig, K.; Kontonikas-Charos, A. Rare earth element behaviour in apatite from the Olympic Dam Cu-U-Au-Ag deposit, South Australia. *Minerals* **2017**, *7*, 135. [CrossRef]
18. Ismail, R.; Ciobanu, C.L.; Cook, N.J.; Teale, G.S.; Giles, D.; Schmidt Mumm, A.; Wade, B. Rare earths and other trace elements in minerals from skarn assemblages, Hillside iron oxide–copper–gold deposit, Yorke Peninsula, South Australia. *Lithos* **2014**, *184–187*, 456–477. [CrossRef]
19. Ehrig, K.; Kamenetsky, V.S.; McPhie, J.; Apukhtina, O.; Ciobanu, C.L.; Cook, N.J.; Kontonikas-Charos, A.; Krneta, S. The IOCG-IOA Olympic Dam Cu-U-Au-Ag deposit and nearby prospects, South Australia. In Proceedings of the 14th SGA Biennial Meeting, Quebec City, QC, Canada, 20–23 August 2017; pp. 823–826.
20. Ehrig, K.; McPhie, J.; Kamenetsky, V.S. Geology and mineralogical zonation of the Olympic Dam iron oxide Cu-U-Au-Ag deposit, South Australia. In *Geology and Genesis of Major Copper Deposits and Districts of the World, a Tribute to Richard Sillitoe*; Hedenquist, J.W., Harris, M., Camus, F., Eds.; Society of Economic Geologists: Littleton, CO, USA, 2012; pp. 237–268.

21. Oreskes, M.; Einaudi, M.T. Origin of Hydrothermal Fluids at Olympic Dam: Preliminary Results from Fluid Inclusions and Stable Isotopes. *Econ. Geol.* **1992**, *87*, 64–90. [CrossRef]

22. Bastrakov, E.N.; Skirrow, R.G.; Davidson, G.J. Fluid evolution and origins of iron oxide Cu-Au prospects in the Olympic Dam District, Gawler Craton, South Australia. *Econ. Geol.* **2007**, *102*, 1415–1440. [CrossRef]

23. BHP Billiton. Available online: https://www.bhp.com/-/media/bhp/documents/investors/annual-reports/2016/bhpbillitonannualreport2016.pdf?la=en (accessed on 8 August 2018).

24. Reeve, J.S.; Cross, K.C.; Smith, R.N.; Oreskes, N. Olympic Dam copper-uranium-gold-silver deposit. In *Geology of the Mineral. Deposits of Australia and Papua New Guinea*; Hughes, F.E., Ed.; Australasian Institute of Mining and Metallurgy Monograph: Melbourne, Australia, 1990; pp. 1009–1035.

25. Creaser, R.A. The Geology and Petrology of Middle Proterozoic Felsic Magmatism of the Stuart Shelf, South Australia. Unpublished Ph.D. Thesis, La Trobe University, Melbourne, Australia, 1989.

26. Kontonikas-Charos, A.; Ciobanu, C.L.; Cook, N.J.; Ehrig, K.; Krneta, S.; Kamenetsky, V.S. Feldspar evolution in the Roxby Downs Granite, host to Fe-oxide Cu-Au-(U) mineralisation at Olympic Dam, South Australia. *Ore Geol. Rev.* **2017**, *80*, 838–859. [CrossRef]

27. Jagodzinski, E.A. The age of magmatic and hydrothermal zircon at Olympic Dam. In Proceedings of the 2014 Australian Earth Sciences Convention (AESC), Sustainable Australia, Newcastle, Australia, 7–10 July 2014; Volume 110, p. 260.

28. Ciobanu, C.L.; Wade, B.; Cook, N.J.; Schmidt Mumm, A.; Giles, D. Uranium-bearing hematite from the Olympic Dam Cu-U-Au deposit, South Australia; a geochemical tracer and reconnaissance Pb-Pb geochronometer. *Precambr. Res.* **2013**, *238*, 129–147. [CrossRef]

29. Courtney-Davies, L.; Zhu, Z.; Ciobanu, C.L.; Wade, B.P.; Cook, N.J.; Ehrig, K.; Cabral, A.R.; Kennedy, A. Matrix-matched iron-oxide laser ablation ICP-MS U-Pb geochronology using mixed solutions standards. *Minerals* **2016**, *6*, 85. [CrossRef]

30. Pasero, M.; Kampf, A.; Ferraris, C.; Pekov, I.V.; Rakovan, J.; White, T. Nomenclature of the apatite supergroup minerals. *Eur. J. Miner.* **2010**, *22*, 163–179. [CrossRef]

31. White, T.J.; Dong, Z.L. Structural derivation and crystal chemistry of apatites. *Acta Cryst.* **2002**, *B59*, 1–16. [CrossRef]

32. Hughes, J.M.; Rakovan, J.F. Structurally robust, chemically diverse: Apatite and apatite supergroup minerals. *Elements* **2015**, *11*, 165–170. [CrossRef]

33. Hughes, J.M.; Rakovan, J.F. The crystal structure of apatite, $Ca_5(PO_4)_3(F,OH,Cl)$. *Rev. Miner. Geochem.* **2002**, *48*, 1–12. [CrossRef]

34. Goldschmidt, V.M. The principles of distribution of chemical elements in minerals and rocks. The seventh Hugo Müller Lecture, delivered before the Chemical Society on 17th March 1937. *J. Chem Soc. (Resumed)* **1937**, 655–673. [CrossRef]

35. Pan, Y.; Fleet, M.E. Compositions of the apatite-group minerals: Substitution mechanisms and controlling factors. *Rev. Miner. Geochem.* **2002**, *48*, 13–49. [CrossRef]

36. Rønsbo, J.G. Coupled substitutions involving REE's and Na and Si in apatites in alkaline rocks from Ilimaussaq, South Greenland, and the petrological implications. *Am. Miner.* **1989**, *74*, 896–901.

37. Hughes, J.M. The many facets of apatite. *Am. Miner.* **2015**, *100*, 1033–1039. [CrossRef]

38. Luo, Y.; Rakovan, J.; Tang, Y.; Lupulescu, M.; Hughes, J.M.; Pan, Y. Crystal chemistry of Th in fluorapatite. *Am. Mineral.* **2011**, *96*, 23–33. [CrossRef]

39. Verdugo-Ihl, M.R.; Ciobanu, C.L.; Cook, N.J.; Ehrig, K.; Courtney-Davies, L.; Gilbert, S. Textures and U-W-Sn-Mo signatures in hematite from the Cu-U-Au-Ag orebody at Olympic Dam, South Australia: Defining the archetype for IOCG deposits. *Ore Geol. Rev.* **2017**, *91*, 173–195. [CrossRef]

40. Morales Ruano, S.; Both, R.A.; Golding, S.D. A fluid inclusion and stable isotope study of the Moonta copper-gold deposits, South Australia: Evidence for fluid immiscibility in a magmatic hydrothermal system. *Chem. Geol.* **2002**, *192*, 211–226. [CrossRef]

41. Schlegel, T.U. The Prominent Hill Iron Oxide-Cu-Au Deposit in South. Australia. A Deposit Formation Model Based on Geology, Geochemistry and Stable Isotopes and Fluid Inclusions. Unpublished Ph.D. Thesis, ETH Zurich, Zürich, Switzerland, 2015.

42. Chiaradia, M.; Banks, D.; Cliff, R.; Marschik, R.; de Haller, A. Origin of fluids in iron oxide–copper–gold deposits: Constraints from $\delta^{37}Cl$, $^{87}Sr/Sr$, and Cl/Br. *Miner. Deposita* **2006**, *41*, 565–573. [CrossRef]

43. Baker, T.; Mustard, R.; Fu, B.; Williams, P.J.; Dong, G.; Fisher, L.; Mark, G.; Ryan, C.G. Mixed messages in iron oxide-copper-gold systems of the Cloncurry district, Australia: Insights from PIXE analysis of halogens and copper in fluid inclusions. *Miner. Deposita* **2008**, *43*, 599–608. [CrossRef]

44. Gleeson, S.A.; Smith, M.P. The sources and evolution of mineralising fluids in iron oxide–copper–gold systems, Norrbotten, Sweden: Constraints from Br/Cl ratios and stable Cl isotopes of fluid inclusion leachates. *Geochim. Cosmochim. Acta* **2009**, *73*, 5658–5672. [CrossRef]

45. Haynes, D.W.; Cross, K.C.; Bills, R.T.; Reed, M.H. Olympic Dam Ore Genesis: A Fluid-Mixing Model. *Econ. Geol.* **1995**, *90*, 281–307. [CrossRef]

46. Cook, N.J.; Ciobanu, C.L.; O'Reilly, D.; Wilson, R.; Das, K.; Wade, B. Mineral chemistry of rare earth element (REE) mineralization, Browns Range, Western Australia. *Lithos* **2013**, *172*, 192–213. [CrossRef]

47. Weng, Z.; Jowitt, S.M.; Mudd, G.M.; Haque, N. A detailed assessment of global Rare Earth Element resources: Opportunities and challenges. *Econ. Geol.* **2015**, *110*, 1925–1952. [CrossRef]

48. GWB Essentials Guide. Available online: https://www.gwb.com/pdf/GWB11/GWBessentials.pdf (accessed on 22 July 2017).

49. Migdisov, A.A.; Williams-Jones, A.E.; Wagner, T. An experimental study of the solubility and speciation of the Rare Earth Elements (III) in fluoride- and chloride-bearing aqueous solutions at temperatures up to 300 °C. *Geochim. Cosmochim. Acta* **2009**, *73*, 7087–7109. [CrossRef]

50. van Dongen, M.; Weinberg, R.F.; Tomkins, A.G. REE-Y, Ti, and P Remobilization in Magmatic Rocks by Hydrothermal Alteration during Cu-Au Deposit Formation. *Econ. Geol.* **2010**, *105*, 763–776. [CrossRef]

51. Schmandt, D.S.; Cook, N.J.; Ciobanu, C.L.; Ehrig, K.; Wade, B.P.; Gilbert, S.; Kamenetsky, V.S. Rare earth element fluorocarbonate minerals from the Olympic Dam Cu-U-Au-Ag deposit, South Australia. *Minerals* **2017**, *7*, 202. [CrossRef]

52. Ciobanu, C.L.; Kontonikas-Charos, A.; Slattery, A.; Cook, N.J.; Ehrig, K.; Wade, B.P. Short-range stacking disorder in mixed-layer compounds: A HAADF STEM study of bastnäsite-parisite intergrowths. *Minerals* **2017**, *7*, 227. [CrossRef]

53. Kontonikas-Charos, A.; Ciobanu, C.L.; Cook, N.J.; Ehrig, K.; Ismail, R.; Krneta, S.; Basak, A. Feldspar mineralogy and rare earth element (re)mobilization in iron-oxide copper gold systems from South Australia: A nanoscale study. *Miner. Mag.* **2018**, *82*, S173–S197. [CrossRef]

54. Kolonin, G.R.; Shironosova, G.P. Influence of acidity–alkalinity of solutions On REE distribution during ore formation: Thermodynamic modeling. *Dokl. Earth Sci.* **2012**, *443*, 502–505. [CrossRef]

55. Shironosova, G.P.; Kolonin, G.R. Thermodynamic modeling of REE partitioning between monazite, fluorite, and apatite. *Dokl. Earth Sci.* **2013**, *450*, 628–632. [CrossRef]

56. Shironosova, G.P.; Kolonin, G.R.; Borovikov, A.A.; Borisenko, A.S. Thermodynamic modeling of REE behavior in oxidized hydrothermal fluids of high sulfate sulfur concentrations. *Dokl. Earth Sci.* **2016**, *469*, 855–859. [CrossRef]

minerals

MDPI

Article

Physicochemical Conditions of Formation for Bismuth Mineralization Hosted in a Magmatic-Hydrothermal Breccia Complex: An Example from the Argentine Andes

Francisco J. Testa [1,2,*], Lejun Zhang [1,2] and David R. Cooke [1,2]

[1] CODES, Centre for Ore Deposit and Earth Sciences, University of Tasmania, Private Bag 79, Hobart, Tasmania 7001, Australia; Lejun.Zhang@utas.edu.au (L.Z.); d.cooke@utas.edu.au (D.R.C.)

[2] TMVC, Transforming the Mining Value Chain, an ARC Industrial Transformation Research Hub, University of Tasmania, Private Bag 79, Hobart, Tasmania 7001, Australia

* Correspondence: F.J.Testa@utas.edu.au; Tel.: +61-3-6226-7211

Received: 3 September 2018; Accepted: 15 October 2018; Published: 26 October 2018

Abstract: The San Francisco de los Andes breccia-hosted deposit (Frontal Cordillera, Argentina) is characterized by complex Bi–Cu–Pb–Zn–Mo–As–Fe–Ag–Au mineralization. After magmatic-hydrothermal brecciation, tourmaline and quartz partially cemented open spaces, followed by quiescent periods where Bi–Cu–Pb–Zn ore formed. Bismuth ore precipitation is characterized by Bi-sulfides, sulfosalts, and tellurosulfide inclusions, which temporally co-exist with Ag-telluride inclusions and chalcopyrite. Three distinct Bi mineralizing stages have been defined based on the following mineral assemblages: (1) Bismuthinite (tetradymite–hessite inclusions); (2) Bismuthinite (tetradymite–hessite inclusions) + cosalite (tetradymite inclusions) + chalcopyrite; and (3) Cosalite (tetradymite inclusions) + chalcopyrite. Overall, Ag-poor bismuthinite hosts both Bi-tellurosulfide and Ag-telluride inclusions, whereas Ag-rich cosalite only hosts tetradymite inclusions.

In this study, we discuss the effects of temperature, pressure, vapor saturation, salinity, acidity/alkalinity, and redox conditions on Bi-rich mineralizing fluids. Evolving hydrothermal fluid compositions are derived from detailed paragenetic, analytical, and previous fluid inclusion studies. Based on trace minerals that co-precipitated during Bi ore formation, mineral chemistry, and quartz geothermobarometry, a thermodynamic model for bismuth species was constructed. Sulfur and tellurium fugacities during Bi-ore precipitation were constrained for the three mineralizing stages at a constant pressure of 1 kbar under minimum and maximum temperatures of 230 and 400 °C, respectively. We infer that Te was transported preferentially in a volatile-rich phase. Given that Te solubility is expected to be low in chloride-rich hydrothermal fluids, telluride and tellurosulfide inclusions are interpreted to have condensed from magmatically-derived volatile tellurium (e.g., $Te_{2(g)}$ or $H_2Te_{(g)}$) into deep-seated, dense, metal-rich brines. Tellurium minerals in the hydrothermal breccia cement provide a direct genetic link with the underlying magmatic system. Though the vertical extent of the breccia complex is unknown, the abundance of Te-bearing minerals could potentially increase with depth and not only occur as small telluride inclusions in Bi-minerals. A vertical zoning of Te-minerals could prove to be important for exploration of similar magmatic-hydrothermal breccia pipes and/or dikes.

Keywords: magmatic-hydrothermal breccias; bismuth mineralization; thermodynamic model; bismuth and tellurium fugacities; San Francisco de los Andes; Frontal Cordillera; San Juan; Argentina

1. Introduction

The San Francisco de los Andes breccia complex (30°50′08′′ S; 69°35′58′′ W) is located on the eastern flank of the Frontal Cordillera, San Juan province, Argentina ([1,2]; Figure 1). The Frontal Cordillera is a N-oriented mountain range that forms part of the Andes and extends from La Rioja to Mendoza province, and it is limited to the west by the Principal Cordillera (Figure 1). The San Francisco de los Andes deposit hosts the largest bismuth concentration in a hydrothermal breccia system in Argentina [3]. It is the most important breccia complex in the mineralized district, characterized by Bi–Cu–Au–As-rich tourmaline-cemented breccias and veins. This district extends up to 30 km north along the Frontal Cordillera, and also contains the Amancay, Cortadera, La Fortuna-El Chorrillo, Mirkokleia, Rodophis, Tres Magos, and Flor de los Andes deposits, among others ([2,3]; Figure 1).

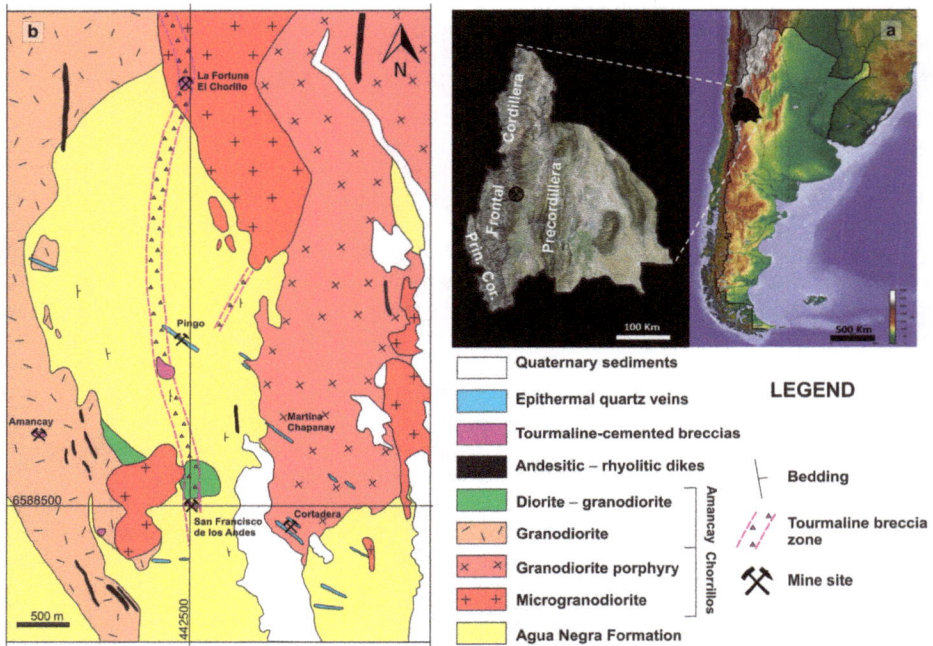

Figure 1. (**a**) Map of Argentina (right) and San Juan province (left). Location of Frontal Cordillera and San Francisco de los Andes mine. (**b**) Geological maps of the southern domain of the San Francisco de los Andes district. Figure based on Llambías and Malvicini (1969), Testa (2017), Testa et al. (2016), and Testa et al. (2018) [1,2,4,5]. Abbreviation: Prin. Cor. = Principal Cordillera.

At a district scale, the study area is characterized by marine sedimentary rocks and granitoids (Figure 1). The oldest unit is the Carboniferous Agua Negra Formation, a sequence of marine sandstones, shales, and siltstones striking north, and dipping steeply eastward ([1,2]; Figure 1). Sedimentary rocks have been intruded by the Tocota Pluton, a Permian intrusive complex that ranges from tonalite to granite in composition ([1,2]; Figure 1). Geological units and hydrothermal alteration assemblages in the San Francisco de los Andes district, along with adjacent epithermal and porphyry districts, have been previously mapped using ASTER Imagery by Testa et al. (2018) [4].

At a deposit scale, the San Francisco de los Andes tourmaline–quartz-cemented breccia complex is spatially and genetically associated with the Tocota Pluton, and has cut the Carboniferous sedimentary host rocks (e.g., [2]; Figure 1). The breccia-hosted deposit is characterized by a complex

Bi–Cu–Au–As–Fe–Zn–Pb–Ag mineral assemblage, which includes hypogene sulfides, sulfosalts, tellurides, and native elements, as well as supergene oxides, hydroxides, arsenates, sulfates, carbonates, and secondary sulfides and sulfosalts [5].

The deposit was mined sporadically between the 1940s and 1980s. A total of 112 tons of bismuth concentrate and 2420 tons of ore with 3–6% Cu and 1.2–4.5% Bi were produced [3,6]. In 1990, the Aguilar Mining Company S.A. conducted geologic studies, including surface, sub-surface, and diamond drill core sampling. Based on a detailed geochemical reconnaissance sampling program of surface exposures and sub-surface samples, resource estimations were generated for ores within 200 m of the surface [7]. Aguilar Mining Company estimated a resource of >0.16 Mt. with 5.4 g/t Au, 77 g/t Ag, 0.9% Cu, 0.15% Bi, 0.4% Pb, and 0.1% Zn for the 35 m thick 'enriched zone'. Although the vertical extent of the breccia complex is unknown, Aguilar Mining Company identified a shallow (20–25 m) hypogene zone comprising sulfide and sulfosalt minerals with average grades of 1.32 g/t Au, 65.5 g/t Ag, 0.55% Cu, 0.02% Bi, 0.38% Pb, and 0.61% Zn [7].

Early comprehensive research of the San Francisco de los Andes deposit was conducted by Llambías and Malvicini (1969) [1]. Supergene arsenate mineral species and supergene luzonite were documented by Bedlivy and Llambías (1969), Malvicini (1969), Bedlivy et al. (1972), and Bedlivy and Mereiter (1982) [8–11]. Preisingerite ($Bi_3(AsO_4)_2O(OH)$) was first documented in the San Francisco de los Andes deposit (type locality), and its crystal structure first described by Bedlivy and Mereiter (1982) [11]. The first occurrence of bismoclite (BiOCl) in a magmatic–hydrothermal breccia complex, and its use as an indicator phase in mineral exploration was reported by Testa et al. (2016) [5]. Bismoclite rarely occurs as a pure phase, but is intimately associated with traces of preisingerite [5].

Despite the available research articles and mining reports focused on the San Francisco de los Andes mine, the evolution of hydrothermal fluids responsible for Bi ore deposition remain mostly undocumented. The current study aimed to resolve that issue through field work, macroscopic sample descriptions, and transmitted–reflected light microscopy, along with detailed analytical work based on Testa (2017) [2]. These data are used to further document an alternative paragenetic sequence of the San Francisco de los Andes breccia complex. Special care has been taken to detect any difference between the NW and SE domain of the figure-8-shaped breccia complex. The major aim of this article is to document the Bi ore and characterize the evolution of hydrothermal fluids responsible for its formation. Based on detailed descriptive and analytical studies of ore minerals and quartz geothermobarometry, a thermodynamic model for bismuth species and co-existing mineral phases was constructed. It constrains S_2 and Te_2 fugacity values for the mineralizing solutions at a constant pressure of 1 kbar under a minimum temperature of 230 °C, and maximum of 400 °C. Along with our thermodynamic model, further genetic implications are proposed for the San Francisco de los Andes magmatic-hydrothermal breccia complex, which aid linking the hydrothermal system with its deep-seated magmatic source.

2. Methods

Eighty-three samples of ore-bearing breccias and host rocks were collected from surface and the main underground mining level at San Francisco de los Andes to represent variations across the breccia complex.

Mineral species were initially determined by macroscopic and microscopic analysis. Mineral percentages, phase relationships, and characteristic textures were recorded using both hand specimens and thin sections. A Nikon E600 POL binocular transmitted/reflected light microscope (Nikon Instruments Inc., Tokyo, Japan) with ×5, ×10, ×20, and ×50 objective lenses, as well as ×10 ocular lenses, was used to identify transparent and opaque minerals. Mineral species were later confirmed by X-ray-diffraction (XRD) analysis. Powder XRD analyses were conducted at the Geology Department, Universidad Nacional del Sur, Argentina, on a Rigaku D-Max III-C automatic powder diffractometer with Cu Kα radiation and graphite monochromator to strip the Kα2 contribution.

Electron microprobe analysis (EMPA) and laser ablation inductively coupled plasma mass spectrometry (LA-ICP-MS) were used for mineral chemical analysis, and to identify microscopic mineral phases in fine-grained intergrowth textures and mineral inclusions. Major element chemical compositions of sulfides, sulfosalts, and tellurides were acquired by EMPA. Carbon-coated mounts were analyzed on a Cameca SX100 electron microprobe housed at the Central Science Laboratory, University of Tasmania, outfitted with five tunable wavelength-dispersive spectrometers (WDS). Backscattered electron images (BSEI) aided further differentiation of mineral species prior to EMP analysis. Minor and trace element compositions of mineral phases were acquired by LA-ICP-MS. Samples were analyzed on a New Wave 213-nm solid-state laser microprobe coupled to an Agilent 4500 quadrupole ICP-MS. All LA-ICP-MS studies were conducted at the CODES LA-ICP-MS analytical facility, University of Tasmania.

Sulfur and Te fugacities during Bi ore deposition were reconstructed based on mineralogical and thermodynamic analysis. Log f_{S_2} vs. log f_{Te_2} diagrams were calculated for Bi-mineralization and co-existing mineral phases from the San Francisco de los Andes breccia complex using available thermodynamic properties for mineral assemblages in equilibrium. Stability limits were calculated using thermodynamic data of minerals from Garrels and Christ (1965), Meyer and Hemley (1967), Robie and Waldbaum (1968), Craig and Barton (1973), Robie et al. (1978), Barton and Skinner (1979), Afifi et al. (1988a,b), Robie and Hemingway (1995), and Krauskopf and Bird (1995) [12–21]. All chemical reactions, log K values, and related thermodynamic properties were based on the aforementioned sources.

3. Results

3.1. Morphology and Spatial Dimensions of the Breccia Complex

The San Francisco de los Andes breccia complex has an elongated NW-trending elliptical shape with a constriction in the middle (Figure 2). Based on underground working, the 'figure 8'-shaped outline in the plan view can be traced for up to 75 m below the surface. The consistent neck in the middle of the structure divides the breccia complex into two domains: the SE and NW domains (Figure 2).

The horizontal dimensions of the breccia body at three levels in the mine can be seen in Figure 2a. The elliptical contour at the surface is markedly elongated WNW, with a maximum length of 67 m (Figure 2a). Fifty meters underground, the SE and NW domains are distinctly larger, and double the area mapped on surface. The 'figure 8'-shaped outline is enhanced, and the maximum axis is 77 m long NW (Figure 2a). Twenty-five meters under the main level, the maximum length of the breccia column is slightly longer (80 m NW), and the area of both domains is wider (Figure 2a). This is the deepest mining level where the outline of the breccia body can be fully documented. A block diagram has been drawn to show three-dimensional variations in the morphology of the breccia complex with depth; Figure 2b is based on the outline of the breccia complex at surface and underground plan views.

Information available from the <200-m-long diamond drill holes also document a gradual increase in length and width with increasing depth, with clear steeply dipping contacts with the host rock (Figure 2b). These features, along with the slightly tilted inwards contacts, are consistent with the upper level of a magmatic-hydrothermal breccia conduit or pipe (e.g., Sillitoe, 1985; 2010 [22,23]). None of the <200-m-long drill holes reached the root of the breccia complex, and the full vertical extent of the breccia column is unknown.

3.2. Paragenetic Sequence

3.2.1. Hydrothermal Cement in the SE Domain

Brecciation was followed by precipitation of silicate minerals from hydrothermal fluids, which partially cemented breccia clasts (Figure 3). Tourmaline, and to a lesser extent quartz, define the earliest cockade banding in the breccia complex (stage SE-a; Figure 3). A second phase of infill followed, which produced major sulfide and sulfosalt deposition (stage SE-b; Figure 3). Based on textural relationship, this stage was subdivided into an early (stage SE-b-I) and late sulfide–sulfosalt phase (stage SE-b-II; Figure 3).

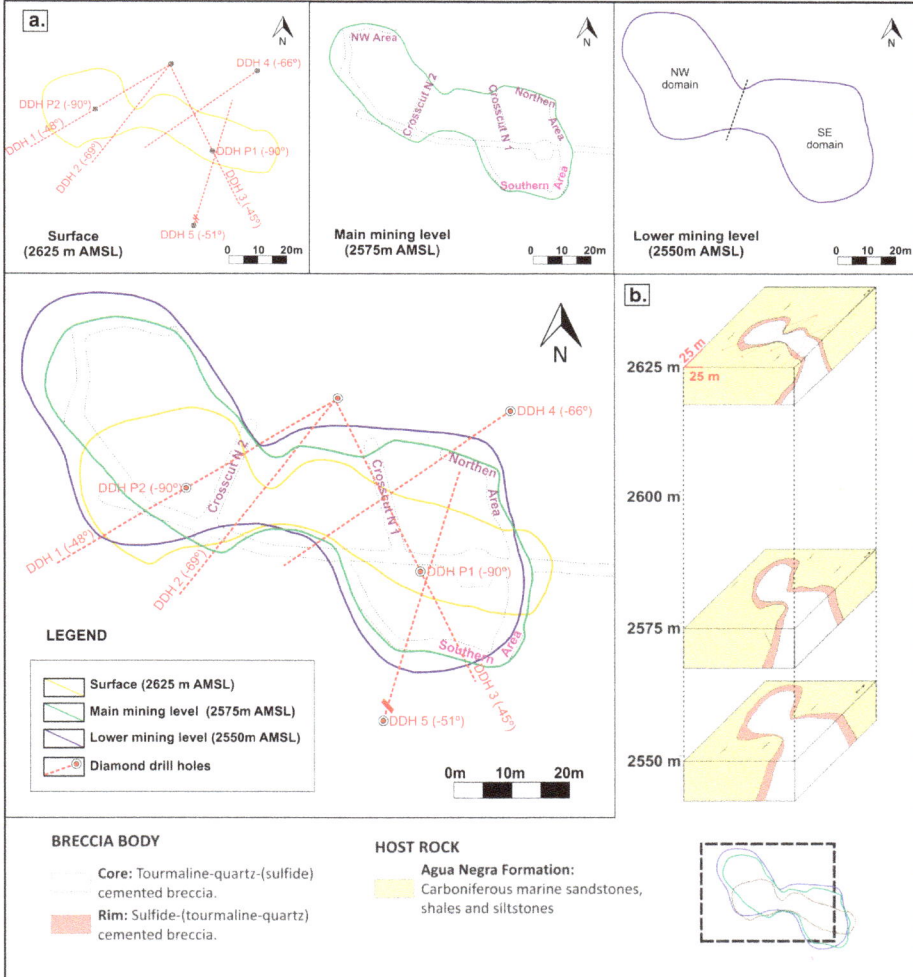

Figure 2. (**a**). Plan view of the San Francisco de los Andes breccia complex at surface, main, and lower mining levels. A fourth panel shows the resulting composite overlay plan view. Locations of drill collars and projected drill holes are shown in red. (**b**). Block diagram of the breccia complex, note the consistent constriction in the middle of the breccia complex. Figure based on Lencinas (1990), Testa et al. (2016), Testa (2017) [2,5,7].

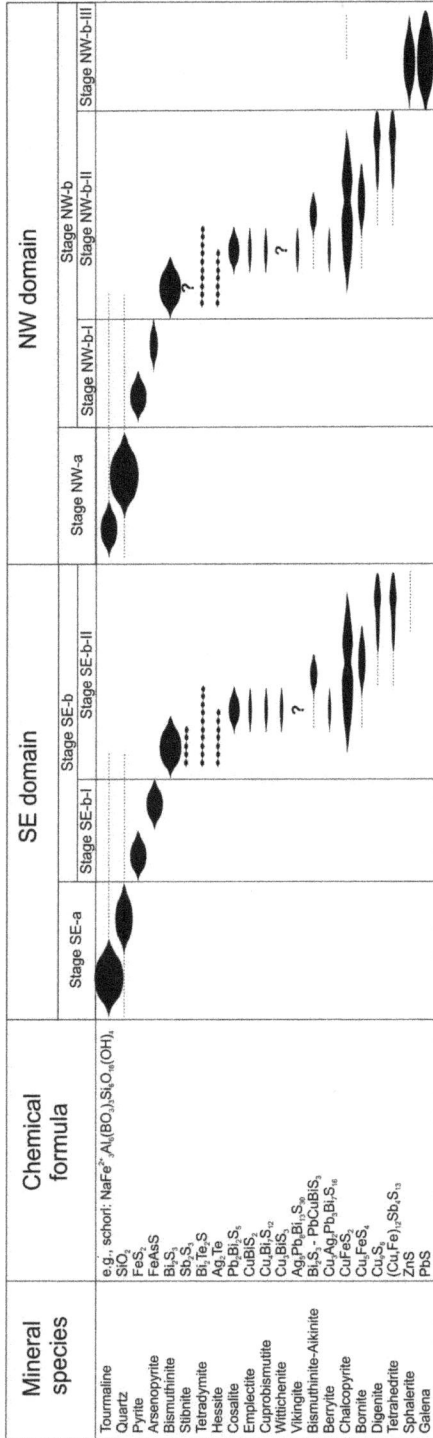

Figure 3. Paragenetic diagram for the SE and NW domains of the San Francisco de los Andes breccia complex. Only the most abundant and distinctive hypogene mineral species are shown.

Coarse- to medium-grained, euhedral tourmaline was the first mineral to precipitate during stage SE-a. It was followed by coarse- to medium-grained euhedral to subhedral quartz, which is intergrown with younger tourmaline (Figures 3 and 4a,b). The abundance of tourmaline and quartz varied during stage SE-a, where abundant tourmaline co-precipitated with minor quartz and vice versa, but in general, abundant tourmaline co-precipitated with minor quartz. Euhedral tourmaline and quartz tended to grow perpendicular to tabular breccia fragments, resulting in cockade textures.

Stage SE-b-I is characterized by euhedral pyrite followed by arsenopyrite (Figure 3). Stage SE-b-I produced more than 95% of pyrite and arsenopyrite in the SE domain of the breccia complex. Pyrite formed euhedral crystals, which now show different degrees of cataclasis (Figure 4c,d). Arsenopyrite exhibits less fractures than pyrite, and it commonly fills fractures in earlier-formed pyrite crystals (Figure 4c,d). Stage SE-b-I pyrite–arsenopyrite is crosscut and overprinted by stage SE-b-II Bi–Cu ore minerals (Figures 3 and 4d).

Stage SE-b-II sulfides and sulfosalts are the main Bi and Cu ores in the SE domain. Bismuthinite was the first Bi mineral to precipitate and was followed by cosalite (Figures 3 and 4e–g). Cosalite co-precipitated and continued to form after bismuthinite stopped (Figure 3). Minor to trace amounts of stibnite formed simultaneously with bismuthinite, whereas minor emplectite, cuprobismutite, wittichenite, and berryite formed synchronous with cosalite (Figure 3). Younger mineral species of the bismuthinite–aikinite series commonly occur at contacts between bismuthinite and cosalite (Figures 3 and 4e), but they are also locally hosted in bismuthinite crystals. Minor amounts of tellurosulfides and tellurides accompanied precipitation of bismuth sulfides and sulfosalts (Figure 3); tetradymite inclusions are hosted in both bismuthinite and cosalite, whereas hessite inclusions are rarer and were only found in bismuthinite (Figures 3 and 4f,g). Stage SE-b-II chalcopyrite crystalized after bismuthinite started to precipitate and began to form before cosalite. Chalcopyrite continued to precipitate well after cosalite stopped (Figure 3; e.g., Figure 4g). Stage SE-b-II paragenetic relationships documented during this study were consistent with the observations of Llambías and Malvicini (1969) [1]. Bornite is a common phase but is less abundant than chalcopyrite. Bornite, together with minor digenite and tetrahedrite formed synchronous with and after stage SE-b-II chalcopyrite began to precipitate (Figure 3). Only minor sphalerite stars and incomplete sphalerite skeletal crystal inclusions occur in stage SE-b-II chalcopyrite crystals (Figures 3 and 4h).

3.2.2. Hydrothermal Cement in the NW Domain

The paragenetic sequence in the NW domain is similar to that in the SE domain, in that silicate minerals (stage NW-a) were followed by sulfides and sulfosalts (stage NW-b; Figure 3). Stage NW-b mineralization can be further subdivided into an early (stage NW-b-I), intermediate (stage NW-b-II) and late phase (stage NW-b-III; Figure 3).

Stage NW-a contains more quartz than tourmaline, which contrasts with stage SE-a (Figure 3). Cockade and comb textures are also common, but the most distinctive textural feature in stage NW-a is the clear to milky cockade quartz, which is not present in stage SE-a (Figure 5a).

Stage NW-b-I is similar to stage SE-b-I in that it is characterized by euhedral pyrite, followed by arsenopyrite (Figure 5b,c). However, arsenopyrite is significantly less abundant in stage NW-b-I than stage SE-b-I (Figure 3).

Identical to the SE domain, stage NW-b-II is characterized by Bi and Cu mineralization (Figure 3). Bismuthinite, cosalite, chalcopyrite, and associated sulfides, sulfosalts, and tellurides are common, but they are less abundant than in stage SE-b-II (Figure 5c,d). Another difference between stages SE-b-II and NW-b-II is the presence of minor vikingite and lack of wittichenite and stibnite from stage NW-b-II (Figure 3).

Stage NW-b-III mineral assemblage is the most distinctive feature of the NW domain of the breccia complex, because it is characterized by abundant galena and sphalerite (Figures 3 and 5e,f). These sulfides are not present in the SE domain, expect for traces of sphalerite as star-like inclusions in stage SE-b-II chalcopyrite (Figure 4h). Stage NW-b-III sphalerite hosts very fine- to medium-grained blebs of chalcopyrite, defining the characteristic chalcopyrite disease texture (Figure 5f; cf., [24]).

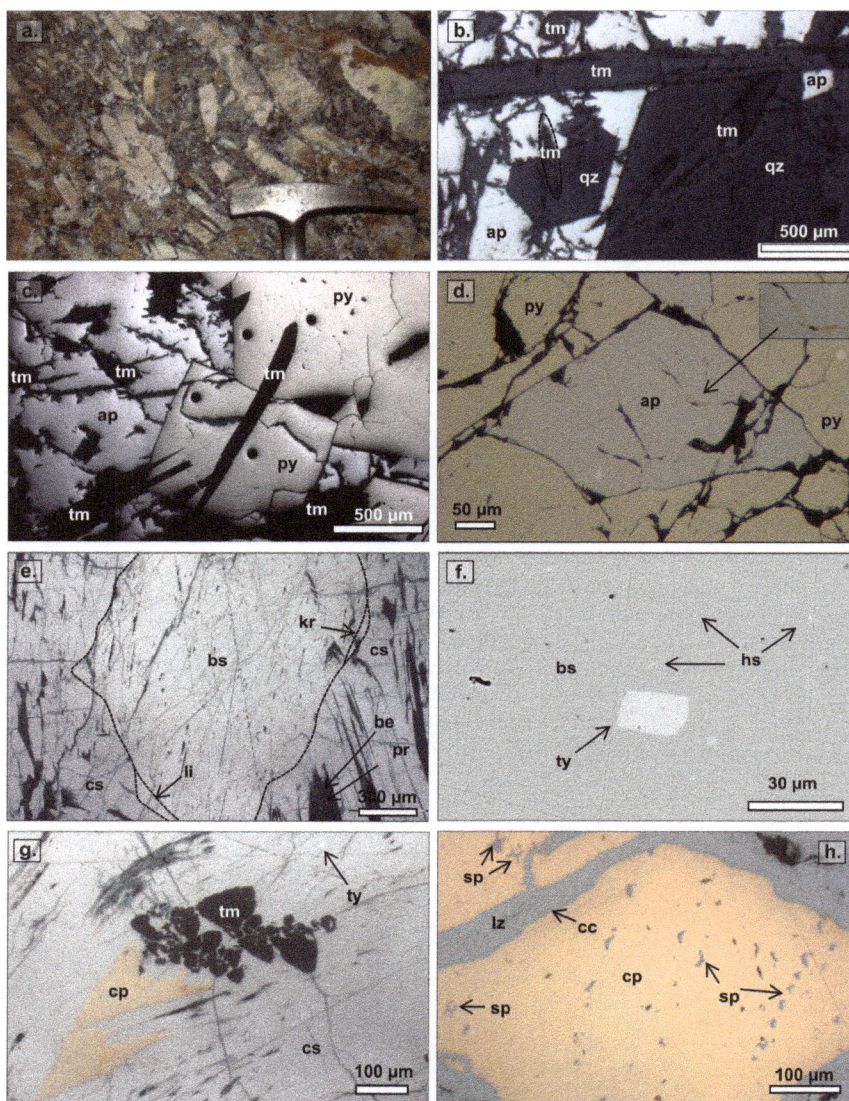

Figure 4. Key mineral phases and assemblages documented in the SE domain. (**a**) Tourmaline > quartz-cemented breccias. (**b**) Stage SE-a tourmaline prisms overgrown by SE-a quartz (hexagonal cross section). Stage SE-b-I arsenopyrite encloses younger silicates (reflected light, parallel nicols). (**c**) Stage SE-a tourmaline overgrown by stage SE-b-I pyrite and younger SE-b-I arsenopyrite (reflected light, parallel nicols). (**d**) Stage SE-b-I arsenopyrite filling fractures in euhedral pyrite. Note that both sulfides are affected by younger fractures, locally cemented by stage NW-b-II chalcopyrite and supergene chalcocite (reflected light, parallel nicols). (**e**) Stage SE-b-II bismuthinite and cosalite. Intermediate members of the bismuthinite–aikinite series formed along the contacts. Cosalite exhibits evident cleavage and parting where the arsenate beudantite and preisingerite, preferentially formed (reflected light, parallel nicols). (**f**) Very fine-grained hessite inclusions are only hosted in bismuthinite, whereas tetradymite inclusions are more abundant and larger (BSEI). (**g**) Stage SE-a tourmaline trigonal prisms overgrown by stage SE-b-II cosalite. Younger SE-b-II chalcopyrite formed along cosalite parting and cleavage (reflected light,

parallel nicols). (**h**) Sphalerite stars and incomplete skeletal sphalerite crystal inclusions in stage SE-b-II chalcopyrite. Supergene chalcocite and luzonite formed along fractures (reflected light, parallel nicols). Abbreviations: ap = arsenopyrite, be = beudantite, bs = bismuthinite, cc = chalcocite, cp = chalcopyrite, cs = cosalite, hs = hessite, kr = krupkaite, li = lindstromite, lz = luzonite, pr = preisingerite, py = pyrite, qz = quartz, sp = sphalerite, tm = tourmaline, and ty = tetradymite.

Figure 5. Key mineral phases and assemblages documented in the NW domain. (**a**) Characteristic milky quartz with cockade texture from the NW domain breccias. (**b**) Stage NW-a tourmaline trigonal prisms overgrown by stage NW-b-I arsenopyrite (reflected light, parallel nicols). (**c**) Stage NW-b-I pyrite moderately fractured and cemented by stage NW-b-II bismuthinite (reflected light, parallel nicols). (**d**) Backscattered electron image of stage NW-b-II bismuthinite and cosalite. Note the presence of tetradymite inclusions hosted in both Bi-bearing minerals, whereas krupkaite formed only along the contacts. (**e**) Stage NW-b-III galena crystal with its characteristic triangular cleavage pits. Stage NW-a tourmaline prisms and NW-b-II chalcopyrite grains are enclosed in the galena crystal (reflected light, parallel nicols). (**f**) Stage NW-b-III sphalerite with chalcopyrite disease texture and supergene chalcocite rims (reflected light, parallel nicols). The different sizes of the chalcopyrite blebs in sphalerite are grouped and aligned along crystallographic planes in sphalerite. Abbreviations: ap = arsenopyrite, bs = bismuthinite, cc = chalcocite, cp = chalcopyrite, cs = cosalite, gn = galena, go = goethite, kr = krupkaite, py = pyrite, sp = sphalerite, tm = tourmaline, and ty = tetradymite.

3.3. Bismuth Mineral Species and Related Phases

In this section, we present two representative samples from our Bi-bearing mineral chemical composition dataset. These samples were selected as they show major chemical variations for each domain. Table 1 lists the average major chemical composition of each species, per domain, analyzed by EMPA. Figure 6 illustrates the theoretical chemical composition for the most relevant species in the Bi–Sb–Cu–Ag–Pb–S system. Only sulfides and sulfosalts were plotted on this ternary diagram.

Table 1. Mean major chemical composition of each Bi-bearing species detected in the NW and SE domains. Average major chemical compositions are based on EMP analyses collected from two samples from opposite domains of the breccia complex: 29-2010 (NW) and 15-2010 (SE).

	NW Domain						SE Domain			
	4b-II Bismuthinite	4b-II Cosalite	Unknown	Unknown	3b-II Bismuthinite	3b-II Tetradymite	3b-II Bismuthinite–Aikinite Series n = 4			
							Gladite–Salzburgite Member	Paarite–Krupkaite Member		Friedrichite
	n = 15	n = 9	n = 1	n = 1	n = 24	n = 7	n = 1	n = 1	n = 1	n = 1
Bi	79.31	43.33	47.56	40.49	79.46	60.81	63.04	59.63	58.52	43.62
Sb	0.93	1.13	0.55	1.68	0.55	0.27	0.53	0.50	0.14	1.08
Pb	0.99	34.74	20.95	26.09	0.84	0.03	13.98	16.88	17.09	28.98
Cu	0.34	2.09	6.71	7.53	0.31	0.08	4.27	5.20	6.53	9.14
Ag	0.01	2.61	7.58	6.56	0.02	0.17	0.00	0.00	0.13	0.10
S	18.34	16.08	16.74	16.62	17.91	5.02	17.44	17.52	17.32	16.78
Se	0.18	0.11	0.19	0.18	0.23	0.31	0.17	0.14	0.18	0.06
Te	0.06	0.14	0.18	0.21	0.30	34.18	0.04	0.07	0.07	0.05
Σ	100.16	100.23	100.46	99.37	99.62	100.89	99.48	99.93	99.98	99.81

n = number of analyses.

Bismuthinite and cosalite are the two major Bi-mineral species identified from the 26 EMP and LA-ICP-MS spots analyzed in the NW domain. Fifteen out of the twenty-six analyses correspond to **bismuthinite**. Its average chemical composition is listed in Table 1 and can be expressed as $Bi_{1.981}Sb_{0.040}Pb_{0.025}Cu_{0.028}Ag_{0.001}(S_{2.986}Se_{0.012}Te_{0.002})$. No apparent chemical variations within the bismuthinite–aikinite solid solution series were detected. Silver telluride inclusions are regularly spread in bismuthinite crystals and can be detected using BSE imagery, and Ag and Te spiky signals in LA-ICP-MS profiles (e.g., Figure 7a). **Cosalite** was detected from nine analyses along the twenty-six-spot-profile. Its average chemical formula can be expressed as $Pb_{1.663}Cu_{0.326}Ag_{0.240}Bi_{2.057}Sb_{0.092}(S_{4.976}Se_{0.013}Te_{0.011})$ based on Table 1. The term 'argentocuprocosalite' can be applied here to refer to Ag- + Cu-bearing cosalite, to explain the shift away from pure end-member cosalite as plotted on Figure 6. No silver telluride inclusion was detected in cosalite in BSE images or in the flat signal of both Ag and Te in LA-ICP-MS profiles (e.g., Figure 7b). The two remaining analyses from the NW domain sample revealed two unknown species (Table 1; Figure 6).

Two mineral phases were found commonly from the 35 EMP and LA-ICP-MS spots analyzed in the SE domain. Twenty-four analyses yielded **bismuthinite**, with no apparent chemical variations in the bismuthinite–aikinite solid solution series (Table 1; Figure 6). Its average major chemical composition can be expressed as $Bi_{2.023}Sb_{0.024}Pb_{0.022}Cu_{0.026}Ag_{0.001}(S_{2.972}Se_{0.015}Te_{0.012})$. Seven out of thirty-five EMPA analyses correspond to **tetradymite** inclusions, which are easily detected by means of BSE imagery and consistent Te and Bi spikes in bismuthinite LA-ICP-MS profiles. Tetradymite's average chemical composition is listed in Table 1 and can be expressed as $Bi_{2.037}Sb_{0.016}Pb_{0.001}Cu_{0.009}Ag_{0.011}(Te_{1.876}S_{1.096}Se_{0.028})$. The remaining four EMP and LA-ICP-MS spots were four different members of the bismuthinite–aikinite solid solution series: (1) **Friedrichite** $Pb_{4.802}Cu_{4.939}Ag_{0.032}Bi_{7.166}Sb_{0.305}(S_{17.961}Se_{0.026}Te_{0.013})$; (2) two species between the **krupkaite–paarite** members with the following chemical composition: $Pb_{0.911}Cu_{1.135}Ag_{0.013}Bi_{3.094}Sb_{0.012}(S_{5.969}Se_{0.025}Te_{0.006})$, and $Pb_{0.891}Cu_{0.894}Ag_{0}Bi_{3.121}Sb_{0.045}(S_{5.975}Se_{0.019}Te_{0.006})$; and (3) a mineral phase between the **salzburgite–gladite** members, $Pb_{1.482}Cu_{1.475}Ag_{0}Bi_{6.624}Sb_{0.096}(S_{11.945}Se_{0.047}Te_{0.008})$.

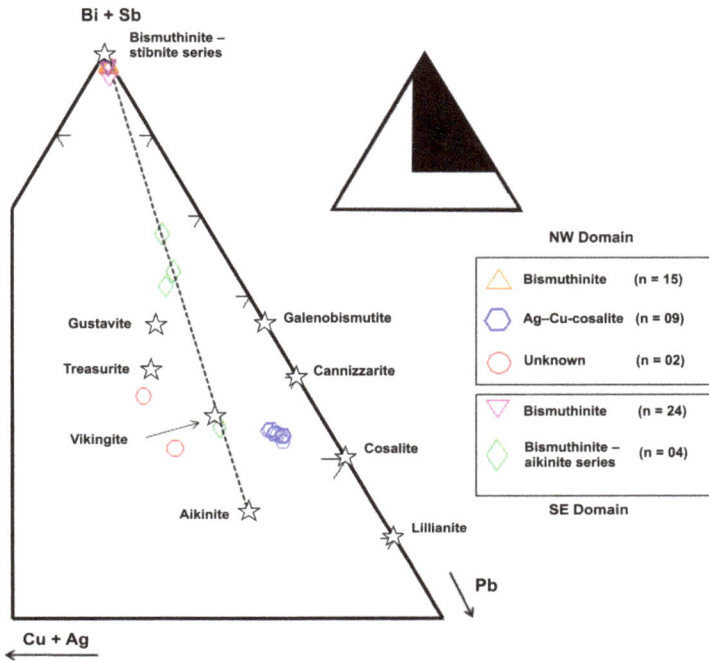

Figure 6. Bismuth-bearing sulfides and sulfosalts plotted on a (Cu + Ag)–(Bi + Sb)–Pb ternary diagram. Major chemical compositions are based on EMP analyses collected from two samples selected from opposite domains of the breccia complex: 29–2010 (NW) and 15–2010 (SE). Abbreviation: n = number of analyses.

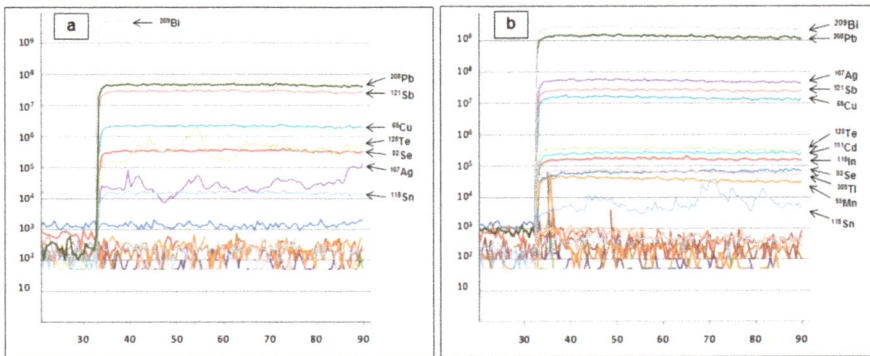

Figure 7. LA-ICP-MS spectra of bismuthinite and cosalite from the NW domain. (**a**) Major and trace elements hosted in the bismuthinite lattice exhibit a flat and smooth LA signal, whereas the Te signal is spiky and mimics the Ag signal as both elements are hosted in tetradymite inclusions. (**b**) Pure cosalite's LA-ICP-MS spectrum with no silver telluride inclusions.

4. Discussion

4.1. Physicochemical Conditions during Bi Ore Deposition

Many physicochemical variables are fundamental to hydrothermal ore transport and deposition; they include temperature, pressure, salinity, acidity/alkalinity, redox, concentrations of sulfur species, and activities of metal species (e.g., Cooke et al., 1996 [25]). Changes in any of these variables can trigger ore deposition. The purpose of this section is to discuss the thermochemical environment during bismuth ore deposition at San Francisco de los Andes, and to constrain fluid physicochemical conditions based on observed bismuth minerals and co-existing minor phases.

4.1.1. Temperature and Pressure

Hydrothermal quartz that cemented the breccias before ore deposition provides an opportunity to constrain temperature and pressure conditions at San Francisco de los Andes. Titanium-in-quartz (TitaniQ) temperatures based on the Thomas et al. (2010) [26] geothermobarometer, calculated for pressure of 1 kbar yielded a temperature ranges of 262–378 °C ([2]; Figure 8). The presence of rutile as an equilibrium phase in the system indicated TiO_2 saturation and allowed Ti activity to be fixed at 1. Titanium-in-quartz geothermobarometry yielded similar T ranges for hydrothermal quartz in the SE domain (i.e., T = 266–378 °C; 1 kbar) and NW domain (i.e., T = 262–364 °C; 1 kbar) of the breccia complex (Figure 8). The lowest limit is based on Ti concentration very close to the detection limit (0.72 ppm; Figure 8a). As seen on the paragenetic diagram in Figure 3, stages SE-a and NW-a quartz are temporally related but earlier than Bi mineralization, which suggests the interval 262–378 °C is the maximum temperature for ore precipitation.

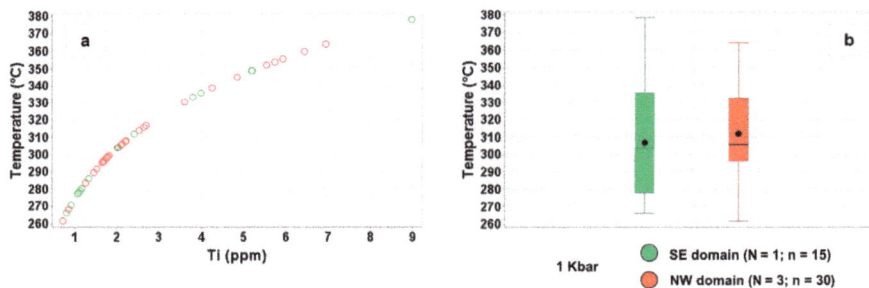

Figure 8. Titanium-in-quartz temperature calculated at 1 kbars after Thomas et al. (2010) [26]. (**a**) Binary plots of Ti contents versus calculated TitaniQ temperatures in quartz from stages SE-a and NW-a. (**b**) Box-and-whisker plot of Ti contents in quartz from stages SE-a and NW-a. Abbreviations: N = number of samples, n = number of analyses.

If fluid/rock ratios are low, minor and trace element contents in minerals can be largely controlled by host rock composition; whereas at high fluid/rock ratios, it is mainly controlled by the composition of hydrothermal fluids. Conduits such as breccia pipes or dikes are a preferred pathway with high fluid fluxes where only minor wall rock interaction occurs; thus it is safe to assume Ti contents in quartz are temperature dependent, where its concentration systematically increases with increasing temperature (e.g., Thomas et al., 2010; Wark and Watson, 2006 [26,27]; Figure 8).

Little has been published on the conditions of formation of the San Francisco de los Andes deposit. The only temperatures based on fluid inclusions were reported by Cardo et al. (2008) [3]. They documented an average homogenization temperature of 228 °C for two-phase, liquid-rich inclusions, and temperatures between 286 and 299 °C for three-phase (liquid + vapor + halite) inclusions hosted in quartz. These homogenization temperatures are comparable with those from hydrothermal quartz in magmatic-hydrothermal tourmaline breccias at Rio Blanco-Los Bronces, Chile [2]. Particularly, in the Sur-Sur breccia, measured homogenization temperature for type ia (two-phase, liquid-rich fluid inclusions) and type iia (three-phases, salt saturated, halite-bearing fluid inclusions) in quartz, range between less than 200 °C and more than 400 °C [28].

Fluid inclusion microthermometry yields a set of homogenization temperatures that provide minimum temperatures of entrapment of a fluid in a mineral. These data are collected in a laboratory under atmospheric pressure, thus pressure corrections should be considered to obtain the real temperatures of entrapment. The temperature of formation of hydrothermal quartz from the San Francisco de los Andes breccia complex can be calculated at 1 kbar based on homogenization temperatures, and ice melting or halite dissolution temperatures provided by Cardo et al. (2008) [3]. Two-phase, liquid-rich fluid inclusions yielded homogenization temperatures between 227 and 229 °C, melting temperature of ice of −10.6 °C (i.e., 14.57% NaCl equivalent) and temperatures of entrapment or formation between 287 and 289 °C. Three-phase (liquid + vapor + halite) fluid inclusions yielded homogenization temperatures between 286 and 299 °C, total homogenization due to dissolution of halite between 367 and 388 °C (i.e., 44.73% and 46.75% NaCl eq.), and temperatures of formation between 359 and 376 °C [29,30]. The estimated temperatures of formation of fluid inclusions in quartz for the pressure prevailing at the time of entrapment were consistent with, and fall in the TitaniQ temperature range under 1 kbar estimated by Testa (2017) [2].

The pressure value of 1 kbar is similar to the conditions of formation of other magmatic- hydrothermal breccia in the Andes, such as Donoso at the Rio Blanco-Los Bronces district (e.g., Skewes et al., 2003 [31]). A previous fluid inclusion study showed that different magmatic fluids were involved in formation of the Donoso breccia: magmatic-hydrothermal fluids cooling under relatively high (>1 kbar) lithostatic pressure (consistent with geologic constraints), and magmatic-hydrothermal fluids cooling at hydrostatic pressure conditions [31]. Both fluids occurred intermittently, as pressure fluctuated between lithostatic and hydrostatic conditions due to sealing and rebrecciation episodes [31].

4.1.2. Phase Separation: Vapor-Rich Phase and Dense Brine Phase

The existence of a vapor-rich phase derived from magmatic-hydrothermal fluids can be inferred from fluid inclusion studies of hydrothermal quartz from San Francisco de los Andes. Testa (2017) [2] documented four types of fluid inclusions in stages SE-a and NW-a quartz from San Francisco de los Andes (Figure 9). He recorded the coexistence of primary type ia (two-phase, liquid-rich) and type ib (two-phase, vapor-rich) fluid inclusions (Figure 9), which traditionally implies boiling. No evidence of leakage, necking or post-entrapment deformation was observed based on detailed petrographic analysis of fluid inclusions. None of these processes have been documented, thus the coexistence of type ia and type ib fluid inclusions cannot be explained by disruption of the original liquid–vapor proportion in individual fluid inclusions. The coexistence of primary type ia and ib fluid inclusions is therefore interpreted to be the product of phase separation, which probably produced a vapor phase and a dense liquid phase during brecciation.

As previously discussed, salinities of 14.57% and 44.73–46.75% NaCl eq. were documented in two-phase and three-phase fluid inclusions, respectively [3]. Testa (2017) [2] documented type iia (four-phase, salt saturated, halite–sylvite-bearing) fluid inclusions and type iib (polyphase, salt saturated, halite–sylvite–mica?-bearing) fluid inclusions from San Francisco de los Andes (Figure 9). Although no microthermometric analyses were conducted, based on the coexistence of large halite (cubic crystals with sharp corners), sylvite (cubic crystals with rounded edges), and unspecified mica (hexagonal sheet-like crystals) in fluid inclusions, salinities above 45 wt % NaCl equivalent are considered likely (Figure 9). Type iia and iib fluid inclusions indicate salinities higher than those reported by Cardo et al. (2008) [3], and evidence that hydrothermal fluids trapped probably contained salts other than NaCl (e.g., KCl, $CaCl_2$, and $MgCl_2$). The elevated salinities in these fluid inclusions were interpreted as evidence of the coexistence of a brine-phase. Stages SE-a and NW-a quartz formed before ore deposition. A deep-seated brine is believed to be the main ore-forming fluid responsible for stage SE-b and NW-b mineralization.

Some of the phase boundaries between liquid and vapor in fluid inclusions hosted in quartz from San Francisco de los Andes appear thick or show apparent concentric patterns (e.g., Figure 9d). These features are artifacts produced by microphotography; no petrographic evidence of non-condensable gas species, such as CO_2, was reported in fluid inclusions by Cardo et al. (2008) [3] or Testa (2017) [2]. Furthermore, no hypogene, C-bearing mineral phases (e.g., carbonates) have been documented from San Francisco de los Andes. The $CO_{2(g)}$ contents of the ore-forming hydrothermal fluids were therefore considered to be significantly lower than 4.4 wt % CO_2, the minimum concentration required to observe visible CO_2 in fluid inclusions at room temperature [32].

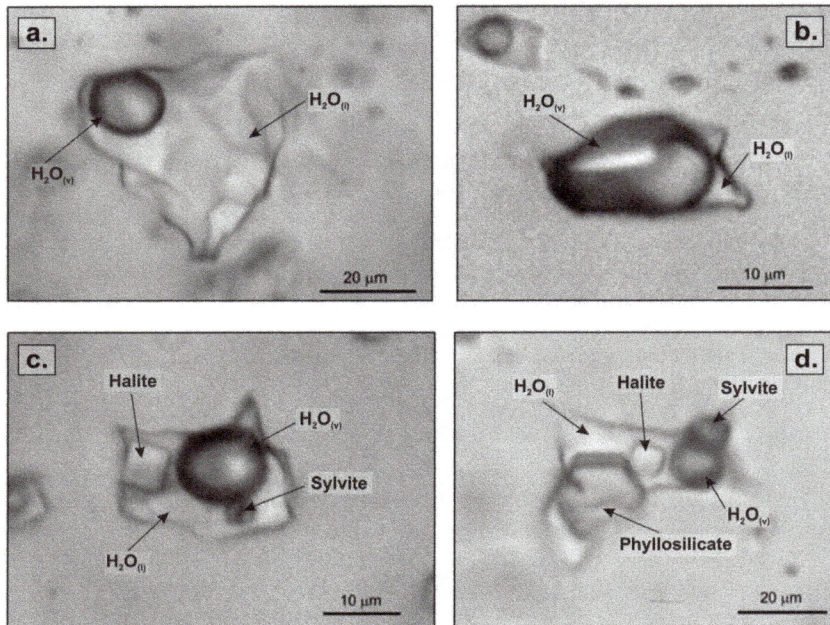

Figure 9. Four types of fluid inclusions hosted in stage SE-a and NW-a quartz from the San Francisco de los Andes Breccia complex. (**a**) Type ia are two-phase, liquid-rich fluid inclusions. (**b**) Type ib are two-phase, vapor-rich fluid inclusions. (**c**) Type iia are four-phase, halite–sylvite-bearing fluid inclusions. (**d**) Type iib are polyphase, halite–sylvite–phyllosilicate-bearing fluid inclusions.

4.1.3. Acidity/Alkalinity

pH values are not always simple to determine and their significance is questionable as the pH of neutral fluids varies with temperature, salinity, etc. pH decreases with increase in temperature, which does not imply that a fluid becomes more acidic at higher temperatures. By definition, a solution is acidic if there is an excess of hydrogen ions over hydroxide ions. At high temperature, pure water is still neutral as the concentration of hydrogen ions is the same as hydroxide ions but pH values are <7. For this reason in this section we prefer to discuss whether fluids were acidic or alkaline, rather than pH values.

The acidity/alkalinity of ore-forming fluids at San Francisco de los Andes can be inferred based on hydrothermal alteration assemblages in the sedimentary host rocks and breccia clasts. Sericite–quartz–tourmaline is the characteristic hydrothermal alteration assemblage adjacent to the breccia complex, whereas epidote–chlorite ± calcite veins are commonly observed distal to the deposit [2]. Based on the presence of illite–quartz, the mineralizing solutions must have been weakly to moderately acidic, evolving outwards from the breccia complex to near-neutral or alkaline conditions by water–rock interaction (epidote–chlorite ± calcite stable; e.g., Cooke et al., 1996; Corbett, and Leach, 1998 [25,33]). The stability of mineral phases in the phyllic alteration assemblage can be constrained as a function of temperature and K^+ and H^+ activities (i.e., <500 °C and intermediate log (a_{K^+}/a_{H^+}) values), with decreasing H^+ metasomatism associated with the propylitic assemblage [34–38]. The absence of potassic alteration associated with the breccia complex indicates that high temperature, alkaline conditions were not attained at the level of exposure of the breccia complex (i.e., K-feldspar stable; >500 °C and high log (a_{K^+}/a_{H^+}); [34–38]). A potassic-altered root is predicted to occur below the current level of exposure, based on comparisons with other well-drilled tourmaline breccia complexes (e.g., Sur-Sur; [28]). The lack of diagnostic minerals characteristic of advanced argillic alteration (e.g., alunite, kaolinite, dickite, pyrophyllite) provides evidence that hydrothermal fluids at the level exposed were only weakly acidic (i.e., moderate log (a_{K^+}/a_{H^+}); [36,37]), preventing formation of hypogene clay minerals; they may have occurred at higher elevations but, if they did, they have been removed by erosion. Scarce clay minerals from intermediate argillic alteration assemblage were locally documented; both chlorite and illite imply weakly acidic conditions of alteration.

4.1.4. Redox Conditions

Magmatic-hydrothermal breccias from the Rio Blanco-Los Bronces district have similarities to the San Francisco de los Andes breccia complex (e.g., pyrite and Cu-sulfide cement), but are different in that oxidized phases (i.e., anhydrite and specularite) were formed at distinctive stages in the Donoso, Sur-Sur, and La Americana magmatic-hydrothermal breccias [2,28,31]. The precipitation of anhydrite during late-stage fluid evolution was interpreted to indicate that hydrothermal fluids evolved to higher oxygen fugacity (i.e., sulfate-stable), and/or higher $a_{Ca^{2+}}$ [28]. The presence of specularite at Rio Blanco-Los Bronces was also consistent with higher oxygen fugacity conditions [28]. At San Francisco de los Andes, the abundance of hypogene sulfide and sulfosalt phases, coupled with an absence of hypogene oxides, sulfates, carbonates or arsenates implies consistently reduced conditions (i.e., H_2S-predominant) and high sulfur fugacities during ore transport and deposition.

4.2. Fluid Evolution and Metal Transport: S and Te Fugacities

Titanium-in-quartz geothermobarometry from San Francisco de los Andes has constrained a temperature range of 262–378 °C at 1 kbar. This temperature probably prevailed during ore deposition. The proposed P–T ranges are comparable with previous studies of fluid inclusions in quartz from magmatic-hydrothermal breccia complexes [3,28,31]. Based on these data and detailed mineralogical observations, two f_{Te_2} vs. f_{S_2} diagrams were calculated (Figure 10).

Thermodynamic models for selected bismuth sulfides, sulfosalts, and tellurides have been proposed for different mineralizing systems, such as orogenic and epithermal deposits (e.g., Cooke et al., 1996; Voicu et al., 1999; Cook and Ciobanu, 2004; Voudouris, 2006 [25,39–41]). Thermodynamic modelling of the San Francisco de los Andes magmatic-hydrothermal breccia complex has been undertaken by assuming a minimum temperature of 230 °C, and a maximum temperature of 400 °C at a constant pressure of 1 kbar. Key mineralogical assemblages in stages SE-b-II and NW-b-II are Bi- and Cu-bearing sulfides and sulfosalts that co-precipitated with tellurides and tellurosulfides, which are used here to constrain both f_{S_2} and f_{Te_2} (Figure 3). The stability fields for Bi sulfides, sulfosalts, tellurosulfides, and co-existing phases have been plotted on f_{S_2}–f_{Te_2} diagrams (Figure 10).

For thermodynamic modeling, paragenetic stages SE-b-II and NW-b-II were divided into three sub-stages. The first stage (SE-b-IIa, NW-b-IIa) was the early phase of bismuthinite (bs) deposition (Figures 3 and 10). During the second stage (SE-b-IIb, NW-b-IIb) bismuthinite–cosalite–chalcopyrite (bs–cs–cp) assemblages were precipitated (Figures 3 and 10). During the third stage (SE-b-IIc, NW-b-IIc), the cosalite–chalcopyrite (cs–cp) assemblage precipitated (Figures 3 and 10).

The chemical behavior of tellurium is one of the keys to understanding the evolution of hydrothermal fluids and ore precipitation at San Francisco de los Andes. Bismuth and Ag tellurosulfides/telluride species occur as inclusions in bismuthinite and cosalite in both domains of the breccia complex (Figure 3). The presence of tellurium-bearing phases may be a key indication of ingress of magmatic volatiles during brecciation and mineralization (e.g., Cooke et al., 1996; Cooke and McPhail, 2001 [25,42]).

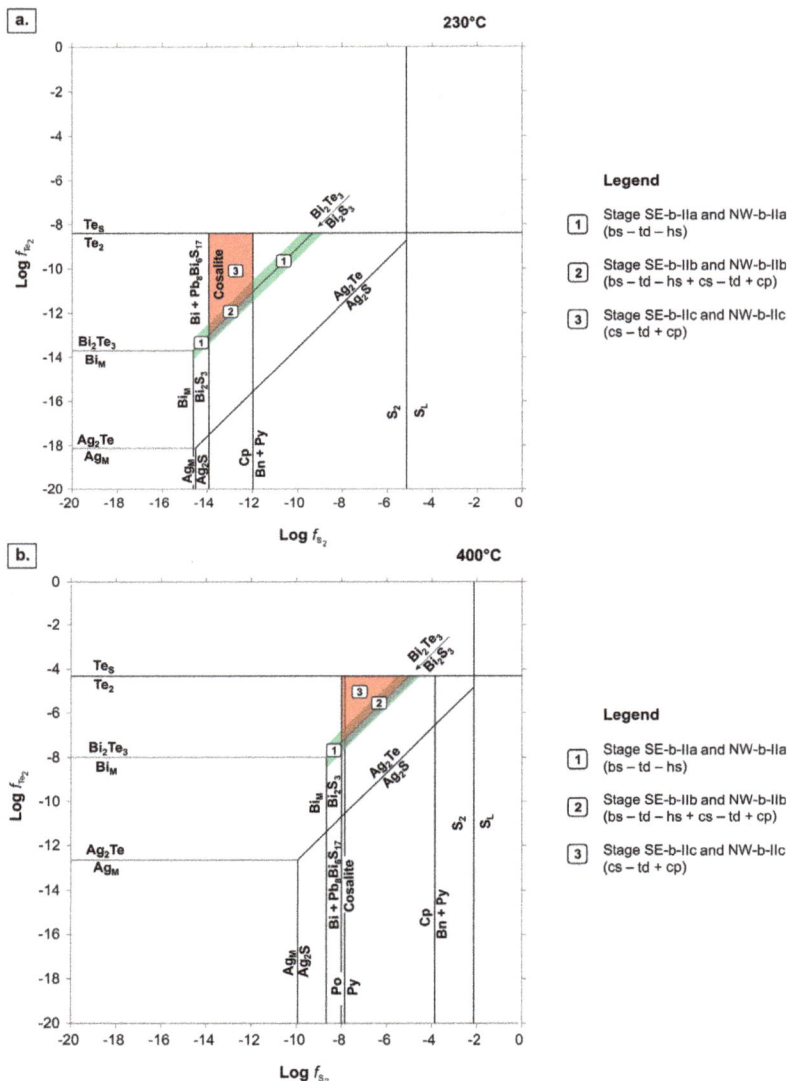

Figure 10. f_{S_2} vs f_{Te_2} diagrams calculated at 230 °C (**a**), and 400 °C (**b**), at 1 kbar pressure. The pyrrhotite–pyrite (po–py) monovariant line was plotted at the higher temperature, but not at 230 °C, as it falls in the stability field of Bi_M. Model abbreviations: Cp = chalcopyrite, Bn = bornite, Py = pyrite, Po = pyrrhotite, L = liquid, S = solid, and M = metallic. Legend abbreviations: bs = bismuthinite, cs = cosalite, cp = chalcopyrite, td = tetradymite, hs = hessite.

Tetradymite (Bi$_2$Te$_2$S; td) inclusions are common and are homogeneously distributed in both stage SE-b-II and NW-b-II bismuthinite and cosalite. Hessite (Ag$_2$Te; hs) inclusions are rare; they were only documented in stage SE-b-II and NW-b-II bismuthinite. Tetradymite (Bi$_2$Te$_2$S; td) can form from tellurobismuthite (Bi$_2$Te$_3$) according to the following reaction:

$$2 \, Bi_2Te_{3(Tellurobismuthite)} + S_{2_{(g)}} \rightleftarrows 2 \, Bi_2Te_2S_{(Tetradymite)} + Te_{2_{(g)}} \qquad (1)$$

Tetradymite can therefore precipitate due to either a slight f_{S_2} increase or a subtle f_{Te_2} decrease. The tetradymite stability space is likely to be similar to that of tellurobismuthite (e.g., Cook and Ciobanu, 2006 [40]). The stability field for the bismuthinite (Bi$_2$S$_3$)–tetradymite (Bi$_2$Te$_2$S) assemblage in Figure 10 is defined in part by the chemical reaction:

$$Bi_2S_{3(Bismuthinite)} + 3/2 \, Te_{2_{(g)}} \rightleftarrows Bi_2Te_{3(Tellurobismuthite)} + 3/2 \, S_{2_{(g)}} \qquad (2)$$

At the minimum temperature of 230 °C, f_{S_2} and f_{Te_2} values for stage SE-b-IIa + NW-b-IIa (bs–td–hs), and stage SE-b-IIb + NW-b-IIb (bs–td–hs + cs–td + cp) are constrained by the equilibrium line defined by the equation $\log f_{Te_2} = \log f_{S_2} + 0.93$; between $\log f_{S_2}$ values of -14.63 and -9.5 (stage SE-b-IIa + NW-b-IIa) and -13.92 and -11.97 (stage SE-b-IIb + NW-b-IIb; Figure 10a). Stage SE-b-IIc + NW-b-IIc (cs $-$ td + cp) stability field is constrained between the line defined by $\log f_{Te_2} = \log f_{S_2} + 0.93$, and $\log f_{Te_2} = -8.3$; and between $\log f_{S_2}$ values of -13.92 and -11.97 (Figure 10a).

For the maximum temperature of 400 °C, stages SE-b-IIa + NW-b-IIa, and SE-b-IIb + NW-b-IIb plot along the equilibrium line $\log f_{Te_2} = \log f_{S_2} + 0.68$, between $\log f_{S_2}$ values of -8.9 and -5 (stage SE-b-IIa + NW-b-IIa), and -8 and -5 (stage SE-b-IIb + NW-b-IIb; Figures 4 and 10b). Stage SE-b-IIc + NW-b-IIc occurs between $\log f_{Te_2} = \log f_{S_2} + 0.68$ and $\log f_{Te_2} = -4.2$ for $\log f_{S_2}$ values between -8 and -5 (Figure 10b).

The position of the bismuthinite–tellurobismuthite monovariant line does not change significantly in the 230–400 °C range. As a result, the f_{S_2} and f_{Te_2} values for stages SE-b-II and NW-b-II a, b, and c markedly depend on the position of the Bi$_M$–Bi$_2$S$_3$, cosalite and cp–(bn + py) monovariant lines and Te$_s$–Te$_2$ equilibrium line (Figure 10). The current model indicates that the stability fields for minerals from the three stages at a given temperature are rather similar, particularly at higher temperatures (Figure 10b).

Tetradymite occurs as inclusions in bismuthinite and cosalite (e.g., Figure 4f,g and Figure 5d). It is inferred to have formed in equilibrium with both the Bi-sulfide and sulfosalt. Conversely, hessite inclusions are only hosted in bismuthinite (e.g., Figure 4f). Bismuthinite has little to no silver in its crystal lattice (Table 1), thus available Ag from hydrothermal fluids has probably bonded with Te to produce the hessite inclusions hosted in bismuthinite. Conversely, cosalite scavenged silver from its surroundings and preferentially partitioned high concentrations of Ag into its crystal structure (Table 1), so pulses of Te-rich fluids could not form hessite inclusions in cosalite. As the hydrothermal fluid evolved, particularly during stages SE-b-IIc and NW-b-IIc, silver concentrations decreased sharply as silver strongly partitioned into cosalite, precluding hessite formation. Tellurium was still available in hydrothermal fluids during stages SE-b-IIc and NW-b-IIc, indicated by the presence of tetradymite inclusions in cosalite (Figure 10). The tetradymite–cosalite assemblage invalidates the possibility of a decrease in f_{Te_2} to explain the lack of hessite inclusions hosted in cosalite (Figure 10).

During stages SE-b-II and NW-b-II, $\log f_{S_2}$ and $\log f_{Te_2}$ were high ($\log f_{S_2} = -14.6$---9.5; $\log f_{Te_2} = -13.7$---8.3; at 230 °C; Figure 10a; and $\log f_{S_2} = -8.9$---5; $\log f_{Te_2} = -8$---4.2; at 400 °C; Figure 10b). Hydrothermal fluids must have had elevated bismuth concentrations to stabilize Bi-sulfosalts as high a$_{Bi}$ is required to form the bismuthinite–cosalite–tetradymite assemblage. Similarly, the a$_{Ag}$ is also interpreted to have been high, based on hessite inclusions hosted in bismuthinite, and silver partitioning into the cosalite structure. Hessite inclusions have not been detected in Ag-rich cosalite, and acanthite has not been documented at San Francisco de los Andes.

At San Francisco de los Andes, formation of bismuth and silver tellurosulfide/telluride inclusions may have been triggered by an increase in tellurium concentrations in the mineralizing fluids (e.g., tellurium-rich magmatic vapor plumes), or less likely, by a slight f_{S_2} decrease. High Te fugacities are interpreted to have been caused by intermittent contributions of magmatic $Te_{2_{(g)}}$ to the mineralizing fluids during stages SE-b-II and NW-b-II. Similar to the mechanism proposed by Cooke et al. (1996) and Cooke and McPhail (2001) [25,42] at Acupan (Philippines), sporadic bursts of $Te_{2_{(g)}}$ and other magmatic volatiles from a deep-seated crystalizing pluton are interpreted to have risen through the breccia column. This process accounts for the local and episodic distribution of tellurosulfide/telluride hosted in Bi sulfides and sulfosalts in the San Francisco de los Andes breccia complex, consistent with a magmatic-hydrothermal origin for the breccia complex. Contemporary viewpoints on the formation of Au and Bi tellurides favor their being magmatically-derived from a Te-rich source (e.g., Cook and Ciobanu, 2006; Cook, 2007 [40,43]). Tellurium is transported either as aqueous or vapor species in hydrothermal fluids, and precipitates as tellurides, triggered by a variety of possible mechanisms (e.g., boiling, mixing, temperature decrease, fluid-rock interaction, etc.; [40,43]). This model explained the formation of most telluride-bearing gold-rich deposits, including epithermal, orogenic, intrusion-related, volcanic-hosted massive sulfide, porphyry Au–(Cu), and skarns [40,43]. Based on literature, the most likely incorporation of tellurium into the breccia column is due to Te-rich magmatic vapor (e.g., Voudouris, 2006; Cooke and McPhail, 2001; Jensen and Barton, 2000 [41,42,44]).

During the very last stage of mineralization in the NW domain at San Francisco de los Andes (i.e., stage NW-b-III), galena and sphalerite are the dominant mineral phases that cement the breccia. The lack of altaite (PbTe), and Pb or Zn-bearing sulfosalts implies low Te_2 and S_2 fugacities throughout stage NW-b-III. A drastic drop in f_{S_2} and particularly f_{Te_2} is consistent with the proposed mechanism of intermittent incorporation of tellurium via magmatic $Te_{2_{(g)}}$-rich plumes.

5. Conclusions

Sulfur and tellurium fugacities during Bi-ore precipitation were constrained by key mineral assemblages formed in equilibrium at a constant pressure of 1 kbar and a minimum temperature of 230 °C, and a maximum temperature of 400 °C. Three mineral assemblages characterize the three main Bi mineralizing stages that provide evidence for S_2 and Te_2 fugacity fluctuations:

1. Bismuthinite (with tetradymite–hessite inclusions): This mineral assemblage formed in equilibrium along the bismuthinite–tellurobismuthite monovariant line. This equilibrium line finishes where it meets the stability fields of Bi_M and Te_S at opposite ends (Figure 10).
2. Bismuthinite (with tetradymite–hessite inclusions) + cosalite (with tetradymite inclusions) + chalcopyrite: This mineral assemblage formed under much more restricted conditions than the previous assemblage (i.e., bismuthinite–tetradymite–hessite); along the bismuthinite–tellurobismuthite equilibrium line where it is constrained in the stability field where cosalite and chalcopyrite co-precipitate (at 230 °C), and where cosalite and Te_2 coexist (at 400 °C; Figure 10).
3. Cosalite (with tetradymite inclusions) + chalcopyrite: This mineral assemblage formed in the stability field constrained by the bismuthinite–tellurobismuthite monovariant line, and the equilibrium lines where cosalite, chalcopyrite, and Te_2 are stable (at 230 °C; Figure 10). At 400 °C, the stability field for the cosalite–tetradymite–chalcopyrite assemblage was defined in the area limited by the equilibrium line that indicates the co-existence of tellurobismuthite, cosalite and Te_2 (Figure 10).

Key findings that have genetic implications for the San Francisco de los Andes breccia complex are:

*   High f_{S_2} and f_{Te_2} conditions prevailed during stages SE-b-II and NW-b-II Bi-ore deposition; hydrothermal fluids must have had high a_{Bi} and a_{Ag} to stabilize Bi-tellurosulfides, sulfosalts, and sulfides, as well as Ag-tellurides and Ag-rich, Bi-sulfosalts.

- Bismuth and Ag telluride/tellurosulfide result from intermittent contributions of magmatic $Te_{2(g)}$ to hydrothermal mineralizing fluids released from the deep-seated crystalizing Tocota Pluton, implying a genetic link between the breccia complex and the underlying magmatic system. Magmatic volatile-rich vapor plumes probably drove fragmentation and buoyantly ascended though the breccia column.

- Abundant galena and sphalerite coupled with the absence of altaite (PbTe) and Pb or Zn-bearing sulfosalts in the NW domain imply lower Te_2 and S_2 fugacities throughout stage NW-b-III; a drastic drop in f_{S_2} and particularly f_{Te_2} is consistent with intermittent incorporation of Te via magmatic $Te_{2(g)}$-rich plumes.

Author Contributions: This article represents a joint effort from the team of authors. F.J.T. conceived and designed the study, interpreted the results, and wrote this article. F.J.T. and L.Z. worked together during data acquisition. D.R.C. supervised this work and was involved during interpretation of thermodynamic data; he provided valuable ideas for the discussion, as well as edited the entire manuscript.

Funding: This research was funded by the AMIRA P1060 project and the ARC Research Hub for Transforming the Mining Value Chain (project number IH130200004).

Acknowledgments: The information published in this article is part of Francisco J. Testa's PhD thesis. The senior author thanks the AMIRA P1060 project, the ARC Research Hub for Transforming the Mining Value Chain (project number IH130200004) and the University of Tasmania—CODES, for providing funding for his PhD thesis. The main author is grateful to CONICET for providing postgrad funding for the early stages of this study. We thank the Central Science Laboratory at the University of Tasmania, the Geology Department at Universidad Nacional del Sur, and the INGEOSUR for allowing the use of their facilities. We gratefully acknowledge the three anonymous reviewers selected by Minerals, and the academic editor who examined the original manuscript; their critical reading and valuable comments greatly improved this article. Last but not least, we would like to express our appreciation to Professor Noel White for his valuable and constructive suggestions during the final revision of this article.

Conflicts of Interest: The authors declare no conflict of interest.

References

1. Llambías, E.J.; Malvicini, L. The geology and genesis of the Bi-Cu mineralized breccia-pipe, San Francisco de los Andes, San Juan, Argentina. *Econ. Geol.* **1969**, *64*, 271–286. [CrossRef]

2. Testa, F.J. Geology, Alteration, Mineralization and Geochemistry of Tourmaline Breccia Complexes in the Andes: Rio Blanco-Los Bronces, Chile and San Francisco de Los Andes, Argentina. Ph.D. Thesis, University of Tasmania, Hobart, Australia, 2017.

3. Cardó, R.; Segal, S.; Korzeniewski, L.I.; Palacio, M.B.; Chernicoff, C. Estudio Metalogenético de brechas hidrotermales portadoras de mineralización de Bi-Au-Cu en el ámbito de Cordillera Frontal, provincia de San Juan. In *Serie Contribuciones Técnicas, Recursos Minerales N° 31; Servicio Geológico Minero Argentino*; SEGEMAR: Buenos Aires, Argentina, 2008; pp. 2–28. (In Spanish)

4. Testa, F.J.; Villanueva, C.; Cooke, D.R.; Zhang, L. Lithological and Hydrothermal Alteration Mapping of Epithermal, Porphyry and Tourmaline Breccia Districts in the Argentine Andes Using ASTER Imagery. *Remote Sens.* **2018**, *10*, 203. [CrossRef]

5. Testa, F.; Cooke, D.; Zhang, L.; Mas, G. Bismoclite (BiOCl) in the San Francisco de los Andes Bi–Cu–Au Deposit, Argentina. First Occurrence of a Bismuth Oxychloride in a Magmatic–Hydrothermal Breccia Pipe and Its Usefulness as an Indicator Phase in Mineral Exploration. *Minerals* **2016**, *6*, 62. [CrossRef]

6. Angelelli, V. *Yacimientos Metalíferos de la República Argentina*; Comisión de Investigaciones Científicas de la Provincia de Buenos Aires: La Plata, Argentina, 1984. (In Spanish)

7. Lencinas, A.N. *Informe sobre Mina San Francisco de los Andes*; Compañía Minera Aguilar S.A.: San Juan, Argentina, 1990. (In Spanish)

8. Bedlivy, D.; Llambías, E.J. Arseniatos de Cu, de Fe, y de Pb de San Francisco de los Andes, Provincia de San Juan, República Argentina. *Rev. la Asoc. Geológica Argentina* **1969**, *24*, 29–40. (In Spanish)

9. Malvicini, L. Luzonita plumbifera de San Francisco de los Andes provincia de San Juan, República Argentina. *Rev. la Asoc. Geológica Argentina* **1969**, *24*, 127–131. (In Spanish)

10. Bedlivy, D.; Llambías, E.J.; Astarloa, J. Rooseveltit von San Francisco de los Andes und Cerro Negro de la Aguadita, San Juan, Argentinien. *Tschermaks Miner. Petrogr. Mitt.* **1972**, *17*, 65–75. [CrossRef]

11. Bedlivy, D.; Mereiter, K. Preisingerite, $Bi_3O(OH)(AsO_4)_2$, a new species from San Juan Province, Argentina: Its description and crystal structure. *Am. Mineral.* **1982**, *67*, 833–840.

12. Garrels, R.M.; Christ, C.L. *Solutions, Minerals, and Equilibria*; Harper & Row: New York, NY, USA, 1965.

13. Meyer, C.; Hemley, J.J. Wall rock alteration. In *Geochemistry of Hydrothermal Ore Deposits*; Barnes, H.L., Ed.; Holt, Rinehart and Winston, Inc.: New York, NY, USA, 1967; pp. 166–232.

14. Robie, R.A.; Waldbaum, D.R. Thermodynamic properties of minerals and related substances at 298.15 K (25.0 °C) and one atmosphere (1.013 bars) pressure and at higher temperatures. In *US Geological Survey Bulletin 1259*; USGS: Washington, DC, USA, 1968; p. 256.

15. Craig, J.R.; Barton, P.B. Thermochemical approximations for sulfosalts. *Econ. Geol.* **1973**, *68*, 493–506. [CrossRef]

16. Robie, R.; Hemingway, B.; Fisher, J. Thermodynamic properties of minerals and related substances at 298.15 K and 1 bar (10^5 pascals) pressure and at higher temperatures. In *US Geological Survey Bulletin 1452*; USGS: Washington, DC, USA, 1978; p. 456.

17. Barton, P.B.J.; Skinner, B. Sulfide mineral stabilities. In *Geochemistry of Hydrothermal Ore Deposits*; Barnes, H.L., Ed.; Wiley Intersci.: New York, NY, USA, 1979; pp. 278–403.

18. Afifi, M.; Kelly, W.C.; Essene, E.J. Phase relations among tellurides, sulfides, and oxides: I. Thermochemical data and calculated equilibria. *Econ. Geol.* **1988**, *83*, 377–394. [CrossRef]

19. Afifi, A.M.; Kelly, W.C.; Essene, E.J. Phase relations among tellurides, sulfides, and oxides: II. Applications to telluride-bearing ore deposits. *Econ. Geol.* **1988**, *83*, 395–404. [CrossRef]

20. Robie, R.A.; Hemingway, B.S. Thermodynamic properties of minerals and related substances at 298.15 K and 1 bar (10^5 pascals) pressure and at higher temperatures. In *US Geological Survey Bulletin 2131*; USGS: Washington, DC, USA, 1995; p. 461.

21. Krauskopf, K.B.; Bird, D.K. *Introduction to Geochemistry*, 3rd ed.; McGraw-Hill: New York, NY, USA, 1995.

22. Sillitoe, R.H. Ore-related breccias in volcanoplutonic arcs. *Econ. Geol.* **1985**, *80*, 1467–1514. [CrossRef]

23. Sillitoe, R.H. Porphyry copper systems. *Econ. Geol.* **2010**, *105*, 3–41. [CrossRef]

24. Barton, P.B.; Bethke, P.M. Chalcopyrite disease in sphalerite: Pathology and epidemiology. *Am. Mineral.* **1987**, *72*, 451–467.

25. Cooke, D.R.; McPhail, D.C.; Bloom, M.S. Epithermal gold mineralization Ancupan Banguio district Philippine: Geology, mineralization, alteration and the thermochemical environmenot of ore deposition. *Econ. Geol.* **1996**, *91*, 243–272. [CrossRef]

26. Thomas, J.B.; Watson, E.B.; Spear, F.S.; Shemella, P.T.; Nayak, S.K.; Lanzirotti, A. TitaniQ under pressure: The effect of pressure and temperature on the solubility of Ti in quartz. *Contrib. Mineral. Petrol.* **2010**, *160*, 743–759. [CrossRef]

27. Wark, D.A.; Watson, E.B. The TitaniQ: A Titanium-in-quartz geothermometer. *Contrib. Mineral. Petrol.* **2006**, *152*, 743–754. [CrossRef]

28. Frikken, P.H.; Cooke, D.R.; Walshe, J.L.; Archibald, D.; Skarmeta, J.; Serrano, L.; Vargas, R. Mineralogical and isotopic zonation in the Sur-Sur tourmaline breccia, Río Blanco-Los Bronces Cu-Mo deposit, Chile: Implications for ore genesis. *Econ. Geol.* **2005**, *100*, 935–961. [CrossRef]

29. Bodnar, R. Revised equation and table for determining the freezing point depression of H_2O-NaCl solutions. *Geochim. Cosmochim. Acta* **1993**, *57*, 683–684. [CrossRef]

30. Lecumberri-Sanchez, P.; Steele-MacInnis, M.; Bodnar, R. A numerical model to estimate trapping conditions of fluid inclusions that homogenize by halite disappearance. *Geochim. Cosmochim. Acta* **2012**, *92*, 14–22. [CrossRef]

31. Skewes, M.A.; Holmgren, C.; Stern, C.R. The Donoso copper-rich, tourmaline-bearing breccia pipe in central Chile: Petrologic, fluid inclusion and stable isotope evidence for an origin from magmatic fluids. *Miner. Depos.* **2003**, *38*, 2–21. [CrossRef]

32. Roedder, E. Fluid inclusions. *Mineral. Soc. Am.* **1984**, *12*, 644.

33. Corbett, G.J.; Leach, T.M. *Southwest Pacific Rim Gold–Copper Systems: Structure, Alteration, and Mineralization*; Special Publication 6; Society of Economic Geologists: Littleton, CO, USA, 1998; p. 238.

34. Burnham, C.W.; Ohmoto, H. Late stage processes of felsic magmatism. *Soc. Min. Geol. Japan* **1980**, *8*, 1–11.

35. Guilbert, J.M.; Park, C.F. *The Geology of Ore Deposits*; Freeman: New York, NY, USA, 1985.

36. Sverjensky, D.A.; Hemley, J.J.; D'Angelo, W.M. Thermodynamic assessment of hydrothermal alkali feldspar-mica-aluminosilicate equilibria. *Geochim. Cosmochim. Acta* **1991**, *55*, 989–1004. [CrossRef]

37. Inoue, A. Formation of clay minerals in hydrothermal environments. In *Origin and Mineralogy of Clays*; Velde, B., Ed.; Springer: Berlin, Germany, 1995; pp. 268–330.

38. Pirajno, F. *Hydrothermal Processes and Mineral Systems*; Springer Science & Business Media: Berlin, Germany, 2009.

39. Voicu, G.; Bardoux, M.; Jébrak, M. Tellurides from the Paleoproterozoic Omai gold deposit, Guiana shield. *Can. Mineral.* **1999**, *37*, 559–573.

40. Cook, N.J.; Ciobanu, C.L. Bismuth tellurides and sulphosalts from the Larga hydrothermal system, Metaliferi Mts, Romania: Paragenesis and genetic significance. *Mineral. Mag.* **2004**, *68*, 301–321. [CrossRef]

41. Voudouris, P. A comparative mineralogical study of Te-rich magmatic-hydrothermal systems in northeastern Greece. *Mineral. Petrol.* **2006**, *87*, 241–275. [CrossRef]

42. Cooke, D.R.; McPhail, D.C. Epithermal Au-Ag-Te mineralization, Acupan, Baguio district, Philippines: Numerical simulations of mineral deposition. *Econ. Geol.* **2001**, *96*, 109–131.

43. Cook, N.J. What makes a gold-telluride deposit? In *GSA Denver Annual Meeting, Session No. 72, Abstracts with Programs, v. 39*; Geological Society of America: Washington, DC, USA, 2007; p. 196.

44. Jensen, E.P.; Barton, M.D. Gold deposits related to alkaline magmatism. *Rev. Econ. Geol.* **2000**, *13*, 279–314.

Article

Nature and Evolution of Paleoproterozoic Sn and Rare Metal Albitites from Central Brazil: Constraints Based on Textural, Geochemical, Ar-Ar, and Oxygen Isotopes

Ana Rita F. Sirqueira [1], Márcia A. Moura [1,*], Nilson F. Botelho [1] and T. Kurt Kyser [2,†]

[1]   Instituto de Geociências, Universidade de Brasília, Campus Darcy Ribeiro, Brasília 70910-900, Brazil; ana_rita_felix@hotmail.com (A.R.F.S.); nilsonfb@unb.br (N.F.B.)

[2]   Department of Geological Sciences and Geological Engineering, Queen's University, 36 Union Street, Kingston, ON K7L 3N6, Canada; kyser@geol.queensu.ca

*   Correspondence: mamoura@unb.br

†   Professor T. Kurt Kyser, Fellow of the Royal Society of Canada and pioneering geochemist, died while teaching in Bermuda on 29 August 2017.

Received: 30 June 2018; Accepted: 3 August 2018; Published: 8 September 2018

Abstract: Economic and subeconomic concentrations of Sn, In, rare earth elements (REE), Ta, and Nb are known in Central Brazil, in the Goias Tin Province. The Sn-P enriched albitites studied in this paper occur in sharp contact with peraluminous granites of the Aurumina Suite (2.0–2.17 Ga) and schists of the Archean to Paleoproterozoic Ticunzal Formation, as dikes or lenses from late-stage magma of the peraluminous magmatism, probably in granite cupolas. Geological, petrological, and isotopic studies were conducted. The albitites consist of albite, quartz, cassiterite, apatite, K-feldspar, and muscovite, and have magmatic texture, such as alignment of albite laths, and snowball texture in quartz, apatite, and cassiterite. They are enriched in Na_2O, P_2O_5, Sn, Ta, and Nb (Ta > Nb), and depleted in CaO, K_2O, TiO_2, MgO, Sr, Ba, Th, and REE. $^{40}Ar/^{39}Ar$ in muscovite gave a plateau age of 1996.55 ± 13 Ma, interpreted as approaching the crystallization age. Oxygen isotope data in albite-cassiterite pairs resulted in an equilibrium temperature of 653–1016 °C and isotopic fluid composition of 8.66–9.72‰. They were formed by crystallization of a highly evolved and sodic granitic magma. This study has implications for Central Brazil's economic potential and offers better understanding of tin behavior in rare, evolved peraluminous granitic magmas.

Keywords: albitite; snowball texture; Tin; rare metals; Ar-Ar age; oxygen isotope; Brazil

1. Introduction

Albitites are uncommon rocks with usually more than 70–80% of albite. Most of the known albitites worldwide have their origin attributed to the action of hydrothermal fluids on granites [1–3]. In this case, quartz and K-feldspar are leached by Na-rich fluids, in an albitization process, which is sometimes related to greisenization from evolved granites [4,5]. More rarely, albitites are formed by direct crystallization from Na-rich magmas, generally related to specialized and rare-metal granites [6–9]. This type can be recognized by typical textures, such as albite inclusions along growth planes in quartz or other mineral, called *snowball texture*, or the alignment of albite laths in the rock matrix, which is interpreted as flow texture. In addition to textural studies, geochemical characterization of associated granites is crucial to distinguish sodic enrichment in the magmatic stage from metasomatic enrichment [7,9–11].

This paper aims to study albitites that occur in a restricted area, i.e., the Goiás Tin Province, located in Central Brazil [12]. Albitites occur as possible dikes or lenses of 1 to 2.5 m thick, cutting peraluminous granites and graphite schists, or as albitite lenses. They have economic and subeconomic tin concentrations. This led to the exploitation by artisanal miners and the interest of mining companies for mineral exploration in the 1980s. However, there was no mine installation in the area.

Geological Context

The study area is located within a region characterized by a geological evolution spanning from rhyacian to cryogenian times. In the region, there are schists and gneisses from the Paleoproterozoic Ticunzal Formation, the Aurumina Granitic Suite (2.17–2.12 Ga) [13–17], rift volcano-sedimentary rocks of the Araí Group, of ca. 1.77 Ga [12,18]; within-plate granites of the Goiás Tin Province, represented by the Pedra Branca Suite [9,12]; and Meso-Neoproterozoic sedimentary rocks of the Paranoá (1.2–0.9 Ga) and Bambuí groups (0.9–0.6 Ga; Figure 1A) [19]. The region was subjected to metamorphism and deformation in the Neoproterozoic (Brasiliano/Pan-African event).

The geological units identified in the study area are the Ticunzal Formation and the Aurumina Suite (Figure 1B). The Ticunzal Formation consists of graphite schist, mica-quartz schist, garnet-mica schist, and biotite gneiss. Its most striking feature is the presence of large amounts of graphite, which suggests a restricted marine sedimentary environment, warm and salt waters, with high biological activity. It is older than 2.17 Ga, the age of the oldest granitic intrusions of Aurumina Suite. Results of Sm-Nd isotopic data indicate model-age between 2.7 and 2.8 Ga for the metasedimentary package [13,20,21]. These lithologies are cut by granitic rocks attributed to the Aurumina Suite, which is generally divided in the following facies: muscovite granite, biotite-muscovite granite, tonalite, biotite granite and turmaline-muscovite granite. Pegmatites, albite granites and rare albitites are locally found. According to Botelho et al. [22], they are syn-collisional and S-type peraluminous Paleoproterozoic granites (2.12–2.17 Ga–zircon U-Pb), but recent data [14] show that the magmas of the Aurumina Suite were generated by the hybridization of mafic magmas and metasedimentary rocks, indicating that these peraluminous rocks are not true S-type granites.

This study was conducted in the Boa Vista and Pelotas artisanal mines. Geological, petrological and isotopic data from albitites and spatially associated-granites were integrated to constrain the genesis and evolution history of albitites. The study also intends to contribute to the literature on the petrogenesis and metallogenesis of evolved granitic systems rich in rare metals.

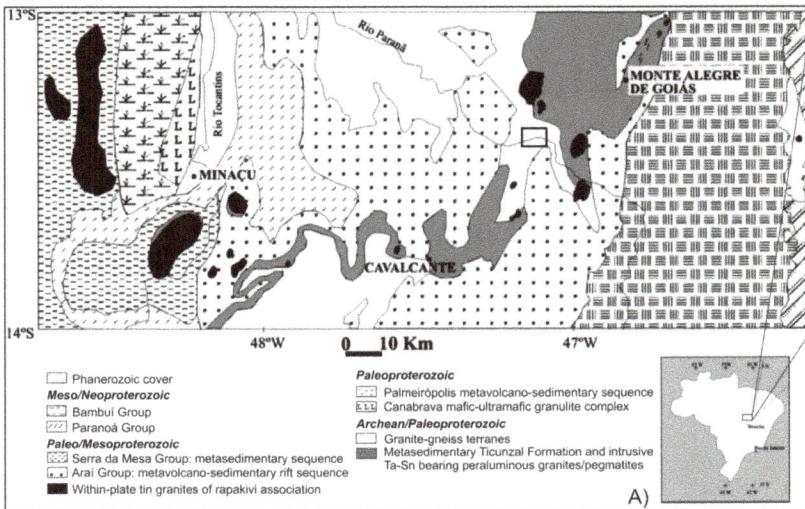

Figure 1. *Cont.*

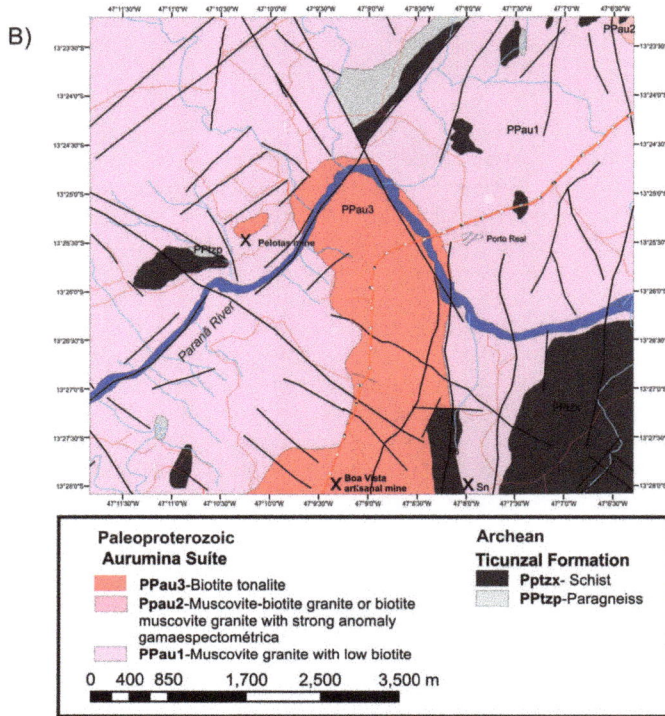

Figure 1. Geological sketch maps of the study area: (**A**) a map of the Goiás Tin Province, with inset showing the approximate region of the study area; and (**B**) a map showing the location of the Pelotas and the Boa Vista artisanal mines [23].

2. Materials and Methods

Two field trips for geological studies and sample collection occurred. Eight drill holes from the Pelotas artisanal mine were sampled. Eighty-six thin sections from selected samples were studied. The petrographic studies were conducted in the Microscopy Laboratory of the Geoscience Institute at the University of Brasilia. The chemical analyses were performed at Acme Analytical Laboratories Ltd. (Vancouver, BC, Canada). Major elements (SiO_2, TiO_2, Al_2O_3, Fe_2O_3, MnO, MgO, CaO, Na_2O, K_2O, P_2O_5) were analysed by ICP-ES (inductively coupled plasma-emission spectrometry). Trace elements (Be, Rb, Cs, Ba, Sr, Ca, V, Sn, W, Ta, Nb, Th, U, Zr, Hf, Y, Sc), including rare earth elements, were analysed by ICP-MS (inductively coupled plasma-mass spectrometry). Cr and Co were analysed by the Leco method.

Mineral chemistry data were obtained from JXA-8230-Jeol electron microprobe (Jeol, Peabody, MA, USA), in the Electronic Microprobe Laboratory of the Geoscience Institute at the University of Brasilia. The standards were commercially supplied by CAMECA.

The $^{40}Ar/^{39}Ar$ geochronological analysis was performed in the Isotope Laboratory at Queen's University, Department of Geological Sciences and Geological Engineering, Ontario (Kingston, ON, Canada). There, 23 temperature steps were performed in primary muscovite from albitite of the Boa Vista Artisanal Mine. Mica was irradiated for 40 h in a McMaster-type nuclear reactor. An 8 W Lexel 3500 specific ion laser (Ar), a MAP 216 mass spectrometer, with a *Baur-Signer* source, and an electron multiplier were used. The measurements of argon isotopes were normalized to atmospheric $^{40}Ar/^{36}Ar$ ratio, using the ratios proposed by [24]. Ages were calculated relative to Hb$_3$Gr (hornblende) with assigned age of 1072 Ma [24] using conventional 40 K decay constants [25,26]. The analytical precision of individual steps for calculation of the plateau was 0.5% at 2σ level.

Stable isotope studies were performed on five samples of albitite, three on albitite from the Boa Vista Artisanal Mine and two on albitite from the Pelotas Artisanal Mine. Cassiterite and albite pairs were considered to be in paragenesis, based on petrographic study. The minerals were manually separated using a binocular loupe in the Geoscience Institute at the University of Brasilia: 5 mg of albite crystals and 3 mg of cassiterite crystals. The oxygen isotope analyses were performed in the laboratory of the Department of Geological Sciences and Geological Engineering at Queen's University (Kingston, ON, Canada). The oxygen isotopic compositions in albite and cassiterite pairs were measured using BrF_5 by the method of Clayton and Mayeda [27]. Measurements of stable isotopes were performed using a Finnigan MAT 252 mass spectrometer. All values are expressed in ‰. Results were presented as $\delta^{18}O$, relatively to the VSMOW standard (Vienna Standard Mean Ocean Water). The analytical precision was of ± 0.3‰ for $\delta^{18}O$ values. The isotopic equilibrium temperature for the quartz-cassiterite and cassiterite-water pairs was calculated according to Zheng [28]. The equation used to calculate the isotopic equilibrium temperature for the albite-water system was proposed by Bottinga and Javoy [29].

3. Geological Setting and Textural Relationships of Albitites

The granitic rocks and albitites described in the studied area are attributed to the Aurumina Suite. The granitic rocks are mainly represented by muscovite-biotite monzogranite and tonalite. The field relationships between the granites could not be defined, as some were identified only in drill holes.

The studied albitites occur as lenses or dikes spatially associated to tonalites and monzogranites, which sometimes contain xenoliths of graphitic schists or even individual graphite lamellae interpreted as xenocrystic inclusions. Schists also occur in contact with both albitites and granitic rocks. They constitute biotite-chlorite-quartz-graphite schist with garnet and graphite-muscovite-chlorite schist (Figure 2A–F).

The muscovite-biotite monzogranite is whitish-gray, medium to coarse-grained, composed of quartz, plagioclase An_{22-1}, orthoclase, muscovite and biotite; chalcopyrite, pyrite, garnet ($Al_{80}Py_{12}Sp_4Gr_2$), and zircon are accessory minerals (Figure 2A,B,D). Tonalite is strongly altered near the albitites. It is whitish-grey to dark grey, composed of quartz (31–38%), altered plagioclase (48–58%), normally resulting in albite composition $An_{10-3\%}$, and microcline (3–5%) as essential minerals; biotite (1–2%) and muscovite (1–5%) as minor minerals (Figure 2E); and pyrite, chalcopyrite, zircon, and monazite as accessory minerals. Fine muscovite and zoisite/clinozoisite are the alteration minerals. Tonalites follow the modal criteria of classification from Streckeisen [30] nomenclature, and also the regional classification adopted by Botelho et al. [16] and Cuadros et al. [14].

Rare pegmatites and albite granite occur in the Pelotas Artisanal Mine, in contact with the granites and schists. They consist of quartz, K-feldspar, plagioclase, and muscovite.

Albitites from both the Boa Vista and the Pelotas artisanal mines are quite similar. In Pelotas, sharp contact between albitite and other rocks is identified in drill-holes (Figure 2F). Albitites are white isotropic rocks, consisting of albite (90–91%), quartz (3–3.5%), cassiterite (1–2%), apatite (1–2%), K-feldspar (0–1%), and primary muscovite (2–4%) (Figure 3A–E).

Albite occurs either as subhedral to euhedral grains, ranging from 0.3 to 1 mm in size, comprising alignment of albite laths, interpreted as flow texture (Figure 3F), or as 0.3 mm euhedral to subhedral laths included in cassiterite and apatite crystals (Figure 3C–E). These Type-2 inclusions of albite crystals arranged almost regularly along growth zones of cassiterite and apatite describe the texture classified as snowball [31] (Figure 3D). Based on textural criteria, albite is interpreted as being of magmatic origin.

Apatite forms subhedral grains 1 to 5.5 mm in size. It occurs as interstitial mineral in most cases. In general, it presents randomly arranged albite lath inclusions. In some crystals, albite is arranged parallel to the apatite border, forming a texture similar to snowball (Figure 3C). It has fluorapatite chemical composition, with low REE content. Cassiterite has euhedral to subhedral habit and ranges from 0.5 to 5.0 mm in size. The crystals are often twinned and the pleochroism varies from reddish brown to yellowish brown. Type-2 albite inclusions are common, often forming snowball texture (Figure 3C,D). There are also inclusions of quartz and, more rarely, of muscovite.

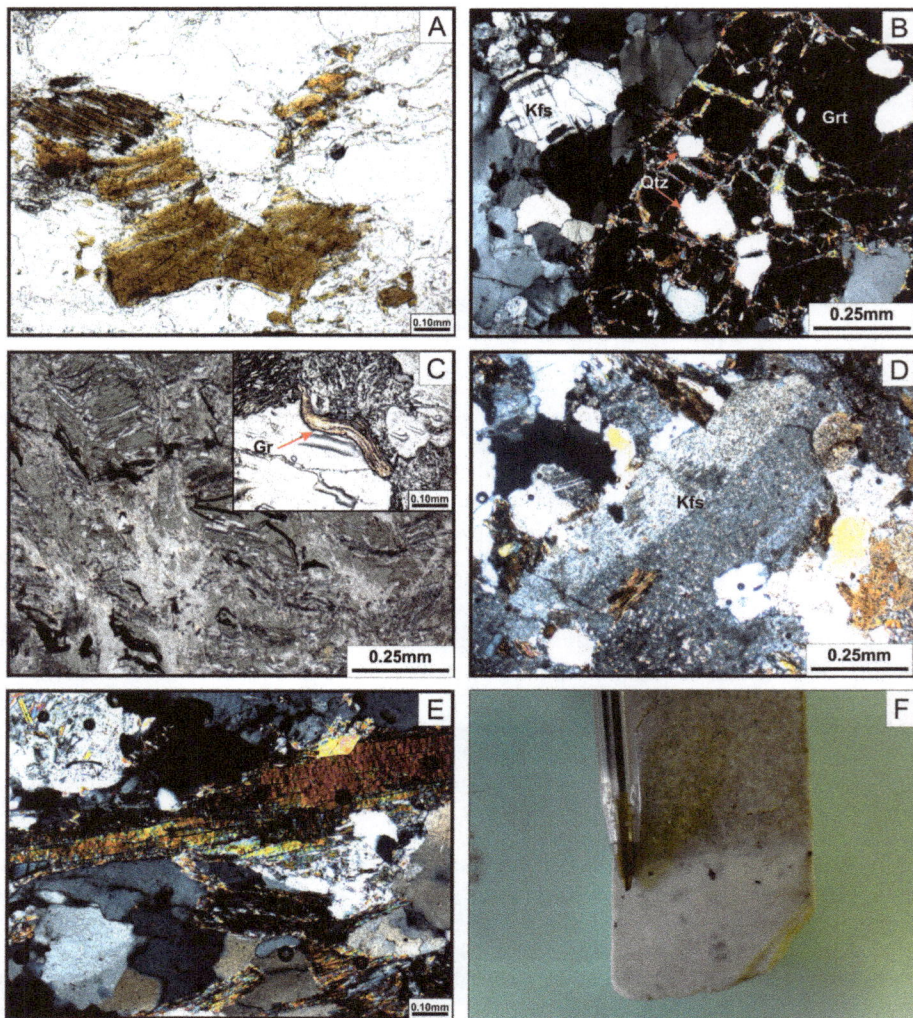

Figure 2. Photographs of representative rocks and features of the study area. (**A**) Coexisting garnet and biotite in muscovite-biotite monzogranite with garnet; (**B**) garnet phenocryst containing quartz inclusion in muscovite-biotite monzogranite with garnet; (**C**) textural aspect of the graphite-chlorite schist in reflected light, showing graphite lamellae; (**D**) altered orthoclase from biotite-muscovite monzogranite; (**E**) magmatic muscovite from biotite-muscovite tonalite; and (**F**) sharp contact between albitite and monzogranite. Kfs: K-feldspar; Grt: garnet; Qtz: quartz; Gr: graphite.

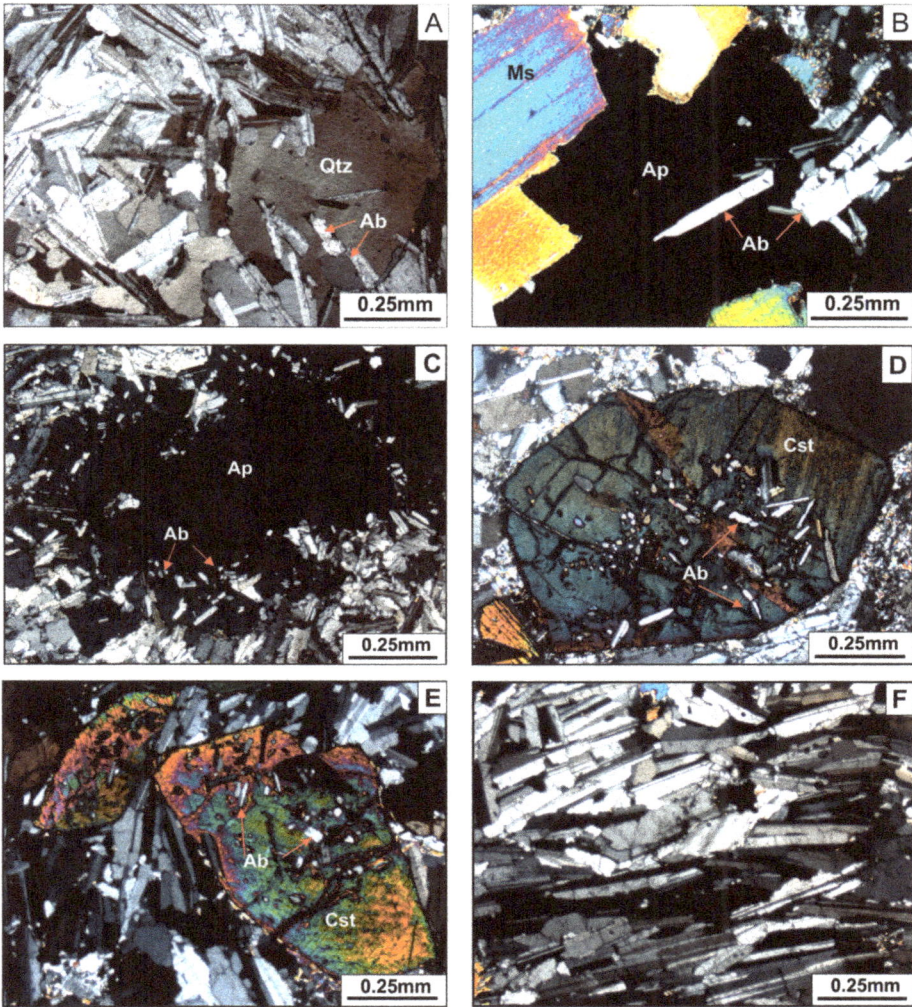

Figure 3. Representative photomicrographs showing textural features of albitites from the Boa Vista and Pelotas artisanal mines. (**A**) Quartz crystals with albite lath inclusions; (**B**) interstitial apatite with albite inclusions in reentrant contact with magmatic muscovite; (**C**) magmatic apatite crystal containing albite inclusions interpreted to be along apatite growth zones; (**D**) cassiterite with albite lath inclusions in the core, forming snowball texture; (**E**) cassiterite with albite lath inclusions and contact relationship with albite from the rock matrix; and (**F**) alignment of albite laths. Ab: albite; Ap: apatite; Cst: cassiterite; Ms: muscovite; Qtz: quartz.

4. Results

4.1. $^{40}Ar/^{39}Ar$ Age

$^{40}Ar/^{39}Ar$ data were obtained for magmatic muscovite of the albitite from Boa Vista Artisanal Mine. In order to estimate age of the crystalization, 23 increment temperature steps were run.

The obtained age spectrum show a staircase pattern. The first step accounts only for 3.5% of [39]Ar released and yields an age of 618.0 ± 36.2 Ma. The high temperature steps yield a plateau-like pattern, which accounts for 47.8% of [39]Ar, with the mean age of 1996.6 ± 13.0 Ma (MSWD = 0.983), near the integrated age of 1957.18 ± 11.15 Ma (Figure 4). Typical criteria to define the plateau age are: (1) the plateau region of age spectrum with at least 70% of total [39]Ar released; (2) at least three steps on the plateau; and (3) the individual ages of these steps must agree with each other within analytical error (e.g., Corsini et al. [32]). The dated sample do not fit two these criteria by the amount of [39]Ar released. Thus, we interpret the age of 1996.6 ± 13.0 Ma as approaching the muscovite crystallization age, but not necessarily reflecting the true crystallization age of albite due to partial resetting at the regional metamorphism during the Brasiliano (Pan-African) event. The lowest age of 618.0 ± 36.2 Ma may directly correspond to this metamorphism. Obviously, the crystallization of the albitite is equal, or older, than the mean age of the plateau-like steps (~2.0 Ga) and may be younger, or equal, to the age of Paleoproterozoic Aurumina granites (~2.12–2.17 Ga).

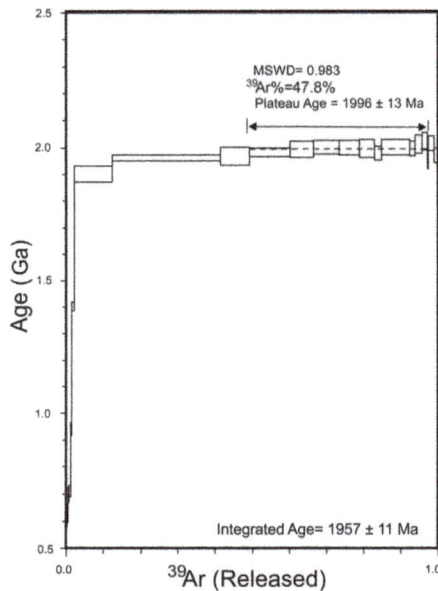

Figure 4. Results of $^{40}Ar/^{39}Ar$ obtained in primary muscovite from the Boa Vista albitite, showing the plateau age, interpreted as approaching the muscovite crystallization age.

4.2. Lithogeochemistry

The chemical representative compositions of the studied rocks are shown in Table 1 and Figures 5 and 6. For comparison, an analysis of a tourmaline-albite granite (TAG) described by Cuadros et al. [14] near the region of the studied albitites was added to Figure 5. According to the Alumina Saturation Index (ASI = $Al_2O_3/CaO + Na_2O + K_2O$ − molar; [33]), albite granites, monzogranites and tonalites are classified as peraluminous rocks (ASI = 1.1 to 1.8), while albitites are classified as metaluminous to peraluminous (ASI = 0.8 to 1.3; Figure 5A). Albitites with lower SiO_2 content show ASI varying from 0.8 to 0.95, and albitites with higher SiO_2 content present ASI ranging from 1.1 to 1.25. These higher ASI values can be explained by the increased amount of muscovite in some samples. Albitite samples richer in muscovite and quartz and poorer in albite and apatite presents higher K_2O and SiO_2 and less Na_2O, CaO, and P_2O_5 contents (Table 1).

Table 1. Representative whole-rock chemical analysis of the studied rocks (wt %).

Rock	Albitite				Tonalite		Schist	Monzogranite							
%															
SiO_2	70.5	71.25	75.9	70.5	74.38	74.9	49.83	75.3	73.6	74.4	75.9	73.5	70.9	75.2	74.1
TiO_2	<0.01	<0.01	0.04	0.01	0.02	0.02	0.8	0.14	0.12	0.13	0.14	0.15	0.3	0.05	0.11
Al_2O_3	17.1	16.83	14.3	15	16.25	15.5	23.58	14.3	14.5	14.3	13.5	14.6	14.4	14.1	14.4
Fe_2O_3	0.1	0.15	0.27	0.22	0.36	0.35	9.19	0.79	1.15	1.21	1.09	1.51	2.45	0.62	1.22
MnO	0.01	0.06	0.01	0.05	<0.01	<0.01	0.27	<0.01	<0.01	0.01	<0.01	0.02	0.02	<0.01	0.01
MgO	<0.01	<0.01	0.06	0.04	0.08	0.1	2.21	0.25	0.24	0.27	0.28	0.34	0.81	0.13	0.26
CaO	1.07	0.53	0.36	2.46	0.71	0.6	2.35	0.46	1.35	1.56	0.89	1.54	1.13	1.12	1.47
Na_2O	9.59	9.86	6.6	8.08	4.79	5.12	2.49	3.76	4.29	3.95	4.03	3.68	3.23	3.66	3.89
K_2O	0.06	0.1	0.88	0.54	2.04	1.73	4.55	3.78	3.58	3.06	2.98	3.43	5.09	4.18	3.29
P_2O_5	0.78	0.43	0.26	1.82	0.07	0.06	0.11	0.07	0.06	0.88	0.09	0.1	0.3	0.09	0.07
LOI	0.7	0.6	0.9	0.8	1.2	1.5	4.3	1.1	0.9	1.0	1.0	1.0	1.1	0.7	1.0
TOTAL	99.9	99.83	99.5	99.6	99.85	99.9	99.72	99.9	99.9	99.9	99.9	99.9	99.8	99.9	99.9
ppb															
Au	<0.5	0.5	<0.5	0.6	1.1	0.8	0.7	<0.5	<0.5	<0.5	1.3	<0.5	0.8	1	1.3
ppm															
Be	31	148	4	4	4	<1	14	4	11	15	4	8	8	<1	8
Rb	1.5	2.5	282	33.4	63.8	57.1	219	157	101	96	115	124	142	114	107
Cs	0.2	2.8	84.6	1.9	10.4	9.5	28.7	9.3	6.4	6.8	9.9	6.8	13.4	2.9	7.7
Ba	3	7	49	31	602	526	701	311	352	259	353	352	1022	394	321
Sr	115	75.3	33.4	139	289.7	265	378.6	78.6	187	152	183	169	145	152	198
Ga	15.7	13.4	15.4	14	13.3	111	28.7	15.7	15.3	17	13	16.8	17	14.9	14.1
V	21	27	21	23	32	33	146	41	36	27	30	31	57	46	46
Sn	8	398	3218	2851	11	11	4	5	2	1	3	5	4	2	1
W	<0.5	<0.5	2	0.9	1.5	1.4	4.1	0.6	<0.5	<0.5	<0.5	<0.5	0.7	<0.5	<0.5
Ta	1.7	18.7	24.2	111	0.1	<0.1	0.9	0.5	0.2	0.2	0.2	0.8	0.6	0.2	0.3
Nb	1.3	12.9	12	72.2	0.6	0.7	11.9	3.8	3.1	3.3	2.2	4.6	5	1.7	3.1
Th	0.5	<0.2	0.9	0.2	2.8	0.4	17.5	8	6.3	7.4	10.5	5.5	6.1	2.8	5.2
U	2.8	0.8	1.5	6.2	<0.1	0.2	4.3	2.4	4	6.3	3.9	5.5	4.4	2.1	2.3
Zr	15.5	21.5	36.3	16.1	1.2	5.5	189.1	70.7	68.6	82.1	114	76.7	90.6	32	58.4
Hf	1.4	2.9	3.3	2.6	<0.1	<0.1	4.6	2.1	2.1	1.9	3.4	2	2.3	1.1	1.4
Y	2	0.1	1.6	4	0.4	0.9	34.9	2.1	2.2	4	3.7	3.4	6.9	3.8	3.1
Sc	<1	<1	<1	<1	<1	<1	23	1	1	2	2	2	3	1	1
La	0.9	0.2	2.7	2.3	3.4	2.7	57.9	16.5	14.9	15.5	22.5	13.4	22	8.2	12.8
Ce	1.7	0.2	4.3	4.5	4.9	5.3	117.7	28.7	26.2	27.1	39	27.2	42.4	15.8	23.7
Pr	0.16	<0.02	0.44	0.46	0.52	0.52	12.96	2.79	2.38	2.51	3.91	2.68	4.36	1.55	2.31
Nd	1.2	<0.3	1.6	0.9	1.2	1.6	46.4	9.2	7.6	8.3	13.3	8.1	15.4	5.6	7.1
Sm	0.21	<0.05	0.26	0.69	0.19	0.24	9.03	1.55	1.42	1.6	2.48	1.81	2.85	1.32	1.41
Eu	0.12	<0.02	0.09	0.27	0.9	0.87	1.72	0.44	0.54	0.44	0.55	0.56	0.74	0.53	0.57
Gd	0.22	<0.05	0.3	0.97	0.2	0.32	7.43	1.23	0.98	1.25	1.8	1.67	2.39	1.1	1.27
Tb	0.06	<0.01	0.05	0.28	0.03	0.03	1.11	0.14	0.14	0.18	0.23	0.21	0.33	0.18	0.16
Dy	0.19	<0.05	0.3	1.11	0.09	0.11	7.08	0.6	0.42	0.83	0.98	0.97	1.37	0.93	0.63
Ho	<0.02	<0.02	0.05	0.09	0.03	0.02	1.37	0.08	0.08	0.21	0.13	0.13	0.28	0.18	0.16
Er	0.05	<0.03	0.06	0.07	<0.03	<0.03	4.09	0.1	0.16	0.27	0.23	0.22	0.44	0.35	0.26
Tm	0.02	<0.01	<0.01	0.01	<0.01	<0.01	0.61	0.01	0.03	0.03	0.03	0.03	0.08	0.05	0.03
Yb	0.07	<0.05	<0.05	<0.05	<0.05	0.09	3.96	0.1	0.09	0.12	0.35	0.24	0.51	0.13	0.17
Lu	<0.01	<0.01	<0.01	<0.01	0.02	0.02	0.59	0.03	0.03	0.04	0.03	0.03	0.06	0.05	0.02
Ni	0.1	0.5	<0.1	<0.1	0.2	0.3	42.1	<0.1	0.6	0.9	0.9	0.6	6.8	<0.1	0.6
Cr	<20	<20	<20	<20	<20	<20	116.3	<20	<20	<20	27.4	<20	20.5	<20	<20
Co	<0.2	<0.2	0.4	0.3	0.7	0.5	21.6	1.1	1.4	1.8	2	2.3	5	1.1	1.4
Cu	1.3	1.7	0.5	1.2	1.2	1.4	73.2	1.7	2.7	0.9	1.9	1.3	9.2	1.4	1.7
Cd	<0.1	<0.1	0.1	2.1	<0.1	<0.1	<0.1	<0.1	<0.1	<0.1	<0.1	<0.1	<0.1	<0.1	<0.1
Zn	10	3	4	14	<1	<1	97	6	19	20	18	18	63	3	13
Pb	4	0.6	3.8	8.5	22.7	25.8	5.5	6.6	9.6	6.3	11.8	29.9	37.8	6.6	5.1
Mo	<0.1	0.1	<0.1	<0.1	<0.1	<0.1	0.7	<0.1	<0.1	<0.1	<0.1	<0.1	0.4	<0.1	<0.1
Ag	<0.1	<0.1	<0.1	<0.1	<0.1	<0.1	<0.1	<0.1	<0.1	<0.1	<0.1	0.2	0.1	<0.1	<0.1
As	<0.5	<0.5	<0.5	<0.5	0.6	<0.5	3	<0.5	<0.5	<0.5	1.3	0.6	1.4	<0.5	<0.5
Sb	<0.1	<0.1	<0.1	<0.1	<0.1	<0.01	<0.1	<0.1	<0.1	<0.1	<0.1	<0.1	<0.1	<0.1	<0.1
Bi	<0.1	<0.1	<0.1	<0.1	1.9	0.8	0.1	<0.1	<0.1	0.1	0.1	0.3	0.3	<0.1	<0.1
Se	<0.5	<0.5	<0.5	<0.5	<0.5	<0.5	<0.5	<0.5	<0.5	<0.5	<0.5	<0.5	<0.5	0	<0.5
Hg	0.03	<0.01	<0.01	<0.01	<0.01	<0.01	<0.01	<0.01	<0.01	<0.01	<0.01	<0.01	<0.01	<0.01	<0.0
Tl	<0.1	<0.1	0.2	<0.1	<0.1	<0.1	0.2	<0.1	0.2	0.1	0.2	0.1	0.4	<0.1	0.2

Albitites display high contents of Na_2O, Al_2O_3, P_2O_5, Sn, Ta, and Nb (Ta > Nb), and low contents of K_2O, TiO_2, Fe_2O_3, MgO, CaO, Ba, Sr, Zr, Th, and Rb. Compared with albitites, albite granites have higher contents of K_2O, Ba, and Sr and lower contents of Na_2O, CaO, P_2O_5, Sn, Ta, Nb, and Zr. Compared with albitites and albite granites, monzogranites are enriched in TiO_2, Fe_2O_3, MgO, K_2O,

Th, Rb, and Zr; depleted in Na_2O, P_2O_5, Al_2O_3, Sn, Nb and Ta (Nb > Ta); and they have intermediate contents of Sr and Ba. The analysed tonalite have intermediate levels of Fe_2O_3, MgO, K_2O, Na_2O ($Na_2O > K_2O$), and CaO; high contents of Al_2O_3, Ba, and Sr; and low contents of TiO_2, P_2O_5, Zr, Rb, Sn, Nb, Ta (Nb > Ta), and Th (Figure 6). The content of U in all the analysed rocks is low (<7 ppm; Table 1). MgO/TiO_2 average ratios are 3.6 and 2.2 for tonalites and monzogranites, respectively (Table 1; Figures 5B,C and 6).

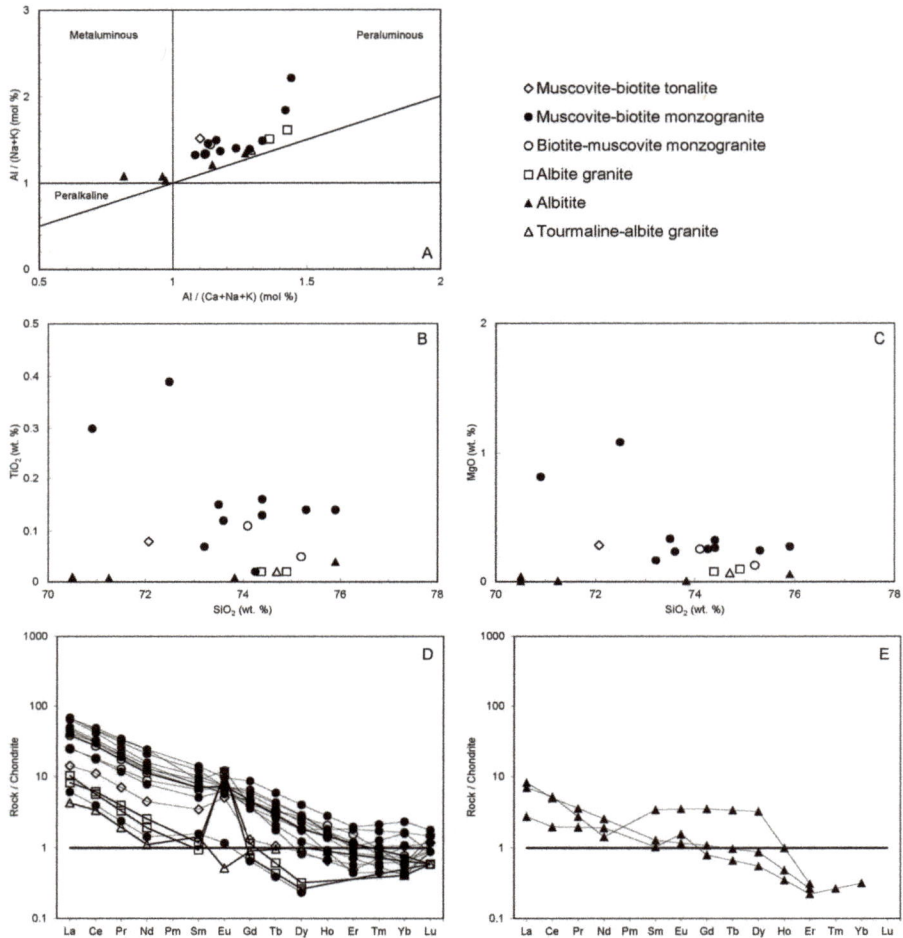

Figure 5. Representative diagrams of compositional characteristics of granites and albitites. (**A**) Classification based on Shand index in Maniar and Piccoli diagram [33], showing the peraluminous character of tonalite, monzogranites and albite granites, and the metaluminous to peraluminous character of albitites from the Pelotas and the Boa Vista artisanal mines; (**B**) $TiO_2 \times SiO_2$ diagram; (**C**) $MgO \times SiO_2$ diagram; (**D**) REE patterns of tonalite, monzogranites and albite granites normalized to chondrite, using Nakamura [34] values; and (**E**) REE patterns of albitites normalized to chondrite using Nakamura [34] values. Tourmaline-albite granite (TAG) sample from Cuadros et al. [14] was added to the diagrams.

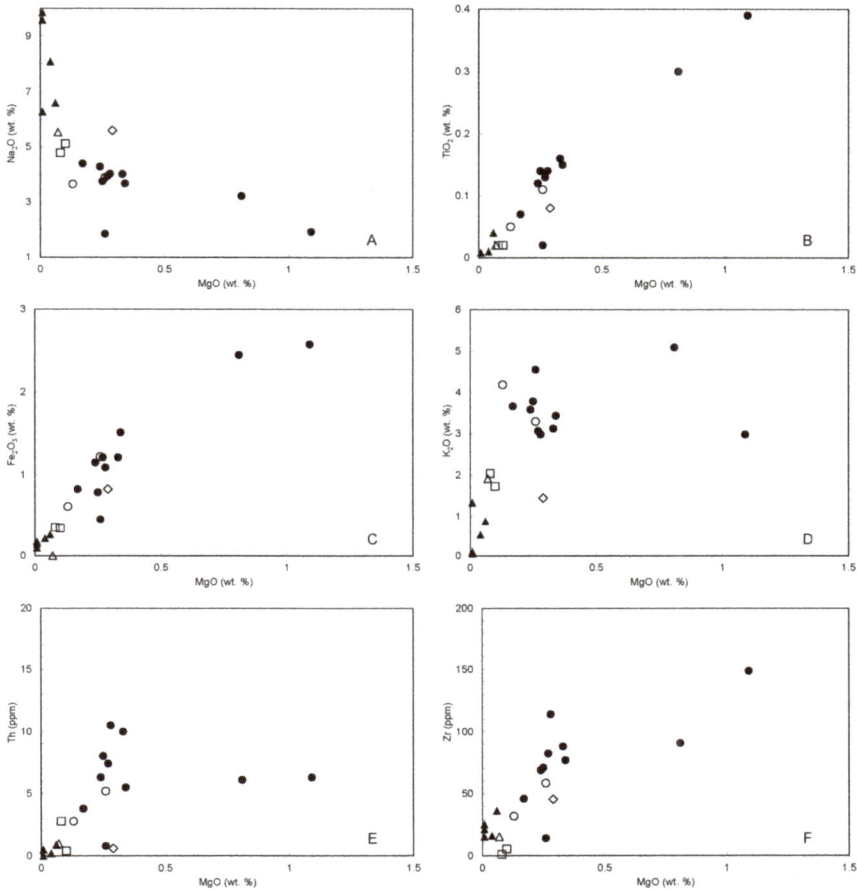

Figure 6. Variation diagrams for monzogranite, tonalite and albitite samples from the Pelotas and the Boa Vista artisanal mines. (**A**) MgO × Na$_2$O diagram; (**B**) MgO × TiO$_2$ diagram; (**C**) MgO × Fe$_2$O$_3$ diagram; (**D**) MgO × K$_2$O diagram; (**E**) MgO × Th diagram (**F**) MgO × Zr diagram. Legend as in Figure 5.

The values of rare earth elements (ΣREE) are low in all felsic rocks, in contrast to the schists values. They range from 0.4 to 11.6 ppm in albitites, from 11.5 to 11.6 ppm in albite granites, from 35–93 ppm in monzogranites, and reach 20.3 ppm in the tonalite (Table 1). The REE patterns normalized to the Nakamura [34] chondrite values for monzogranites (Figure 5D) are enriched in light rare earth elements (LREE) and depleted in heavy rare earth elements (HREE). Eu anomalies range from slightly positive to negative (0.99 < Eu/Eu* < 1.44). The patterns are slightly to strongly fractionated (La$_n$/Yb$_n$ = 15.6 to 110.7). For albite granites, the REE patterns are enriched in LREE as opposed to HREE, whose concentrations are below or near the detection limit of the anlytical method (Table 1). They have a pronounced positive Eu anomaly (Eu/Eu* = 2–14), and their REE patterns are strongly fractionated (La$_n$/Yb$_n$ ranges from 20 to 106; Figure 5D). Albitites do not follow the same REE pattern (Figure 5E). There is a slight enrichment in LREE compared to HREE (La$_n$/Yb$_n$ ranges from 2.6 to 30). The Eu anomalies are absent or very discrete, and slightly positive (Eu/Eu* = 1 to 1.7).

4.3. Mineral Chemistry

Plagioclase chemical composition in the analysed rocks varies from albite to albite-oligoclase (Figure 7A). Albite ($Ab_{90-99}An_{10-0,6}Or_{0,4}$) occurs in albitites and muscovite-biotite tonalite facies. It contains Na_2O values between 12% and 13 wt % in albitites and between 10.5% to 11.9% in the muscovite-biotite tonalite facies. CaO contents are between 0.1% and 0.6 wt %, and 0.5% and 2.8% in albitites and tonalites, respectively.

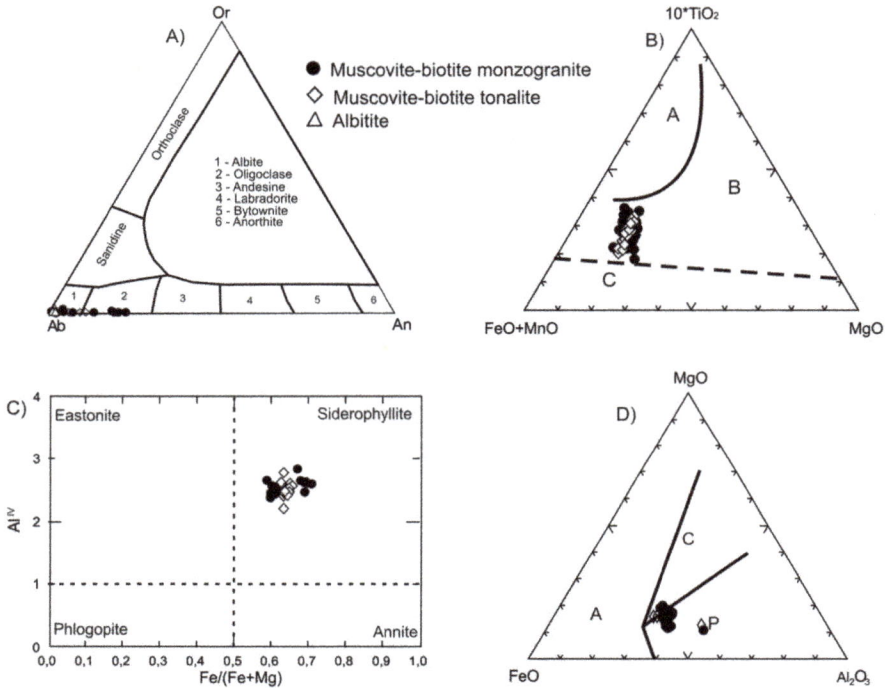

Figure 7. Chemical composition diagrams of plagioclase and biotite of tonalite, monzogranite, and albitite. (**A**) Classification diagram of the analysed plagioclase; (**B**) (FeO + MnO) − (10*TiO_2) − MgO diagram of Nachit et al. [35], showing the compositional variations of the analysed biotite (A = field of primary magmatic biotite; B = field of magmatic reequilibrated biotite; C = field of secondary biotite); (**C**) Al^{IV} vs. Fe/(Fe + Mg) diagram, proposed by Deer et al. [36]; and (**D**) the relationship between MgO-Al_2O_3, $FeO_{(t)}−Al_2O_3$, MgO−$FeO_{(t)}$, and MgO−$FeO_{(t)}−Al_2O_3$ for the biotite from monzogranites and tonalites in the diagram proposed by Abdel-Rahman [37] (A = alkali granites; C = calc-alkaline granites; P = peraluminous granites).

Albite from albitites has a more sodic composition, with Ab = 96–99%. The albite molecule in albite of muscovite-biotite tonalite has Ab = 90–97%, revealing little less sodic compared to albitites.

The muscovite-biotite monzogranite has plagioclase with albite-oligoclase composition ($Ab_{77-99}An_{22-1}Or_1$; Figure 7A). It presents Na_2O values between 9.2% and 12.3% and CaO values between 0.1 and 4.8%. In grains where there is a compositional variation, the core is more calcic than the rim.

According to the triangular diagram 10*TiO_2 vs. FeO + MnO vs. MgO [35], the different biotite types are positioned in the reequilibrated magmatic biotite field (Figure 7B). The TiO_2 content is between 0.8 and 1.7%, MgO content generally ranges from 6 to 9.5%, while the values of Al_2O_3, from 17 to 19% (Table 2). The biotite of monzogranite and tonalite is classified as siderophyllite in the Mg/(Mg + Fe) vs. Al^{IV} diagram proposed by Deer et al. [36]; Figure 7C). The Fe/(Fe + Mg) ratio is between 0.58 to 0.71 apfu, and the content of Al^{IV} varies from 2.1 to 3.3 apfu. The average chemical formula can be defined as: $(K_{1.82},Na_{0.02},Ca_{0.01})_{1.85}$ $(Fe_{3.06},Mg_{1.75},Al^{VI}_{0.69},Ti_{0.14},Mn_{0.03})_{5.71}$ $Si_{5.44}Al^{IV}_{2.56}O_{20}(OH_{3.7}F_{0.3})_4$.

Table 2. Representative analyses of muscovite, biotite, and cassiterite from the studied rocks (wt %).

	Muscovite				Biotite					Cassiterite			
	Albitite		Tonalite		Tonalite		Monzogranite			Boa Vista Albitite		Pelotas Albitite	
	S	P	S	P						N	B	N	B
SiO$_2$	47.70	48.40	48.77	47.36	36.04	35.61	35.84	36.05	SiO$_2$	0.03	0.00	0.00	0.10
TiO$_2$	0.02	0.07	0.15	0.28	1.10	1.20	1.41	0.96	Al$_2$O$_3$	0.00	0.00	0.00	0.02
Al$_2$O$_3$	33.79	33.67	33.60	34.67	17.65	17.47	17.91	17.76	FeO	0.10	0.25	0.12	0.24
FeO	0.93	0.86	2.50	1.85	23.04	23.45	22.92	22.51	MnO	0.02	0.03	0.00	0.03
MnO	0.06	0.02	1.19	0.12	0.29	0.10	0.15	0.25	WO$_3$	0.00	0.00	0.00	0.00
MgO	0.02	0.11	0.05	1.06	7.48	8.11	8.08	8.08	As$_2$O$_5$	0.00	0.00	0.00	0.00
CaO	0.06	0.00	0.04	0.00	0.02	0.07	0.30	0.05	Ta$_2$O$_5$	0.45	0.96	0.48	0.73
Na$_2$O	0.88	0.68	0.04	0.27	0.05	0.08	0.04	0.21	Sb$_2$O$_5$	0.38	0.46	0.35	0.36
K$_2$O	10.56	10.23	10.67	10.60	9.65	9.56	9.28	9.47	SO$_3$	0.00	0.01	0.00	0.00
SrO	0.11	0.00	0.00	0.05	0.03	0.00	0.02	0.00	Bi$_2$O$_3$	0.00	0.00	0.00	0.00
BaO	0.02	0.00	0.06	0.00	0.26	0.00	0.01	0.00	Nb$_2$O$_5$	0.11	0.38	0.22	0.66
F	0.00	0.00	0.38	0.32	0.62	0.27	0.30	0.29	In$_2$O$_3$	0.17	0.13	0.12	0.17
Cl	0.00	0.00	0.01	0.00	0.03	0.03	0.03	0.04	SnO$_2$	99.32	98.30	99.31	97.74
H$_2$O *	4.46	4.49	4.38	4.39	3.61	3.77	3.79	3.78	UO$_2$	0.00	0.02	0.00	0.00
Total	98.61	98.58	101.90	100.86	99.66	99.65	99.98	99.40	CuO	0.00	0.01	0.02	0.00
									ZnO	0.00	0.00	0.00	0.01
FORMULE ON THE BASIS OF 22O									TOTAL	100.57	100.55	100.63	100.05
Si	6.409	6.470	6.406	6.249	5.524	5.466	5.453	5.510					
Al^{iv}	1.591	1.530	1.594	1.751	2.476	2.534	2.547	2.490	FORMULE ON THE BASIS OF 2O				
Site T	8	8	8	8	8	8	8	8	Si	0.001	0.000	0.000	0.002
Al^{vi}	3.760	3.775	3.609	3.640	0.712	0.628	0.666	0.710	Al	0.000	0.000	0.000	0.000
Ti	0.002	0.007	0.015	0.028	0.126	0.138	0.162	0.110	Fe	0.002	0.005	0.003	0.005
Fe	0.105	0.096	0.274	0.204	2.953	3.011	2.916	2.878	Mn	0.001	0.001	0.000	0.001
Mn	0.003	0.003	0.132	0.014	0.038	0.013	0.013	0.014	W	0.000	0.000	0.000	0.000
Mg	0.007	0.021	0.011	0.208	1.708	1.857	1.832	1.840	As	0.000	0.000	0.000	0.000
Site M	3.876	3.907	4.040	4.094	5.538	5.647	5.594	5.570	Ta	0.003	0.007	0.003	0.005
Ca	0.008	0.001	0.006	0.000	0.003	0.012	0.049	0.007	Sb	0.003	0.004	0.003	0.003
Na	0.229	0.176	0.068	0.076	0.016	0.024	0.013	0.063	Nb	0.001	0.004	0.003	0.008
K	1.810	1.744	1.788	1.784	1.886	1.872	1.802	1.847	In	0.002	0.001	0.001	0.002
Sr	0.009	0.000	0.000	0.004	0.003	0.000	0.001	0.000	Sn	0.987	0.977	0.986	0.973
Ba	0.001	0.000	0.003	0.000	0.015	0.000	0.001	0.000	U	0.000	0.000	0.000	0.000
Site I	2.057	1.921	1.866	1.864	1.926	1.911	1.867	1.924	Cu	0.000	0.000	0.000	0.000
OH *	4.000	4.000	3.842	3.866	3.691	3.862	3.851	3.853	Zn	0.000	0.000	0.000	0.000
F	0.000	0.000	0.157	0.134	0.301	0.130	0.142	0.138	TOTAL	1.000	0.999	1.000	1.000
Cl	0.000	0.000	0.001	0.000	0.009	0.008	0.006	0.009					
Site A	4	4	4	4	4	4	4	4					
TOTAL	31.867	31.656	31.812	31.915	35.905	35.940	35.823	35.964					

* Calculated values; S = Secondary mineral; P = Primary mineral; N = Nucleus; B = Border.

In tectonic discrimination diagrams proposed by Abdel-Rahman [37], the biotite analyses of monzogranite and tonalite are located predominantly in the biotite field of peraluminous granitic suites (Figure 7D).

Muscovite with textural characteristics of primary muscovite has variation in the chemical composition between biotite-muscovite tonalite and albitite (Table 2). The muscovite from biotite-muscovite tonalite has the following composition: SiO$_2$ = 45.43–49.50%, Al$_2$O$_3$ = 32.58–38.75%, FeO (t) = 0.54–3.70%, TiO$_2$ = 0.0–1.15%, MgO = 0.04–1.07%, MnO = 0.0–1.8%, K$_2$O = 10.11–11.24%, Na$_2$O = 0.20–0.69%, and CaO = 0.0–0.08%. The muscovite from albitites has the following content: SiO$_2$ = 45.26–48.40%, Al$_2$O$_3$ = 26.28–33.99%, FeO (t) = 0.20–3.40%, TiO$_2$ = 0.0–0.07%, MgO = 0.0–0.93%, MnO = 0.0–0.15%, K$_2$O = 9.20–10.89%, Na$_2$O = 0.26–0.96%, and CaO = 0.0–0.17%. It has mostly no fluorine and chlorine.

Primary muscovite of tonalite differs from the secondary one for having higher Al, Mg, and Na contents and lower Fe and Si contents (Table 2; Figure 8). In relation to Ti concentrations, usually used to distinguish primary (high Ti) from secondary (low Ti) muscovite [38,39], analyses of magmatic muscovite of tonalite displayed different contents of titanium, apparently contradicting the textural interpretation. On the other hand, the muscovite petrographically interpreted as secondary always has low TiO_2 (0.0–0.02 apfu) contents (Figure 8A). With the exception of Ti content, the chemical composition of primary muscovite of tonalite follows the conclusions of Miller et al. [38], according to whom crystals that have the textural aspect of primary muscovite have typically higher Ti, Na, and Al contents and lower Si and Mg contents compared with secondary muscovite.

All analyses of muscovite of albitites had almost zero Ti contents, and low Mg and Fe contents. In contrast, both the primary and the secondary muscovite of albitite have high Na, Al and Si contents (Table 2; Figure 8).

The results made it possible to conclude that the combination of textural and compositional characteristics should be used to distinguish primary from secondary muscovite. Similar conclusions were obtained by Zane and Rizzo [40] and Tao et al. [41]. The chemical composition of the studied rocks is interpreted to have acted as a control factor on the composition of primary muscovite, especially in Na_2O and TiO_2 contents.

Cassiterite with textural characteristics of primary mineral from albitites of the Boa Vista and Pelotas artisanal mines were analysed. In general, the reddish-brown portions have higher SnO_2 contents and lower FeO, Ta_2O_5, and Nb_2O_5 contents (Ta > Nb) for both areas. The opposite occurs in bands with light brown pleochroism (Figure 9). In cassiterite from Boa Vista, this pattern is more evident than in cassiterite from the Pelotas mine. In crystals without zonation, the nucleus is generally richer in Sn and poorer in Ta, Nb, and Fe in comparison with the border (Table 2).

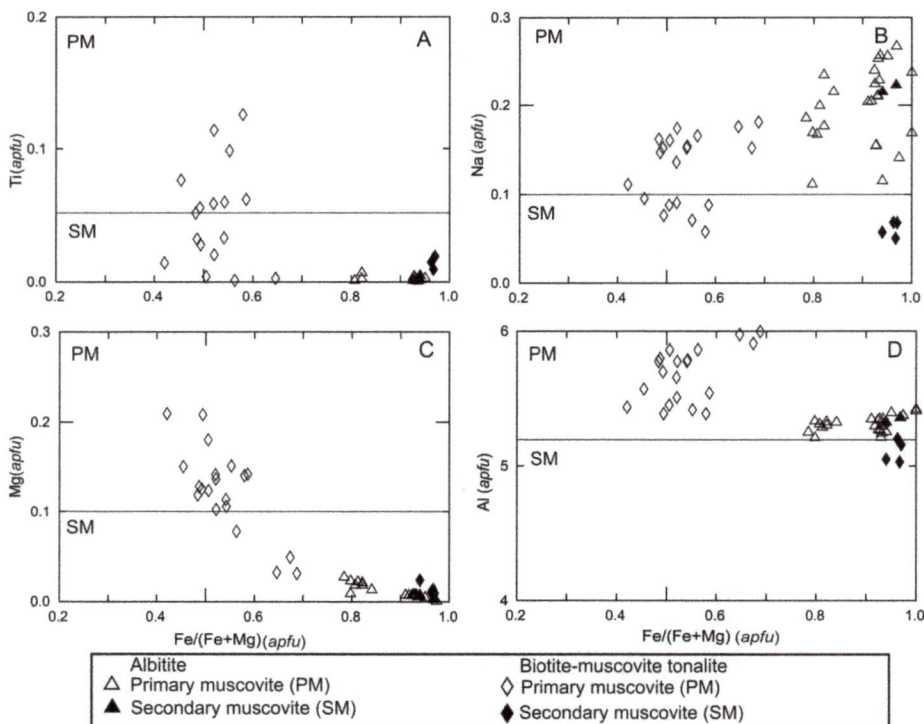

Figure 8. Geochemical discrimination diagram for the primary and secondary muscovite. (**A**) Ti vs. Fe/(Fe + Mg) diagram; (**B**) Na vs. Fe/(Fe + Mg) diagram; (**C**) Mg vs. Fe/(Fe + Mg) diagram; and (**D**) Al (t) vs. Fe/(Fe + Mg) diagram, based on Miller et al. [38]. *apfu = atoms per formula unit.*

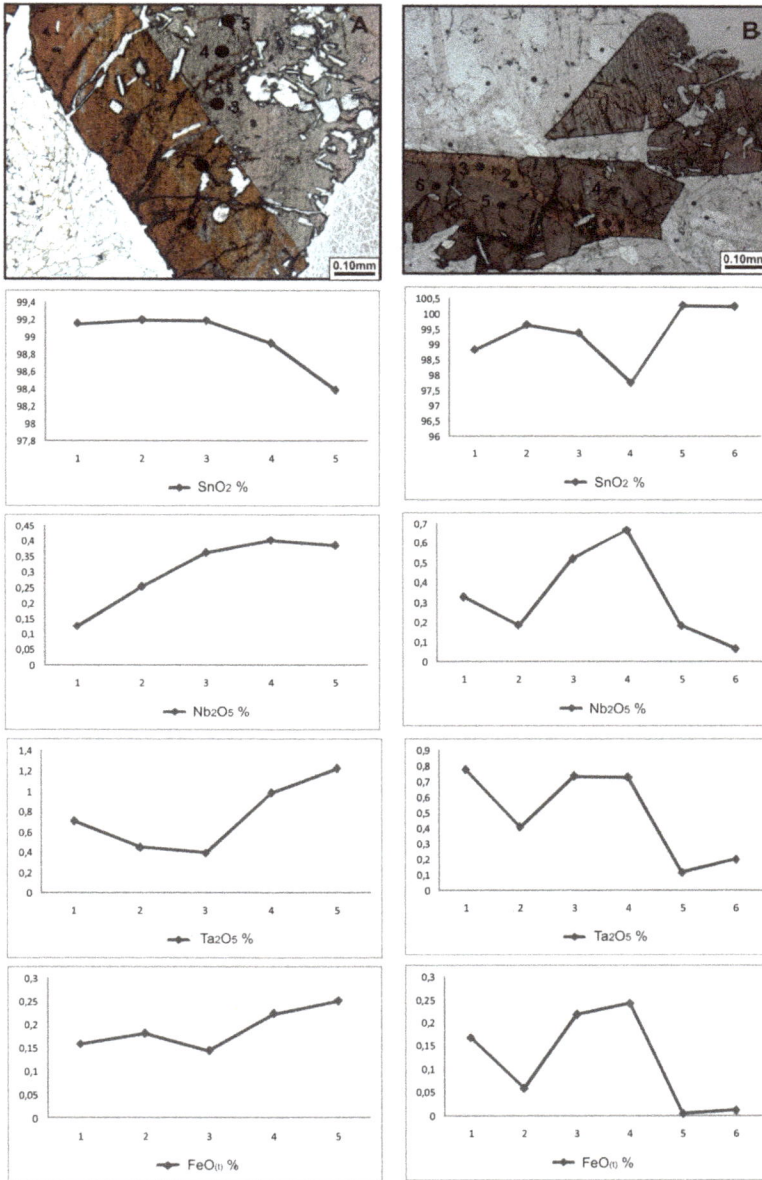

Figure 9. Compositional variation in different cassiterite crystals. (**A**) Cassiterite from albitites in Boa Vista Artisanal Mine; and (**B**) cassiterite from albitites in the Pelotas Artisanal Mine.

The compositional variation is due to the substitution of Sn for Ta, Nb and Fe in the cassiterite structure. Some possible substitution equations are [42]:

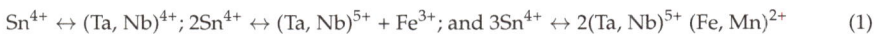

$$Sn^{4+} \leftrightarrow (Ta, Nb)^{4+}; 2Sn^{4+} \leftrightarrow (Ta, Nb)^{5+} + Fe^{3+}; \text{ and } 3Sn^{4+} \leftrightarrow 2(Ta, Nb)^{5+} (Fe, Mn)^{2+} \quad (1)$$

Similar results were obtained by Costi et al. [43] in magmatic cassiterite of within-plate granites in the Pitinga Province, which has a high Nb_2O_5 (0.063 to 0.6%) and Ta_2O_5 (0.1 to 1.3%) contents. However, cassiterite from hydrothermal origin showed lower Nb_2O_5 (0.0–0.1%) and Ta_2O_5 (0.0–0.1%) contents. Higher Ta_2O_5 values in relation to Nb_2O_5 were also identified in other magmatic cassiterites from peraluminous granites of the Aurumina suite, while the opposite (Nb > Ta) occurs in hydrothermal cassiterite from within-plate granites of the Goiás Tin Province [44].

4.4. Oxygen Isotopes

Analyses of oxygen isotopes were performed in pairs of albite and cassiterite interpreted petrographically as to be in paragenesis. The obtained values were $\delta^{18}O_{VSMOW}$ = 6.7 to 9.3‰ and $\delta^{18}O_{VSMOW}$ = 5.3 to 6.6‰, respectively (Table 3).

Table 3. Oxygen isotope values ($\delta^{18}O_{VSMOW}$) of albite and cassiterite from Pelotas and Boa Vista albitite samples, and calculated isotopic equilibrium temperature and isotopic composition of fluid.

| Deposit | Mineral (V-SMOW) | | Isotopic Equilibrium Temperature (°C) [a] | Isotopic Composition of the Fluid (‰) |
	Albite (‰)	Cassiterite (‰)		
Pelotas	9.3	5.3	653 °C	9.35
Pelotas	6.7	6.6	1163–1319 °C	9.17
Boa Vista	7.7		-	9.40 [b]
Boa Vista	7.8	6.0	943 °C	9.39
Boa Vista	7.9	6.5	1016 °C	9.72

[a] Temperature calculated considering the analytical error (±0.3‰); [b] Composition estimated using the isotopic equilibrium temperature average for the Boa Vista artisanal mine.

The calculated temperature of isotopic equilibrium of the albite-cassiterite pair from albitites varied from 653 to 1319 °C. The highest temperature, however, was achieved for albite and cassiterite with relative isotopic fractionation (Δ_{xy}) of 0.1‰. Since the difference in the minerals' isotopic composition is smaller than the analytical error (0.3‰), the temperature of isotopic equilibrium estimated for this pair must be considered with caution. Considering the margin of analytical error, the temperature obtained for this pair is 1163 to 1319 °C (Table 3).

The isotopic composition of albitite fluids from the two artisanal mines is very close regardless of the difference in the temperature of isotopic equilibrium: it varies from 9.17‰ to 9.72‰ (Table 3).

Data of the isotopic fluid composition obtained for albitites are consistent with water of magmatic origin. According to Taylor Jr. [45], they show $\delta^{18}O$ between 5.5‰ and 10.0‰. Despite the wide range of values for the isotopic equilibrium temperature, the results are consistent with petrological data and with the interpretation of albite crystallization in magmatic conditions and at high temperature.

The calculated values of the isotopic fluid composition and isotopic equilibrium temperature are different from those obtained for greisens mineralized in tin described in the region: 3.4–3.9‰ and 285–370 °C for Li-muscovite greisen, and 4.8–7.9‰ and 520–660 °C for zinnwaldite greisen and quartz-topaz rock [5].

As Kalamarides [46] stated, the plutonic rocks generally crystallize in several stages and experience changes of $\delta^{18}O$ in the subsolidus phase until they reach the isotopic closure temperature. Consequently, the isotopic equilibrium temperatures obtained for these rocks commonly lie below 800 °C. In contrast, the $\delta^{18}O$ values of volcanic rocks and their minerals tend to represent the final product of igneous processes that generated them, as crystallization occurs instantaneously. Considering the obtained data and the mode of occurrence of the studied albitites, it is reasonable to suggest elevated isotopic closure temperature and the existence of subsolidus reequilibrium.

5. Discussion

5.1. Tectonic Setting and Regional Geological Context

Field, petrographic, and chemical characteristics obtained from the studied monzogranites and tonalites are consistent with the interpretation that these are peraluminous granites, generated in a syn-collisional environment. Diagnostic primary minerals of peraluminous granites are identified in tonalites and granites, such as muscovite and garnet.

Chemical compositions of biotite from tonalite and monzogranite, where TiO_2 varies from 0.8% to 1.7%, MgO from 4.9% to 11.6%, and Al_2O_3 from 15.9% to 22.5%, are similar to biotite from peraluminous granitic suites [35,37].

In the Rb vs. (Y + Nb) and Rb vs. (Ta + Yb) diagrams elaborated by Pearce et al. [47], samples of monzogranites from the Pelotas Artisanal Mine are located near the boundary between volcanic arc granites (VAG) and syn-collisional (syn-COLG) granites. The plot of tonalite and some monzogranite samples in the field of VAG is in agreement with the interpretation of Cuadros et al. [14] that these rocks are derived from hybrid magmas (Figure 10A).

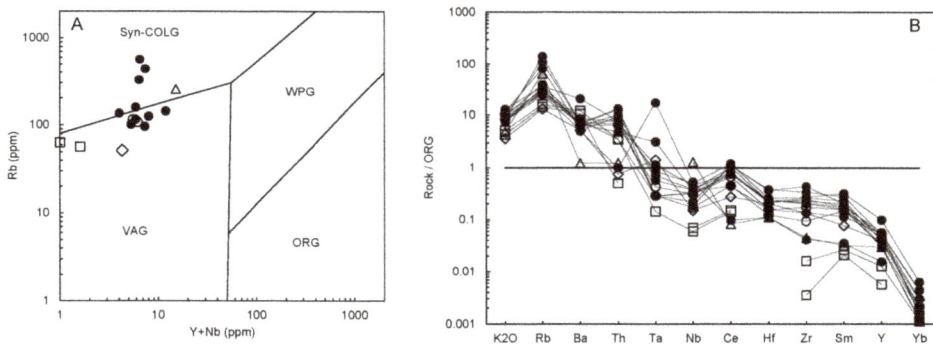

Figure 10. Tonalite, monzogranite, and albite granite samples in Pearce et al. [47] diagrams. (**A**) Rb − (Y + Nb) discriminant diagram; (**B**) ocean ridge granite (ORG) normalized geochemical patterns for representative analyses. syn-COLG: syn-collision granites; VAG: volcanic arc granites; WPG: within-plate granites. Legend as in Figure 5.

In the multi-element diagram normalized to ocean ridges granite (ORG), with values of Pearce et al. [47], positive anomalies of Rb, Th, Ce, and Sm are observed in monzogranite and albite granite samples (Figure 10B). Pearce et al. [47] attributed the enrichment of Rb and Th with respect to Nb and Ta, and Ce and Sm in regard to their adjacent elements, to crustal involvement. The pattern is described by the authors as crust-dominated. Patterns show that the studied monzogranites and albite granites are similar to those presented by Pearce et al. [47] for syn-COLG granites of Tibet.

The Sn-mineralized albitites are in contact with peraluminous granites and schists, attributed to the Aurumina granite suite and the Ticunzal Formation, respectively [12,16]. The $^{40}Ar/^{39}Ar$ age obtained for muscovite from albitite of Boa Vista Artisanal Mine (1996.55 ± 13 Ma) is interpreted as approaching the crystallization age, which suggests that the albitites were formed during the Paleoproterozoic. They probably comprise, together with tourmaline-bearing pegmatites and granites, the younger phases of the Aurumina Suite peraluminous magmatism, whose granites are considered to have crystallized between 2.12 and 2.17 Ga (zircon U-Pb) [14,22]. These Ar-Ar data are similar to the results obtained by Sparrenberger and Tassinari [17] in peraluminous pegmatites situated at about 40 km north of the studied area. According to the authors, coarse-grained muscovite from pegmatites yielded K-Ar ages of 2129 ± 26 Ma and 2006 ± 24 Ma. After the albitites' formation, the $^{40}Ar/^{39}Ar$ data registered an event at 1901.2 ± 30 Ma, which caused the release of 47% of [39]Ar. It is suggested that this release of [39]Ar could be related to the first crust perturbations due to the continental rifting experienced by the region in the Paleoproterozoic. In this geotectonic setting, the intrusion of within-plate granites occurred about 1.8 Ga [48,49]. Another heating event was recorded at 618.0 ± 36.2 Ma by [40]Ar/[39]Ar data, interpreted as related to the Brasiliano tectono-metamorphic event (Pan-African) described in the region and registered as incipient deformation in the studied rocks.

5.2. Genesis and Evolution of Albitites and Associated Granitic Rocks

Albitites in the Pelotas and the Boa Vista artisanal mines have textures interpreted as magmatic, such as snowball texture in quartz, apatite and cassiterite. Albite inclusions have crystallized during its host mineral growth. Albite also has flow texture, with alignment of equigranular albite laths in the matrix (Figure 3). Field relationships suggest that, at least in the Pelotas Artisanal Mine, the studied albitite occurs as dikes or lenses 1 to 2.5 m thick. They occur in sharp contact with peraluminous monzogranites and tonalites, assigned to the peraluminous Aurumina Granitic Suite (2.15–2.0 Ga) by Botelho et al. [16], and with schists of the Ticunzal Formation, which was intruded by the Aurumina Suite. Magmatic albitite, regarded as the end stage of magmatic crystallization, has also been reported in other regions of the world [7,8,50].

The obtained data allow interpretation of the snowball texture in the studied rocks following the authors who consider those textural relationships as a characteristic of evolved granites that crystallized from a residual melt. These granites occur in the apical part of subvolcanic granite complexes and are commonly associated with rare-metal mineralization [51].

The oxygen isotope data showed that the isotopic fluid composition in equilibrium with albitite, from 8.68 to 9.72, is consistent with magmatic signature without influence of meteoric water. This reinforces the interpretation of magmatic origin for the Sn-mineralized albitite of the two studied areas. The estimated albitite isotopic equilibrium temperatures are also consistent with direct crystallization from highly specialized magma rich in Na_2O, P, and B, and depleted in fluorine, with subsolidus reequilibrium.

The studied albitites are probably related to tourmaline-albite granites and tourmaline pegmatites recently described by Cuadros et al. [14] in the region and considered as the most evolved rocks of the Aurumina Suite. In the Boa Vista and Pelotas area we find only tourmaline-bearing pegmatites and tourmaline-free albite granites. We interpret the albitite as primary in origin. One possibility is its formation by the accumulation of albite crystals, which separated from an evolved granitic liquid that gave rise to the tourmaline-albite granite and associated pegmatitic facies. The albite granite would represent a residual liquid. The albitite is similar to the Sn-Ta-bearing albitite described in Slovakia by Breiter et al. [52]. However, at least considering the actual level of investigation in the area, a granite cupola does not occur. The albitite and pegmatite occur as isolated bodies into the granite-tonalite terrain. The albitite is composed of more than 90% albite instead of the compositional variation in Slovakia. The observed flow textures and the pockets of massive albite allow suggest an accumulation of albite laths by separation of the liquid. These interpretations are reinforced by the oxygen isotope data and the flow textures described above.

5.3. Tin Transport and Concentration

Crystallization and economic concentration of cassiterite have been reported as a product of direct crystallization from evolved granitic magma or, more commonly, as the result of hydrothermal alteration. In the latter case, tin was leached from granite and/or country rocks and crystallized as cassiterite.

Cassiterite in the studied artisanal mines is interpreted as having been formed by crystallization from an evolved tin-rich granitic magma. When cassiterite occurs as a magmatic mineral, tin must be incompatible with the whole history of melt crystallization. According to Linnen [53], tin must be partitioned into silicate liquid, or the vapor fraction should be sufficiently low so that the tin concentration in the liquid will continue to increase with fractionation. Magmatic cassiterite can be crystallized in highly evolved systems because, in granitic intrusions with low chlorine content, tin is partitioned into the granitic melt [53]. The results of Linnen et al. [54] and Bhalla et al. [55] showed that Al plays an important role in tin solubility in evolved granitic magmas. SnO_2 solubility in peraluminous magmas will also depend on oxygen fugacity and temperature. Oxygen fugacity increase and temperature decrease tend to reduce tin solubility [31,54–56]. Linnen et al. [55] uses the strong oxygen fugacity dependence of tin solubilities to explain the existence of some magmatic tin deposits. The presence of volatiles in magma is also an important factor in Sn concentration: B, F, Cl, and P contents are generally concentrated in the final stages of evolved magmas and contribute to tin solubility [55,57,58].

The studied granitic rocks have features typical of LCT [59], peraluminous and syn-orogenic systems. Albitites have primary apatite due to high contents of phosphorous. The less-evolved system in the region contains high boron, expressed as magmatic tourmaline. Granitic rocks do not possess cassiterite on their mineral assemblages and have low Sn, Ta, and Nb contents (Table 1). The existing data suggested that tin remained in solution in the peraluminous granitic magma until it precipitated in albitites. Probably, Sn transport occurred in low oxygen fugacity conditions, in P-rich peraluminous magma. Tantalum and niobium precipitation occurred along with Sn and tantalite was recovered together with cassiterite in some artisanal mines

6. Conclusions

The main conclusions of this research are:

1. The studied albitites have a snowball texture in quartz, apatite, and cassiterite and flow texture, with alignment of albite laths in the matrix, interpreted as magmatic texture. These rocks are interpreted as magmatic and occur as dikes or lenses in monzogranite, peraluminous tonalite, and graphite schist.

2. Albitites have high Na_2O, Al_2O_3, P_2O_5, Sn, Ta, and Nb (Ta > Nb) contents and represent cumulates separated from evolved Na-rich peraluminous magmas related to the Aurumina suite tourmaline-bearing rocks. The albite granite represents a residual liquid from the late stages of magmatism.

3. Biotite from monzogranite and tonalite have compositions similar to those from peraluminous granitic suites. Tin in cassiterite is replaced by Fe, Ta, and Nb, with Ta > Nb contents. The chemical composition of the studied rocks acted as a control factor on the composition of primary muscovite, especially in Na_2O and TiO_2 contents. Primary muscovite from tonalite has higher Al, Mg, and Na, and lower Fe and Si contents than secondary ones. While secondary muscovite always has low TiO_2 (0.0–0.07%) contents, magmatic muscovite of tonalite has variable Ti contents. Both primary and secondary muscovite from albitites and albite granites have virtually no Ti and are high in Na, Al, and Si.

4. The isotopic fluid composition in equilibrium with albitites varies from 8.68‰ to 9.72‰ and is consistent with the interpretation of magmatic origin. Although the calculated isotopic equilibrium temperatures are elevated for the evolved peraluminous granite system in place, they also demonstrate the absence of hydrothermal influence on albitite crystallization.

5. $^{40}Ar/^{39}Ar$ data in muscovite suggest that albitites crystallized around 1996 ± 13 Ma and they can be correlated to the final stages of the Aurumina Suite (2.12–2.17 Ga). Argon loss in 618.0 ± 36.2 Ma is interpreted as related to the Brasiliano (Pan-African) tectono-metamorphic event.

6. In addition to containing hydrothermal tin mineralization hosted in greisens and within-plate granite magmatism, of approximately 1.7 Ga, the Goiás Tin Province has magmatic tin economic concentrations hosted in igneous albitite of about 2.0 Ga. These results, therefore, extend the possibilities of a tin source in the Goiás Tin Province. They have implications for the province's

economic potential and also help understand solubility and tin concentration in peraluminous granitic systems highly evolved and very rich in sodium.

Author Contributions: M.A.M. and N.F.B. conceived and coordinated the research; A.R.F.S. and M.A.M. took part to the field campaigns; A.R.F.S. prepared and analyzed the samples and the data; T.K.K. was responsible for the isotopic data analyses; and A.R.F.S., M.A.M. and N.F.B. took part in the discussion and wrote the paper.

Funding: This study was financially supported by CNPq (Brazilian National Council for Scientific and Technological Development) (process no. 133302/2012-1) and CAPES (Coordenação de Aperfeiçoamento de Pessoal de Nível Superior) (scholarship for the first author).

Acknowledgments: The authors would like to thank CNPq (Brazilian National Council for Scientific and Technological Development) for financial support. Verena Mineração Ltd. is thanked for giving permission for this study and for providing access to drill holes and technical information. We sincerely thank Valmir da Silva Souza (University of Brasilia) for providing the 40Ar/39Ar data. We also thank University of Brasilia and Queen's geological laboratories used in this research.

Conflicts of Interest: The authors declare no conflict of interest

References

1. Charoy, B.; Pollard, P.J. Albite-rich, silica-depleted metasomatic rocks at Emuford, Northeast Queensland: Mineralogical, geochemical and fluid inclusion constraints on hydrothermal evolution and tin mineralization. *Econ. Geol.* **1989**, *84*, 1850–1874. [CrossRef]
2. Castorina, F.; Masi, U.; Padalino, G.; Palomba, M. Constraints from geochemistry and Sr–Nd isotopes for the origin of albitite deposits from central Sardinia (Italy). *Miner. Depos.* **2006**, *41*, 323–338. [CrossRef]
3. Mohammad, Y.O.; Maekawa, H.; Lawa, F.A. Mineralogy and origin of Mlakawaalbitite from Kurdistan region, Northeastern Iraq. *Geosphere* **2007**, *3*, 624–645. [CrossRef]
4. Kovalenko, V.I. The genesis of rare metal granitoids and related ore deposits. In *Metallization Associated with Acid Magmatism Czech Geological Survey*; Stemprok, M., Burnol, L., Tischendorf, G., Eds.; Geological Survey: Prague, Czech Republic, 1978; Volume 3, pp. 235–247.
5. Moura, M.A.; Botelho, N.F.; Olivo, G.R.; Kyser, K.; Pontes, R.M. Genesis of the Proterozoic Mangabeira tin–indium mineralization, central Brazil: Evidence from geology, petrology, fluid inclusion and stable isotope data. *Ore Geol. Rev.* **2014**, *60*, 36–49. [CrossRef]
6. Cuney, M.; Marignac, C.; Weisbrod, A. The Beauvoir topaz-lepidolite albite granite (Massif Central, France); the disseminated magmatic Sn-Li-Ta-Nb-Be mineralization. *Econ. Geol.* **1992**, *87*, 1766–1794. [CrossRef]
7. Schwartz, M.O. Geochemical criteria for distinguishing magmatic and metasomatic albite-enrichment in granitoids: Examples from the Ta–Li granite Yichun (China) and the Sn–W deposit Tikus (Indonesia). *Miner. Depos.* **1992**, *27*, 101–108. [CrossRef]
8. Costi, H.T.; Dall'Agnol, R.; Pichavant, M.; Rämö, O.T. The peralkaline tin-mineralized Madeira cryolite albite-rich granite of Pitinga, Amazonian craton, Brazil: Petrography, mineralogy and crystallization processes. *Can. Miner.* **2009**, *47*, 1301–1327. [CrossRef]
9. Wang, G.; Wang, Z.; Zhang, Y.; Wang, K. Zircon geochronology and trace element geochemistry from the Xiaozhen Copper Deposit, North Daba Mountain: Constraints on albitites petrogenesis. *Acta Geol. Sin. Engl.* **2014**, *88*, 113–127. [CrossRef]
10. Lenharo, S.L.R.; Moura, M.A.; Botelho, N.F. Petrogenetic and mineralization processes in Paleo-to Mesoproterozoic rapakivi granites: Example from Pitinga and Goiás, Brazil. *Precambr. Res.* **2002**, *119*, 277–299. [CrossRef]
11. Lenharo, S.L.R.; Pollard, P.J.; Born, H. Petrology and textural evolution of granites associated with tin and rare-metals mineralization at the Pitinga mine, Amazonas, Brazil. *Lithos* **2003**, *66*, 37–61. [CrossRef]
12. Marini, O.J.; Botelho, N.F. A Província de Granitos Estaníferos de Goiás. *Rev. Bras. Geociênc.* **1986**, *16*, 119–131.
13. Cuadros, F.; Botelho, N.F.; Fuck, R.A.; Dantas, E.L. The Ticunzal Formation in central Brazil: Record of Rhyacian sedimentation and metamorphism in the western border of the São Francisco Craton. *J. S. Am. Earth Sci.* **2017**, *79*, 307–325. [CrossRef]

14. Cuadros, F.; Botelho, N.F.; Fuck, R.A.; Dantas, E.L. The peraluminous Aurumina Granite Suite in central Brazil: An example of mantle-continental crust interaction in a Paleoproterozoic cordilleran hinterland setting? *Precambr. Res.* **2017**, *299*, 75–100. [CrossRef]

15. Alvarenga, C.J.S.; Botelho, N.F.; Dardenne, M.A.; Lima, O.N.B.; Machado, M.A. *Monte Alegre de Goiás—SD.23-V-C-III, Escala 1:100,000: Nota Explicativa Integrada com Nova Roma e Cavalcante*; CPRM: Brasília/Goiás, Brazil, 2007; 67p.

16. Botelho, N.F.; Alvarenga, C.J.S.; Menezes, P.R.; D'El Rey Silva, L.J.H. Suíte Aurumina: Uma suíte de granitos paleoproterozóicos, peraluminosos e sin-tectônicos na Faixa Brasília. In Proceedings of the 7th Simpósio de Geologia do Centro-Oeste, Brasília, Brasil, 14–19 November 1999; p. 17.

17. Sparrenberger, I.; Tassinari, C.C.G. Subprovíncia do Rio Paraná (GO): Um exemplo de aplicação dos métodos de datação U-Pb e Pb-Pb em cassiterita. *Rev. Bras. Geociênc.* **1999**, *29*, 405–414. [CrossRef]

18. Pimentel, M.M.; Fuck, R.A.; Botelho, N.F. Granites and the geodynamic evolution of the Neoproterozoic Brasilia belt, central Brazil. *Lithos* **1999**, *46*, 463–483. [CrossRef]

19. Dardenne, M.A. The Brasília fold belt. In Proceedings of the 31st International Geological Congress, Rio de Janeiro, Brazil, 6–17 August 2000; pp. 231–236.

20. Marini, O.J. Nova unidade litostratigráfica do Pré-Cambriano do estado de Goiás. In Proceedings of the 30th Congresso Brasileiro de Geologia, Recife, Brazil, November 1978; pp. 126–127.

21. Pimentel, M.M.; Jost, H.; Fuck, R.A. O embasamento da Faixa Brasília e o Arco Magmático de Goiás. In *Geologia do Continente sul Americano: Evolução da Obra de Fernando Flávio Marques de Almeida*; Mantesso-Neto, V., Bartorelli, A., Carneiro, C.D.R., Brito-Neves, B.B., Eds.; Beca: São Paulo, Brazil, 2004; pp. 356–368.

22. Botelho, N.F.; Fuck, R.A.; Dantas, E.L.; Laux, J.H.; Junges, S.L. The Paleoproterozoic peraluminous Aurumina Granite Suite, Goiás and Tocantins, Brazil: Geological, whole rock geochemistry and U-Pb and Sm-Nd isotopic constraints. In *The Paleoproterozoic Record of the São Francisco Craton, Proceedings of the IGCP 509 Annual Meeting, Bahia/Minas Gerais, Brazil, 9–21 September 2006*; p. 92. Available online: https://earth.yale.edu/sites/default/files/files/IGCP/IGCP%20Brazil.pdf (accessed on 3 August 2018).

23. Instituto de Geociências/UnB. Mapa Geológico do Projeto Nova Roma—Porto Real. Bachelor's Thesis, University of Brasilia, Brasília, Brazil, 2005.

24. Roddick, J.C. High precision intercalibration of ^{40}Ar/^{39}Ar standards. *Geochim. Cosmochim. Acta* **1983**, *47*, 887–898. [CrossRef]

25. Steiger, R.H.; Jäger, E. Subcommission on geochronology: Convention on the use of decay constants in geo- and cosmo-chronology. *Earth Planet. Sci. Lett.* **1977**, *36*, 359–362. [CrossRef]

26. Dalrymple, G.B.; Alexander, E.C., Jr.; Lanphere, M.A.; Kraker, G.P. *Irradiation of Samples for ^{40}Ar/^{39}Ar Dating Using the Geological Survey TRIGA Reactor*; U.S. Geological Survey Professional Paper 1176; U.S. Government Printing Office: Washington, DC, USA, 1981; p. 1176.

27. Clayton, R.N.; Mayeda, T.K. The use of bromine pentafluoride in the extraction of oxygen from oxides and silicates for isotopic analysis. *Geochim. Cosmochim. Acta* **1963**, *27*, 43–52. [CrossRef]

28. Zheng, Y.F. Calculation of oxygen isotope fractionation in metal oxides. *Geochim. Cosmochim. Acta* **1991**, *55*, 2299–2307.

29. Bottinga, Y.; Javoy, M. Comments on oxygen isotope geothermometry. *Earth Planet. Sci. Lett.* **1973**, *20*, 250–265. [CrossRef]

30. Streckeisen, A. To each plutonic rocks its proper name. *Earth Sci. Rev.* **1976**, *12*, 1–33. [CrossRef]

31. Schwartz, M.O.; Rajah, S.S.; Askury, A.K.; Putthapiban, P.; Djaswadi, S. The Southeast Asian Tin Belt. *Earth Sci. Rev.* **1995**, *38*, 95–293. [CrossRef]

32. Corsini, M.; Figueiredo, L.L.; Caby, R.; Feraud, G.; Ruffet, G.; Vauchez, A. Thermal history of the Pan-African/Brasiliano Borborema Province of northeast Brazil deduced from ^{40}Ar/^{39}Ar analysis. *Tectonophysics* **1997**, *285*, 103–117. [CrossRef]

33. Maniar, P.D.; Piccoli, P.M. Tectonic discrimination of granitoids. *Geol. Soc. Am. Bull.* **1989**, *101*, 635–643. [CrossRef]

34. Nakamura, N. Determination of REE, Ba, Fe, Mg, Na and K in carbonaceous and ordinary chondrites. *Geochim. Cosmochim. Acta* **1974**, *38*, 757–775. [CrossRef]

35. Nachit, H.; Ibhi, A.; Abia, E.H.; Ohoud, M.B. Discrimination between primary magmatic biotites, reequilibrated biotites and neoformed biotites. *C. R. Geosci.* **2005**, *337*, 1415–1420. [CrossRef]

36. Deer, W.A.; Howie, R.A.; Zussman, J. *Rock Forming Minerals: Sheet Silicates*; Longman Green and Co.: London, UK, 1963.

37. Abdel-Rahman, A.M. Nature of biotites from alkaline, calc-alkaline, and peraluminous magmas. *J. Petrol.* **1994**, *35*, 525–541. [CrossRef]

38. Miller, C.F.; Stoddard, E.F.; Bradfish, L.J. Composition of plutonic muscovite: Genetic implications. *Can. Miner.* **1981**, *19*, 25–34.

39. Koh, J.S.; Yun, S.H. The compositions of biotite and muscovite in the Yuksipryong two-mica granite and its petrological meaning. *Geosci. J.* **1999**, *3*, 77–86. [CrossRef]

40. Zane, A.; Rizzo, G. The compositional space of muscovite in granitic rocks. *Can. Mineral.* **1999**, *37*, 1229–1238.

41. Tao, J.; Li, W.; Cai, Y.; Cen, T. Mineralogical feature and geological significance of muscovites from the Longyuanba Indosinian and Yanshannian two-mica granites in the eastern Nanling Range. *Sci. China Ser. D* **2014**, *57*, 1150–1157. [CrossRef]

42. Möller, P.; Dulski, P.; Szacki, W.; Malow, G.; Riedel, E. Substitution of tin in cassiterite by tantalum, niobium, tungsten, iron and manganese. *Geochim. Cosmochim. Acta* **1988**, *52*, 1497–1503. [CrossRef]

43. Costi, H.T.; Horbe, A.M.C.; Borges, R.M.K.; Dall'agnol, R.; Rossi, O.R.R.; Sighnolfi, G.P. Mineral chemistry of cassiterites from Pitinga Province, Amazonian Craton, Brazil. *Rev. Bras. Geociênc.* **2000**, *30*, 775–782. [CrossRef]

44. Pereira, A.B. Caracterização dos Granitos e Pegmatitos Peraluminosos, Mineralizados em Sn-Ta, de Monte Alegre de Goiás. Master's Thesis, Instituto de Geociências, Universidade de Brasília, Brasília, Brazil, 2002.

45. Taylor, H.P., Jr. The application of oxygen and hydrogen isotope studies to problems of hydrothermal alteration and ore deposition. *Econ. Geol.* **1974**, *69*, 843–883. [CrossRef]

46. Kalamarides, R.I. High-temperature oxygen isotope fractionation among the phases of the Kiglapait Intrusion, Labrador, Canada. *Chem. Geol.* **1986**, *58*, 303–310. [CrossRef]

47. Pearce, J.A.; Harris, N.W.; Tindle, A.G. Trace element discrimination diagrams for the tectonic interpretation of granitic rocks. *J. Petrol.* **1984**, *25*, 956–983. [CrossRef]

48. Pimentel, M.M.; Heaman, L.; Fuck, R.A.; Marini, O.J. U/Pb zircon geochronology of Precambrian tin-bearing continental type acid magmatism in central Brazil. *Preccambr. Res.* **1991**, *52*, 321–335. [CrossRef]

49. Pimentel, M.M.; Botelho, N.F. Sr and Nd isotopic characteristics of 1.77–71.58 Ga rift-related granites and volcanics of the Goiás Tin Province, central Brazil. *Anais Acad. Bras. Ciênc.* **2001**, *73*, 263–276. [CrossRef]

50. London, D. Magmatic-hydrothermal transition in the Tanco rare-element pegmatite: Evidence from fluid inclusions and phase-equilibrium experiments. *Am. Miner.* **1986**, *71*, 376–395.

51. Müller, A.; van den Kerkhof, A.M.; Behr, H.-J.; Kronz, A.; Koch-Müller, M. The evolution of late-Hercynian granites and rhyolites documented by quartz—A review. *Earth Environ. Sci. Trans. R. Soc. Edinb.* **2009**, *100*, 185–204. [CrossRef]

52. Breiter, K.; Broska, I.; Uher, P. Intensive low-temperature tectono-hydrothermal overprint of peraluminous rare-metal granite: A case study from the Dlhá dolina valley (Gemericum, Slovakia). *Geol. Carpath.* **2015**, *66*, 19–36. [CrossRef]

53. Linnen, R.L. Depth of emplacement, fluid provenance and metallogeny in granitic terranes: A comparison of western Thailand with other tin belts. *Miner. Depos.* **1998**, *33*, 461–476. [CrossRef]

54. Linnen, R.L.; Pichavant, M.; Holtz, F. The combined effects of fO_2 and melt composition on SnO_2 solubility and tin diffusivity in haplogranitic melts. *Geochim. Cosmoch. Acta* **1996**, *60*, 4965–4976. [CrossRef]

55. Bhalla, P.; Holtza, F.; Linnen, R.L.; Behrens, H. Solubility of cassiterite in evolved granitic melts: Effect of T, fO_2, and additional volatiles. *Lithos* **2005**, *80*, 387–400. [CrossRef]

56. Linnen, R.L.; Pichavant, M.; Holtz, F.; Burgess, S. The effect of fO_2 on the solubility, diffusion, and speciation of tin in haplogranitic melt at 850 °C and 2 kbar. *Geochim. Cosmoch. Acta* **1995**, *59*, 1579–1588. [CrossRef]

57. Bea, F.; Fershlater, G.; Corretgé, L.G. The geochemistry of phosphorus in granite rocks and the effect of aluminium. *Lithos* **1992**, *29*, 43–56. [CrossRef]

58. Thomas, R.; Förster, H.-J.; Heinrich, W. The behaviour of boron in a peraluminous granite-pegmatite system and associated hydrothermal solutions: A melt and fluid-inclusion study. *Contrib. Mineral. Petrol.* **2003**, *144*, 457–472. [CrossRef]

59. Černý, P. Fertile granites of Precambrian rare-element pegmatite fields: Is geochemistry controlled by tectonic setting or source lithologies. *Precambr. Res.* **1991**, *51*, 429–468. [CrossRef]

minerals

MDPI

Article

Geochronology and Genesis of the Xitian W-Sn Polymetallic Deposit in Eastern Hunan Province, South China: Evidence from Zircon U-Pb and Muscovite Ar-Ar Dating, Petrochemistry, and Wolframite Sr-Nd-Pb Isotopes

Jingya Cao [1,2,3], Qianhong Wu [1,*], Xiaoyong Yang [2,*], Hua Kong [1], Huan Li [1], Xiaoshuang Xi [1], Qianfeng Huang [1] and Biao Liu [1]

[1] Key Laboratory of Metallogenic Prediction of Nonferrous Metals and Geological Environment Monitoring (Central South University), Ministry of Education, Changsha 410083, China; jingyacao@126.com (J.C.); konghua2006@126.com (H.K.); lihuan@csu.edu.cn (H.L.); xiyiyun@sina.com (X.X.); hqianfeng@gmail.com (Q.H.); liubiaoznuedu@163.com (B.L.)

[2] CAS Key laboratory of Crust-Mantle Materials and Environments, University of Science and Technology of China, Hefei 230026, China

[3] Geological Survey of Anhui Province, Hefei 230001, China

* Correspondence: qhwu19@163.com (Q.W.); xyyang@ustc.edu.cn (X.Y.)

Received: 6 February 2018; Accepted: 24 February 2018; Published: 8 March 2018

Abstract: The recently explored Xitian tungsten-tin (W-Sn) polymetallic ore field, located in Hunan province, South China, is one of the largest ore fields in the Nanling Range (NLR). Two major metallogenic types appeared in this ore field, skarn- and quartz vein-type. They are distributed within Longshang, Heshuxia, Shaiheling, Hejiangkou, Goudalan, and so on. Hydrothermal zircons from two altered granites yielded U-Pb ages of 152.8 ± 1.1 Ma, and 226.0 ± 2.8 Ma, respectively. Two muscovite samples from ore-bearing quartz vein yielded $^{40}Ar/^{39}Ar$ plateau ages of 156.6 ± 0.7 Ma, 149.5 ± 0.8 Ma, respectively. Combined with the geological evidence, two metallogenic events are proposed in the Xitian ore field, with skarn-type W-Sn mineralization in Late Triassic (Indosinian) and quartz vein/greisen type W-Sn mineralization in Late Jurassic (Yanshanian). The relatively low Ce/Ce* ratios and high Y/Ho ratios in zircons from two altered granites indicate that the hydrothermal fluids of two metallogenic events are characterized by low oxygen fugacities and enrichment in F. The similar chondrite-normalized patterns between the skarn and Xitian Indosinian granites and Sr-Nd-Pb isotopic compositions of wolframite suggest that the metal sources for both types W-Sn mineralization are derived from a crustal source.

Keywords: zircon U-Pb; muscovite Ar-Ar; wolframite Sr-Nd-Pb isotopes; Xitian W-Sn deposit; Eastern Hunan

1. Introduction

Tungsten (W) and tin (Sn) are important metals in many aspects of industrial manufacture. Accompanied by the greater demand for W-Sn, the study and exploitation of W-Sn deposits have long been a hot topic [1–13]. China holds the largest resources of W and Sn in terms of production and reserves, and their reserves have accounted for ca. 58% and ca. 31% in the world, respectively [14]. In China, more than 83% of the W and 63% of the Sn reserves are in the Nanling region [15]. The Nanling region is famous for its large-scale and multi-stage magmatism and abundant W, Sn and other rare-metal resources and reserves [2,3,15–22]. Previous studies have revealed the presence of many large W-Sn polymetallic deposits in this region, such as Shizhuyuan, Dajishan, Xianghualing,

Xihuashan, and Xitian (Figure 1; [17–19,23–25]. Furthermore, they are closely related to the Mesozoic granitic intrusions, on both temporal and spatial scales [12,23,26–28]. Since the 1990s, a considerable amount of high-precision data of rock- and ore-forming ages have been obtained from the Nanling range with the help of progressive dating technologies, such as zircon U-Pb, molybdenite Re-Os, and mica ^{40}Ar-^{39}Ar dating methods, and most of these ages show that these deposits were formed in late Mesozoic (Yanshanian), such as Shizhuyuan (149 ± 2 Ma; [29]), Xianghualing (156 ± 4 Ma; [24]), Xihuashan (157.8 ± 0.9 Ma; [18]), Dajishan (161.1 ± 1.3 Ma; [30]), Taoxikeng (154.4 ± 3.8 Ma; [17]) and Xitian (151.8 ± 1.4 Ma; [19]). Recently, some new data of metallogenic age for the W-Sn deposits in this area are proven to be early Mesozoic (Indosinian), such as Wangxianling (220.6 ± 1.1 Ma; [31]), Hehuaping (224.0 ± 1.9 Ma; [32]), and Limu (214.1 ± 1.9 Ma; [33]). This evidence demonstrated that two periods of metallogenetic events existed in the Nanling region. However, further studies on the mineral genetic epoch for the deposits are required, especially for these deposits with multiple phase-intrusive activities.

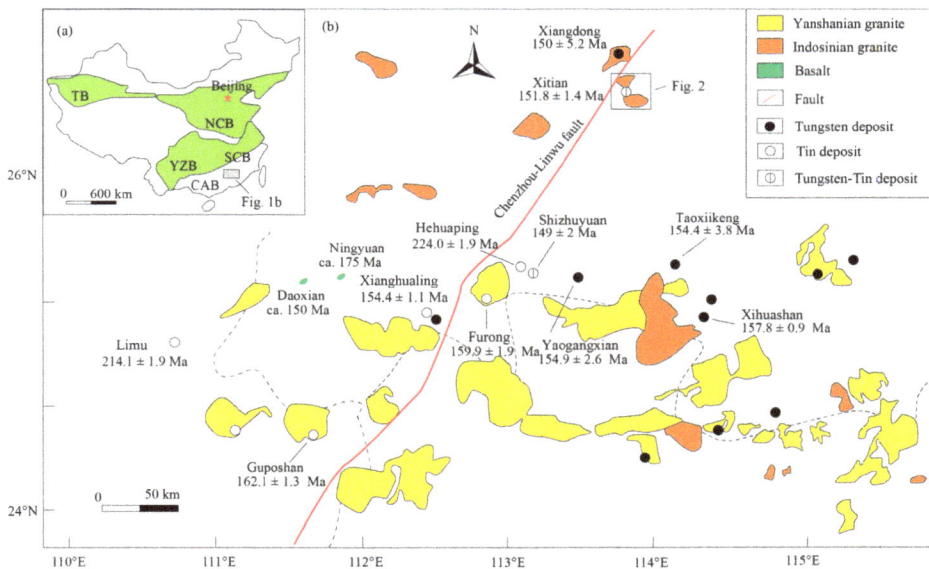

Figure 1. (**a**) Geological sketch map of China; (**b**) Geological sketch map of the Nanling region (modified from [2]), showing the distribution of granitic plutons, basalts, and related W-Sn deposits and their geochronological data compiled from [17–19,21,29,31–37]. TB: Tarim block; CAB: Cathaysian Block; NCB: North China Block; SCB: South China Block; YZB: Yangtze Block.

The Xitian W-Sn polymetallic ore field, located in the middle of the Nanling region, is one of the largest newly discovered ore fields in recent years (Figure 2). A large number of studies have been carried out in the Xitian area by geochemical and isotopic methods, and these studies have shown that the formation of this deposit is genetically related to the Xitian pluton [19,28,38,39]. Many dating technologies have been applied to study the emplacement age of this pluton and metallogenic age of this deposit, including LA-ICP-MS, and ion probe by either SHRIMP or CAMECA zircon U-Pb, mica and cassiterite ^{40}Ar-^{39}Ar, and molybdenite Re-Os isotopic techniques [19,40–42]. These precise data provide detailed chronological constraints on the emplacement age of the Xitian pluton and for the time interval between W-Sn mineralization of the Xitian deposit. The majority of these chronological data show that the Xitian pluton could be subdivided into the Late Triassic (Indosinian) granites (230–220 Ma) and the Late Jurassic (Yanshanian) granites (160–140 Ma), and the time interval for W-Sn mineralization is 160–150 Ma. From the above-mentioned evidence, it seems that the formation of this ore field is attributed to Yanshanian magmatic activity. Deng et al. [43] obtained a molybdenite Re-Os age of 225.5 ± 3.6 Ma from altered granites in the Indosinian granitic batholith, indicating a possibility for the Indosinian mineralization event in the Xitian ore field; however, this age may not represent the age of the large-scale skarn-type W-Sn mineralization in the Xitian ore field. It was proposed that the skarn occurred in the contact zone between the Yanshanian granites and the Devonian dolomitic limestone [19]. However, recent studies show that the granites belong to Indosinian granites rather than Yanshanian granites [34]. Therefore, is the skarn type ore body related to the Indosinian granites, rather than to the Yanshanian granites? Ore-forming age of some deposits in the Xitian ore field is still in doubt; for example, the Hejiangkou deposit. In this study, we display the results of zircon U-Pb dating, zircon compositions, muscovite ^{40}Ar-^{39}Ar dating, skarn geochemistry, and wolframite Sr-Nd-Pb isotopic compositions, with the aims of constraining the time interval between mineralization and the emplacement of associated granitic rocks, outlining the genetic relationship between two episodes of granitic magmatism and two types of W-Sn mineralization, and probing into the genesis of the two types of W-Sn mineralization in the Xitian ore field.

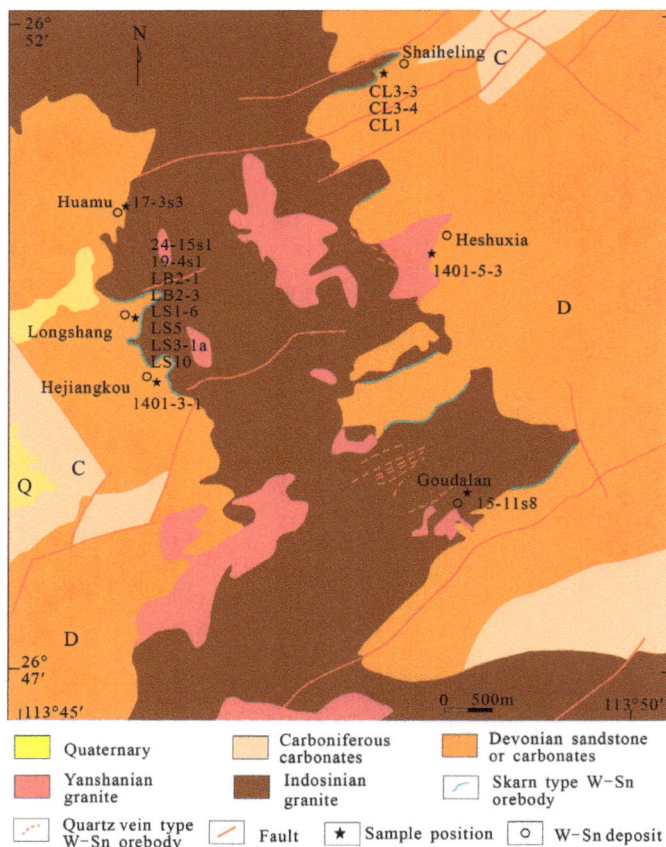

Figure 2. Schematic geological map of the Xitian W-Sn ore field showing the location of samples (modified from [44]).

2. Geological Setting

The South China Block (SCB) was formed by the amalgamation of the Yangtze Block (YZB) to the northwest and the Cathaysian Block (CAB) to the southeast at ca. 820 Ma (Figure 1a; [45]). The Nanling region, located in the central section of SCB, is comprised of Guangxi, Guangdong, Hunan, and Jiangxi province occupying an area of 170,000 km^2 [46]. This region has undergone several significant tectonic-magmatic events, the most famous of which are the Indosinian and Yanshanian tectonic events during the Mesozoic [47–53]. Due to the superior metallogenic geotectonic setting, it is characterized by widespread igneous rocks and numerous large-scale W-Sn polymetallic deposits [2,3,16,20,54,55]. The basement of the Nanling region consists of weakly metamorphosed Precambrian, late Paleozoic sedimentary strata which are mainly Devonian and Carboniferous carbonate rocks, and lesser amounts of Upper Triassic to Tertiary sandstone and siltstone [2,46]. The regional fault is the NE-trending Chenzhou-Linwu fault which controls the spatial distribution of the granitic intrusions and numerous W-Sn polymetallic deposits associated with the granitic magma activities (Figure 1b; [56,57]). Numerous granitic intrusions were emplaced in this region, most of which are Indosinian and Yanshanian pluton, and these granites are mostly peraluminous, calc-alkaline and remelted granites [12,23,27,28,58,59].

The Xitian ore field, located in Chaling, Hunan province, is characterized by intensive and widely distributed granitoids associated with numerous non-ferrous and rare-metal minerals of Mesozoic age [19,39,44,60].

2.1. Sedimentary Rocks

The strata outcropping in the Xitian area are Ordovician, Devonian, Carboniferous, and Cretaceous sedimentary rocks, among which the middle to upper Devonian and Carboniferous rocks are dominant (Figure 2). Lying unconformably on Ordovician metasedimentary rocks, the Devonian strata can be subdivided into the Middle Devonian Tiaomajian and Qiziqiao Formations, and the upper Devonian Shetianqiao and Xikuangshan Formations [61]. The Tiaomajian formation, 35–42 m in thickness, consists of conglomerate-bearing quartzite. The Qiziqiao formation, over 200 m thick, comprises impure carbonate rocks and arenaceous shale, and is the typical ore reservoirs of the Xitian ore field. The Shetianqiao formation is up to 500 m thick, composed of quartz sandstone and argillaceous siltstone. The Xikuangshan formation, 110–130 m in thickness, are mainly quartz sandstone, arenaceous shale and nodular limestones. The Carboniferous Yanguan formation which is about 275 m in thickness, mainly consists of sandshale and siltstone.

2.2. Structure

The Xitian ore field is located to the east of the NE-trending Chenzhou-Linwu deep fault (Figure 1b), which is considered to be the boundary between the Yangtze Block and Cathaysian Block [56,57,62]. The tectonic framework of this region is controlled mainly by two trends of faults which are approximately NE-, nearly SN- and NW-trending. The NE-trending faults are the larger in scale, and some of these faults are truncated by the NW- and/or SN trending-faults (Figure 2). The NE-trending faults are the main ore-controlling faults in Xitian ore field, with 2–13 km in outcropped length, 60–70° in angle of trend, and 60–85° in angle of dip [60]. The nearly SN-trending faults are also important ore-bearing structures, including a series of NNW-, SN- and NNE-trending small faults [60]. The NW-trending faults are about 1.5–8.0 km in outcropped length, with a dip of NNE and large inclined angle [60].

2.3. Igneous Rocks

The Xitian pluton, occupying an area of ~240 km^2, are intruded into Paleozoic rocks which are mainly Devonian and Carboniferous carbonate and sandstone. Previous studies have recorded three stages of granitic magmatic activities in this area: Indosinian (230–220 Ma; [38], early Yanshanian

(160–150 Ma; [28]) and late Yanshanian (141 Ma; [63]). The Indosinian granites, outcropped as intrusive stock, are mainly coarse-grained porphyritic biotite granites, with K-feldspar (~40%), plagioclase (~25%), quartz (~20%), and biotite (~15%) as the main minerals and zircon, apatite, sphene, and magnetite as the accessory minerals [34]. The early Yanshanian granites are mainly composed of fine-grained two mica granites as dykes, with K-feldspar (28–30%), quartz (28–38%), plagioclase (25–30%), and mica (5–12%, including biotite and muscovite) as the main minerals and magnetite, tourmaline, apatite, topaz and zircon as the accessory minerals [28]. The late Yanshanian granites are exposed rarely, which are mainly muscovite granite [63].

3. Geology of the Ore Deposits

Previous studies have revealed that four types of W-Sn polymetallic ore bodies were exploited in the Xitian ore field consisting of skarn-, quartz vein-, greisen- and structurally altered rock-types [19,60]. The skarn-type ore bodies, occurred mainly in Longshang, Hejiangkou, and Shaiheling, are characterized by W-Sn mineralization (Figure 2). The quartz vein- and greisen-type ore bodies are also characterized by W-Sn mineralization, distributed in Longshang, Hejiangkou, Heshuxia, and Goudalan (Figure 2). The structurally altered rock-type ore bodies are mainly found in Shaiheling featured by Lead (Pb)-Zinc (Zn) mineralization (Figure 2). The morphology of ore body, specimen and micrographs were presented in Figures 3 and 4, respectively.

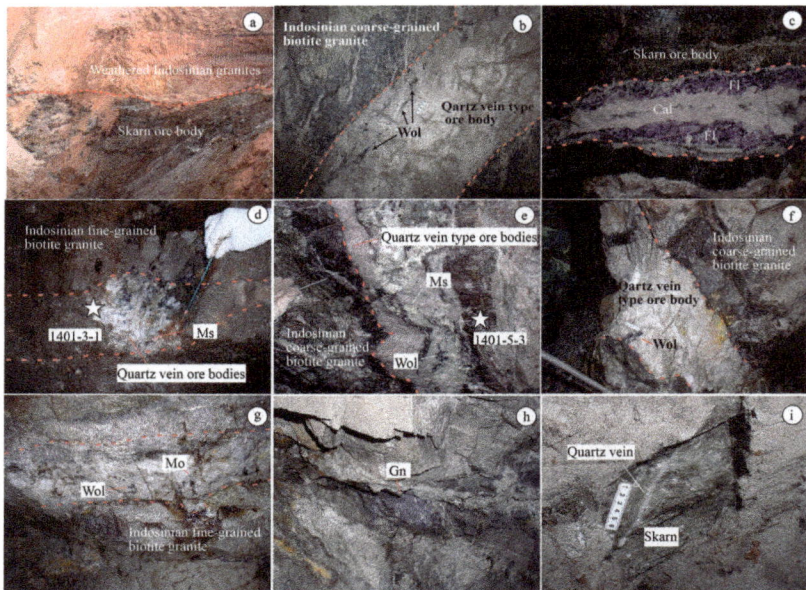

Figure 3. Photographs of the related ore bodies showing the morphology of skarn- and quartz vein-type ore body. (**a**) The stratiform and stratoid ore bodies occurring in the contact zone between the Indosinian granites and the Devonian dolomitic limestone in Longshang ore district; (**b**) Quartz vein type of ore body in Longshang ore district; (**c**) Skarn-type ore bodies are superimposed by calcite- and fluorite-veins in Longshang ore district; (**d**) Quartz vein type of ore body in Hejiangkou ore district; (**e**) Quartz vein type of ore body in Heshuxia ore district; (**f**) Quartz vein type of ore body in Heshuxia ore district; (**g**) Quartz vein type of ore body in Goudalan ore district; (**h**) The structurally altered rock-type of Pb-Zn ore body in Shaiheling ore district; (**i**) The stratiform and stratoid ore bodies occurring in the contact zone between the Indosinian granites and the Devonian dolomitic limestone in Shaiheling ore district; Cal: calcite; Fl: fluorite; Gn: galenite; Mo: molybdenite; Ms: muscovite; Wol: wolframite.

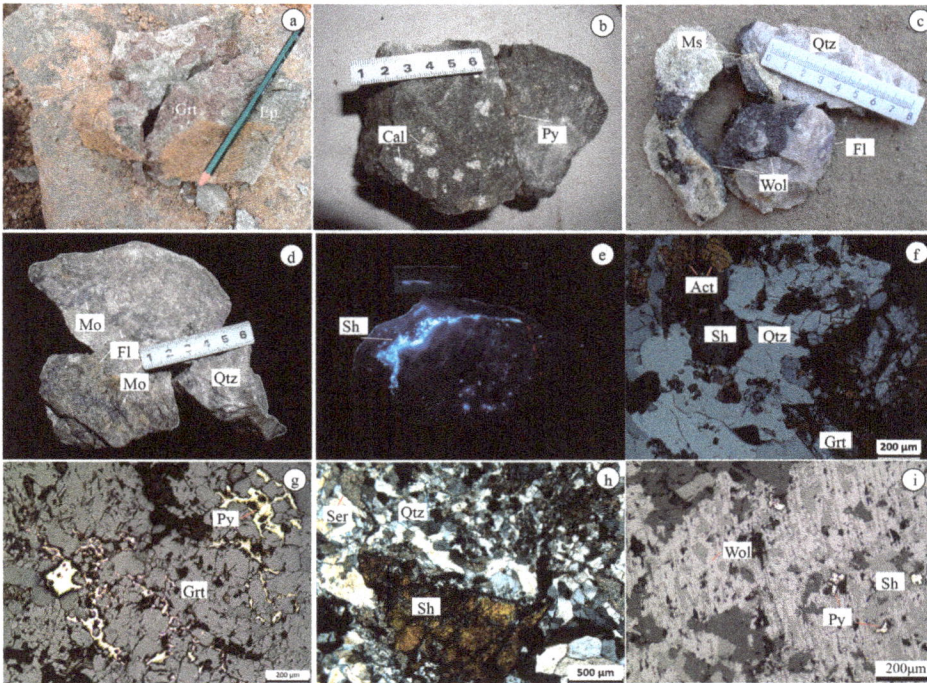

Figure 4. Photographs showing representative mineral assemblages and textural features of the skarn- and quartz vein type ore. (**a**) Hand specimen of the skarn showing the skarn mineral assemblages of garnet and epidote; (**b**) Hand specimen of the skarn type ore showing mineral assemblages of pyrite and calcite; (**c**) Wolframite-bearing quartz ore in Heshuxia ore district; (**d**) Molybdenite-bearing quartz ore in Heshuxia ore district; (**e**) Scheelite-bearing quartz ore in Heshuxia ore district (under a tungsten lamp); (**f**) Scheelite coexisting with quartz, garnet, and actinolite; (**g**) The xenomorphic pyrite coexisting with the garnet; (**h**) Scheelite with quartz and sericite; (**i**) Scheelite coexisting with wolframite and pyrite, and replacing the wolframite. Act: actinolite; Cal: calcite; Ep: epidote; Fl: fluorite; Grt: garnet; Py: pyrite; Mo: molybdenite; Ms: muscovite; Qtz: quartz; Ser: sericite; Sh: scheelite; Wol: wolframite.

3.1. Longshang

The Longshang deposit, located in the western part of the Xitian ore field, is the largest W-Sn deposit in this area (Figure 2). Two types of ore body are exploited in this deposit, which are skarn- and quartz vein type. The skarn-type W-Sn ore bodies are stratiform, stratoid and lentoid, mainly occurring at the endo- or exo-contact zone between the Xitian Triassic granites and the Devonian dolomitic limestone (Figure 3a). In addition, the N–S-trending and E-dipping skarn-type ore bodies are 2700 m long and 4.5–33.1 m thick with ore grade of 0.14–0.77% Sn and 0.038–0.83% WO_3 [35]. The quartz vein-type ore bodies, occurring in the Xitian Triassic granites, are E–W-trending and S-dipping with ore veins of 3–20 cm in thickness (Figure 3b). A complex skarn-vein type W-Sn ore bodies, newly found in the Longshang deposit, have relatively high grade with 2.3% WO_3 and 1.7% Sn in some parts of the ore bodies (Figure 3c).

The major ore minerals of skarn-type ore bodies comprise scheelite, pyrite, and some other minerals, with gangue mineral mainly consisting of garnet, epidote, idocrase, quartz, and other skarn minerals (Figure 4a,b,f,g). The major ore minerals in quartz vein-type ore bodies are wolframite, cassiterite, scheelite, pyrite, and some other minerals, with gangue mineral mainly consisting of quartz, fluorite, calcite, and sericite.

3.2. Hejiangkou

The Hejiangkou deposit is also located in the western part of the Xitian ore field (Figure 2). Similar to the Longshang deposit, Hejiangkou is also characterized by skarn-type Sn-Cu mineralization and quartz vein type W-Sn mineralization. The stratiform and lentoid skarn-type ore bodies are the mainly mining target in this deposit, which occur in the endo- or exo-contact zone between the Xitian Triassic granites and the Devonian limestone, with 500–1200 m long and 30–50 m thick. The quartz vein-type W-Sn ore veins, hosted in the Xitian Triassic granites and Devonian sandstone, are NEE and/or NNW-trending and N-dipping with ore veins of 50–200 m long and 0.1–0.8 m thick (Figure 3d).

Ore minerals of the skarn-type ore bodies are mainly cassiterite, chalcopyrite, scheelite, sphalerite, and pyrite, with gangue minerals consisting of quartz, calcite, garnet, epidote, idocrase and other skarn minerals. Ore minerals of the quartz vein-type ore bodies are mainly composed of wolframite, cassiterite, scheelite, and pyrite, and gangue minerals are quartz, fluorite, and muscovite.

3.3. Heshuxia

The Heshuxia deposit is in the eastern part of the Xitian ore field (Figure 2). Unlike the Longshang and Hejiangkou deposits in the western part of the Xitian ore field, the Heshuxia deposit is mainly characterized by quartz vein type of W mineralization. The NW-trending and SW-dipping ore veins are mainly hosted in the Xitian Triassic granites, with length of 100–1000 m, thickness of 30–100 cm, and ore grade of 0.172–0.700% WO_3 (Figure 3e,f). Ore minerals are mainly wolframite, with a small quantity of scheelite, molybdenite, pyrite, and chalcopyrite. Gangue minerals are mainly composed of quartz, fluorite, and muscovite (Figure 4c,d,e,h,i).

3.4. Goudalan

The Goudalan deposit, located in the southeast part of the Xitian ore field, is also characterized by quartz vein type of W-Sn mineralization. The NEE trending ore veins are also hosted in the Xitian Triassic granites with length of 100–1000 m, thickness of 0.1–1 m and ore grade of 0.086–0.762% WO_3 (Figure 3g). Ore minerals mainly consist of wolframite, with a small quantity of scheelite, molybdenite, pyrite, and chalcopyrite. Gangue minerals are mainly composed of quartz, fluorite, and muscovite.

3.5. Shaiheling

The Shaheling deposit, located in the northeast part of the Xitian ore field, are characterized by the structurally altered rock-type of Pb-Zn mineralization and skarn-type of W-Sn mineralization. The structurally altered rock-type of Pb-Zn ore bodies are mainly hosted in the fracture zone of the skarn and/or carbonate formations, with NW trending and NE dipping (Figure 3h). The skarn-type W-Sn ore bodies are stratiform, stratoid and lentoid, mainly occurring at the endo- or exo-contact zone between the Xitian Triassic granites and the Devonian dolomitic limestone as in the Longshang and Hejiangkou deposits, with average length of 1.4 km and thickness of 3.3 m (Figure 3i). The ore minerals of the structurally altered rock-type ore mainly comprise of sphalerite, galenite, scheelite, pyrite and chalcopyrite, with gangue minerals consisting of quartz, feldspar, chlorite, and so on. Ore minerals of the skarn-type W-Sn ore are mainly composed of scheelite, cassiterite, pyrite and chalcopyrite, with gangue minerals consisting of quartz, garnet, epidote, and so on.

4. Sampling and Analytical Methods

The analyzed samples were collected from underground mines (Figure 2). Zircon grains used for LA-ICP-MS U-Pb dating were separated from a sericitic coarse-grained biotite granite (sample No. 19-4s1, Figure 5a,b) and a sericitic fine-grained two mica granite (sample No. 24-15s1, Figure 5c,d), which were collected from the Longshang deposit. The muscovite used for Ar-Ar dating were extracted from quartz vein-type ore in Hejiangkou (Figure 3d, sample 1401-3-1) and Heshuxia (Figure 3e, sample 1401-5-3). The skarn used for geochemical analyses were all endo-skarn, some of which contained sulfides, collected from the Longshang, Shaiheling and Huamu deposits. The wolframite, separated from the quartz vein type ore and used for Sr-Nd-Pb analysis, were collected from Longshang, Goudalan and Heshuxia. The sampling locations were marked in Figure 2.

Figure 5. Hand specimen and micrographs of the altered granite. (**a**) Hand specimen of the altered Indosinian coarse-grained biotite granite; (**b**) The K-feldspar is altered and replaced by sericite; (**c**) Hand specimen of the altered Yanshanian fine-grained two-mica granite; (**d**) Almost the feldspars are altered and replaced by sericite. Kfs: K-feldspar; Pl: plagioclase; Ser: sericite.

4.1. In Situ *LA-ICP-MS Zircon U-Pb Dating and Trace Element Compositions*

Zircon grains from these samples were separated by conventional magnetic and heavy liquid techniques before they were hand-picked under a binocular microscope. They were then mounted into epoxy resin blocks and polished to obtain flat surfaces. Cathodoluminescence (CL) imaging technique was used to visualize the internal structures of individual zircon grains, with a scanning electron microscope (TESCAN MIRA 3 LMH FE-SEM, TESCAN, Brno, Czech Republic) at the Sample Solution Analytical Technology Co., Ltd., Wuhan, China. Zircon U-Pb dating was undertaken with an Agilent 7700 inductively coupled plasma-mass spectrometer (ICP-MS, Agilent, Santa Clara, CA, USA), combined with a Coherent 193 laser ablation (LA) system at Sample Solution Analytical Technology Co., Ltd., Wuhan, China. Two zircon standards, 91500 (1062 ± 4 Ma; [64] and GJ-1 (610.0 ± 1.7 Ma; [65], were used as external standards for dating. Standard silicate glass (NIST SRM610) was used for external standardization for trace element analysis, and ^{29}Si was used for internal standardization (32.8% SiO_2 in zircon). The standard protocol correction method was used in analyzing the 91500 and

GJ-1 standard zircons twice and once, respectively, after every five analyses. The raw ICP-MS data were processed using ICPMSDataCal software [66], and common Pb was corrected following [67]. Concordia diagrams and weighted mean calculations were processed using Isoplot (version 3.0; [68]).

4.2. Muscovite ^{40}Ar-^{39}Ar Dating

The Muscovite grains were carefully handpicked using a binocular microscope from the crushed sample to ensure purity up to 99.9%, then these grains were washed repeatedly in an ultrasonic bath using deionized water and acetone. Aliquots of approximately 10 mg were wrapped in Al foil and stacked in quartz vials. After samples had been stacked, the sealed quartz vials were put in a quartz canister, which was wrapped with cadmium foil (0.5 mm in thickness) to act as a slow neutron shield thereby preventing interface reactions during irradiation. The irradiation procedure was put the samples in channel B4 of Beijing 49-2 reactor for 50 h at the Chinese Academy of Nuclear Energy Sciences. During irradiation, the vials were rotated at a speed of two cycles per minute to ensure uniformity of the irradiation. The biotite standard ZBH-2506 (132.5 Ma; [69]) was used to monitor the neutron flux. $^{40}Ar/^{39}Ar$ stepwise heating analyses were performed at the Key Laboratory of Tectonics and Petroleum Resources, China University of Geosciences, Wuhan, China. Analyses were carried out using an Argus VI mass spectrometer combined with Coherent 50 W CO_2 laser system. The time of heating was 60 s for every single stage with a laser beam diameter of 2.5 mm, and the time of gas purification was 400 s with two Zr-Al scavenger. The detailed analytical procedures were given by [70]. K_2SO_4 and CaF_2 crystals were analyzed to calculate Ca, K correction factors: $(^{39}Ar/^{37}Ar)_{Ca} = 8.984 \times 10^{-4}$, $(^{36}Ar/^{37}Ar)_{Ca} = 2.673 \times 10^{-4}$, $(^{40}Ar/^{39}Ar)_K = 5.97 \times 10^{-3}$. The data-processing software and diagrams of plateau age we used was the ArArCALC 2.52 software by [71].

4.3. Skarn Major and Trace Elements Analysis

The skarn samples were crushed in a milling machine to 200 mesh before elemental analyses were conducted. The major and trace element compositions of skarn were analyzed at ALS Chemex, Guangzhou, China. The major element contents were measured using a Panalytical Axios Max X-ray fluorescence (XRF, Panalytical, Almelo, The Netherlands) instrument, with analytical accuracy of about 1–5%. Trace element compositions were measured using ICP-MS (Perkin Elmer Elan 9000, Perkin, Waltham, MA, USA), with analytical accuracy of better than 5%.

4.4. Wolframite Sr-Nd-Pb Isotopic Composition Analysis

Sr-Nd-Pb isotopic analyses were carried out at the Key Laboratory of Crust-Mantle Materials and Environments, Chinese Academy of Sciences, University of Science and Technology of China, Hefei, China, using a Finnigan MAT-262 multicollector thermal ionization mass spectrometer (MC-TIMS). Rb-Sr was separated and purified using conventional cation exchange (AG50W-X12, 200–400 resin), whereas Sm and Nd were separated and purified using Teflon and a Power resin, respectively. The correction for mass fractionation of the Sr-Nd isotopic ratio was undertaken by normalizing to $^{86}Sr/^{88}Sr = 0.1194$ and $^{146}Nd/^{144}Nd = 0.7219$. The Sr standard (NBS987, $^{87}Sr/^{86}Sr = 0.710249 \pm 0.000012$ (2σ)) and the Nd standard (La Jolla, $^{143}Nd/^{144}Nd = 0.511869 \pm 0.000006$ (2σ)) were used as the standard solution in this study. The analytical accuracy of the Sr and Nd isotope data are superior to 0.003%. The $^{208}Pb/^{206}Pb$, $^{207}Pb/^{206}Pb$, and $^{204}Pb/^{206}Pb$ ratios of the Pb standard (NBS981) are 2.1681 ± 0.0008 (2σ), 0.91464 ± 0.00033 (2σ), and 0.059042 ± 0.000037 (2σ), respectively. The analytical accuracy of the Pb isotope data is better than 0.01%. Specific procedures of the Sr-Nd-Pb isotopic analytical techniques are given by [72].

5. Results

5.1. Zircon U-Pb Dating

Most of zircon grains of sample 19-4s1 are euhedral or subhedral in shape and black in CL imaging. The length of zircons ranges from 80 to 180 μm with length-to-width ratios from 1:1 to 3:1. CL images show that these zircons are with weak internal oscillatory zoning, and/or irregular, patchy to granular internal structures, and growth zonings can be found in some zircons (zircon No. 8, 9, 12, and 15, Figure 6a), indicating that they might not be the typical magmatic zircons [73]. Th and U contents of these zircon grains vary from 509 ppm to 1495 ppm (mean = 931 ppm) and 663 ppm to 3129 ppm (mean = 1705 ppm), respectively. The Th/U ratios are variable ranging from 0.18 to 1.73 (mean = 0.66). Several isotopic data of the analyzed zircon grains have relatively big errors, which are eliminated in the process of dating calculation. The ^{206}Pb/^{238}U ages of thirteen zircons ranges from 216.4 Ma to 233.5 Ma which plot on or near the concordia curve (Table S1 of Supplementary Materials), yielding a weighted mean ^{206}Pb/^{238}U age of 226.0 ± 2.8 Ma (MSWD = 2.1, Figure 6b).

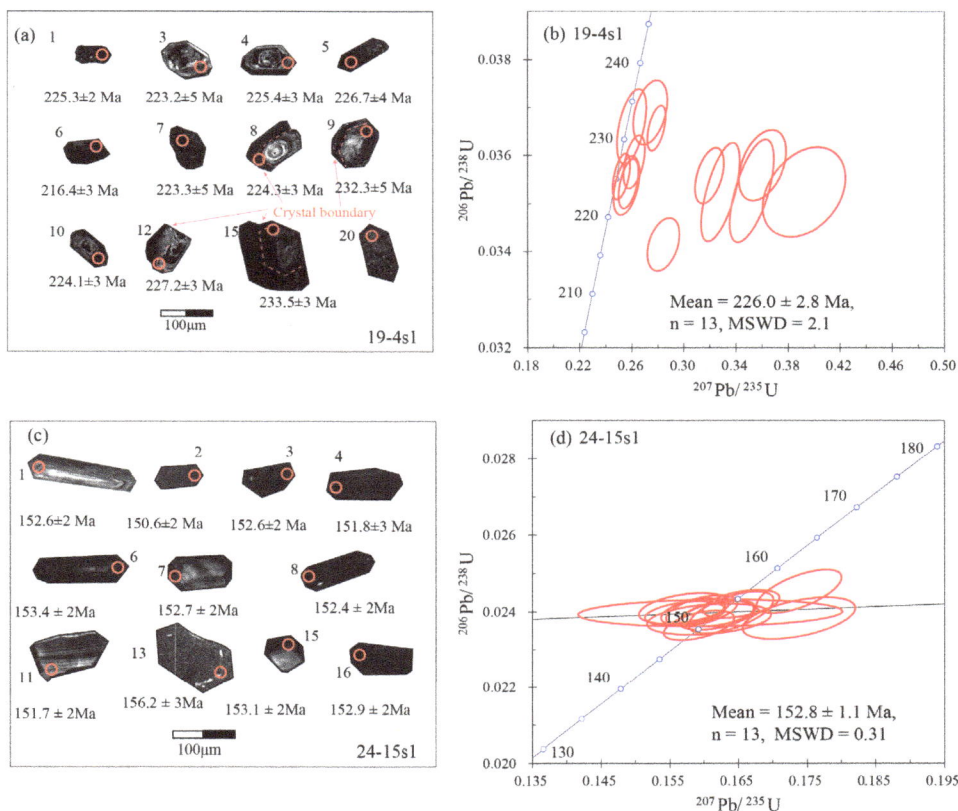

Figure 6. Cathodoluminescence (CL) images of representative zircon grains and concordia diagrams of zircon U-Pb geochronological data for the samples taken from the Xitian ore field. (**a**) Cathodoluminescence (CL) images of the zircons from altered Indosinian granites; (**b**) Concordia diagram of zircon U-Pb data for the altered Indosinian granites; (**c**) Cathodoluminescence (CL) images of the zircons from altered Yanshanian granites; (**d**) Concordia diagram of zircon U-Pb data for the altered Yanshanian granites.

Most of the zircon grains from sample 24-15s1 are xenotopic, with small amounts of idiomorph, and the aspect ratios are ranging from 1:1 to 4:1 with lengths of 50–200 μm. CL imaging indicates that most of the zircon grains are black in color with weak internal oscillatory zoning, and/or granular internal texture (Figure 6c). These grains show abnormally high contents of U (3277–59,113 ppm; mean = 30,823 ppm), and Th (3347–14,922 ppm; mean = 6389 ppm, Table S1 of Supplementary Materials), which is much higher than the granites without alteration in Xitian pluton with U (354–7047 ppm) and Th (192–1257 ppm) contents [28]. Their relatively low Th/U ratios (0.209–0.67, mean = 0.25) indicate a hydrothermal origin [73]. The ^{206}Pb/^{238}U ages of 13 zircon grains from this sample range from 150.6 Ma to 156.2 Ma (Table S1 of Supplementary Materials) and are plotted on or close to the concordia curve, with a weighted mean ^{206}Pb/^{238}U age of 152.8 ± 1.1 Ma (MSWD = 0.31; Figure 6d).

5.2. Trace Element Compositions of Zircons

The zircon grains used for trace element analysis are the same as those which were dated in this study. Their trace element compositions and related parameters are given in Table S2 of Supplementary Materials.

Zircon grains of sample 19-4s1 are characterized by high contents of the heavy rare earth elements (HREEs) and relatively low contents of light rare earth elements (LREEs), with LREE/HREE ratios ranging from 0.04 to 0.24 (mean = 0.1). They also have relatively variable and high contents of the rare earth elements (REEs) with ΣREE ranging from 623 ppm to 2058 ppm (mean = 1071 ppm). Chondrite normalized REE patterns of these zircon grains are characterized by steep slopes, elevated heavy rare earth elements (HREEs), positive Ce anomalies (most of the Ce/Ce* ratios range from 1.61 to 74.14, with average = 10.84), and negative Eu anomalies (Eu/Eu* = 0.17–0.41, mean = 0.28, Figure 7a).

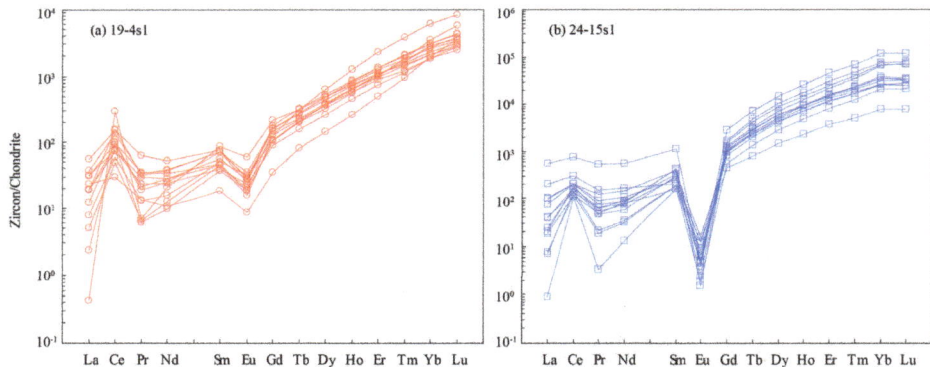

Figure 7. Chondrite-normalized REE chemistry of zircon grains for the samples taken from the Xitian ore field, with normalizing factors from [74]. (**a**) Chondrite-normalized REE chemistry of zircon grains for altered Indosinian granites; (**b**) Chondrite-normalized REE chemistry of zircon grains for altered Yanshanian granites.

Zircon grains from sample 24-15s1 are also characterized by high contents of the heavy rare earth elements (HREEs) and relatively low contents of light rare earth elements (LREEs), with LREE/HREE ratios ranging from 0.01 to 0.13 (mean = 0.02). The REEs contents are tremendously high with ΣREE ranging from 7073 ppm to 39,062 ppm (mean = 15,241 ppm) and Y ranging from 3995 ppm to 43,198 ppm (mean = 18,180 ppm). Chondrite normalized REE patterns that are also characterized by steep slopes, elevated heavy rare earth elements (HREEs), relatively positive Ce anomalies (most of the Ce/Ce* ratios range from = 1.33–8.88, mean = 3.68), and significant negative Eu anomalies (Eu/Eu* = 0.01–0.04, mean = 0.01, Figure 7b).

In addition, both of these zircon grains of two samples have relatively high contents of La, low ratios of $(Sm/La)_N$ and Ce/Ce*, and most of the zircon grains are plotted in the hydrothermal field in the diagram of La versus $(Sm/La)_N$ and $(Sm/La)_N$ versus Ce/Ce*, indicating a hydrothermal origin of these samples (Figure 8a,b; [75]).

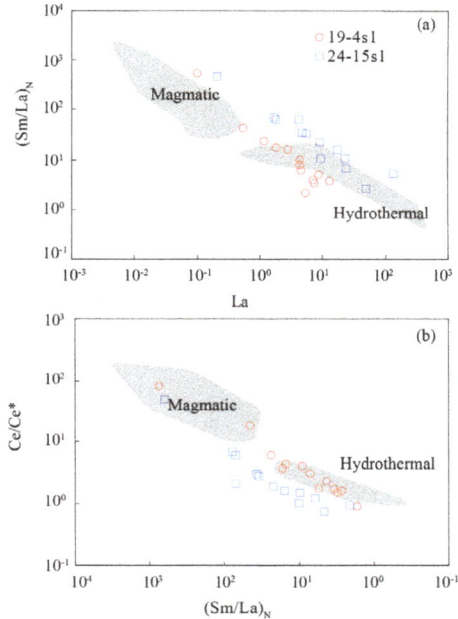

Figure 8. (**a**) La versus $(Sm/La)_N$; and (**b**) $(Sm/La)_N$ versus Ce/Ce* plots of samples from the Xitian ore field, modified from [75].

5.3. Muscovite ^{40}Ar-^{39}Ar Dating

The Ar-Ar isotopic data of two muscovite samples are given in Table S3 of Supplementary Materials. Thirteen laser-heating stages were carried out for sample 1401-3-1, and stages from 6 to 12 had been obtained the flat age spectrum. This sample yields a plateau age of 156.6 ± 0.7 Ma (MSWD = 0.51, Figure 9a) with 60% release of ^{39}Ar, consistent with the inverse isochron age of 156.0 ± 0.7 Ma (MSWD = 0.50, Figure 9b).

Seventeen laser-heating stages were carried out for sample 1401-5-3, and stages from 4 to 16 had been obtained the flat age spectrum. This sample yields a well-defined plateau age of 150.0 ± 0.6 Ma (MSWD = 0.39 Ma, Figure 9c) with 97% release of ^{39}Ar, consistent with the inverse isochron age of 149.5 ± 0.8 Ma (MSWD = 0.17, Figure 9d).

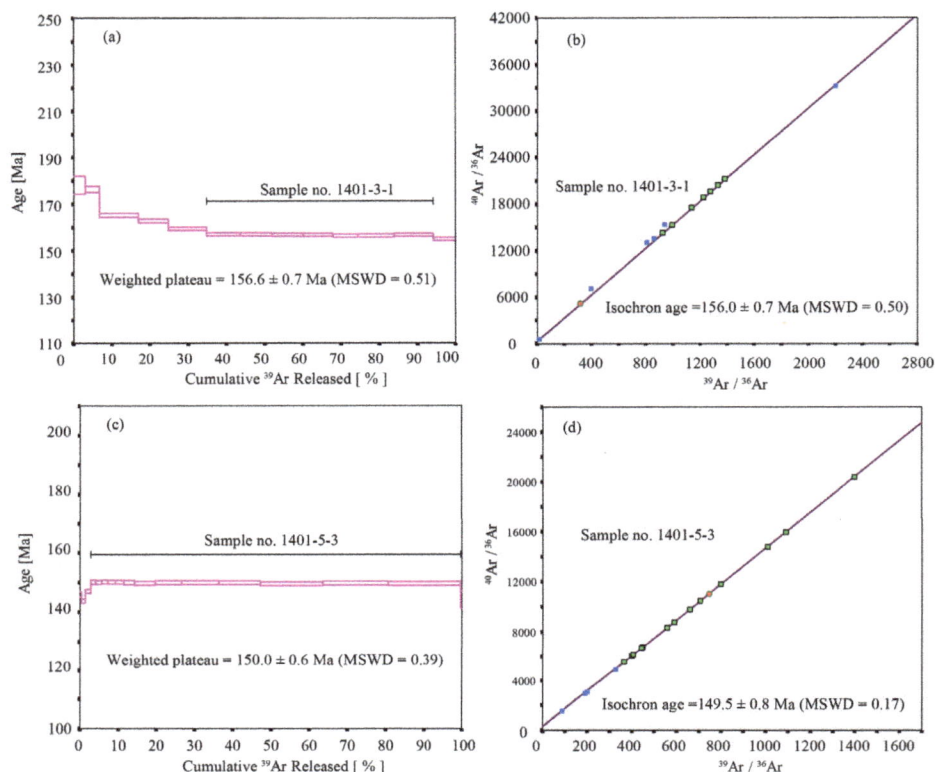

Figure 9. ^{39}Ar-^{40}Ar age spectra and isochron for muscovite samples from the Xitian ore field. (**a**) Diagram of the plateau age for the 1401-3-1 muscovite; (**b**) Diagram of isochron age for the 1401-3-1 muscovite; (**c**) Diagram of the plateau age for the 1401-5-3 muscovite; (**d**) Diagram of isochron age for the 1401-5-3 muscovite.

5.4. Skarn Major and Trace Elements Analysis Results

The representative bulk chemical compositions (major elements) of skarns in the Xitian ore field are presented in Table S4 of Supplementary Materials. These samples have relatively wide ranges of chemical compositions, with SiO_2 = 30.1–51.2%, TiO_2 = 0.07–0.56%, Al_2O_3 = 2.33–12.50%, TFe_2O_3 = 2.12–23.96%, MnO = 0.07–0.82%, MgO = 1.86–12.85%, CaO = 14.80–34.8%, Na_2O = 0.03–0.28%, K_2O = 0.04–4.68%, and SO_3 = 0.03–10.20%.

The trace and rare earth elements (REEs) contents of skarns are given in Table S4 of Supplementary Materials. The total REE (ΣREE) contents range from 40.4 to 184.5 ppm (mean = 103.2 ppm), with high $(La/Yb)_N$ ratios (5.2–13.2), and significant negative Eu anomalies (Eu/Eu* = 0.36–0.82). The REE distribution patterns display right-dipping V-type curves, which are exactly similar to those of the Indosinian granites in Xitian (Figure 10a). These samples also show similar primitive-mantle-normalized trace element patterns to those of Indosinian granites in Xitian, characterized by enrichment in Rb, K, U, Zr, Hf and REE, and depletion in Ti, P, Sr, Ba, and Nb, which are also similar to the Indosinian granites in the Xitian ore field (Figure 10b).

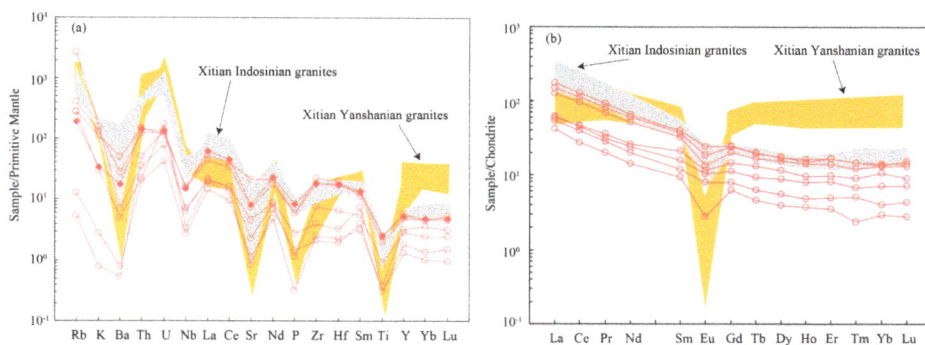

Figure 10. (**a**) Chondrite-normalized REE chemistry; and (**b**) primitive-mantle-normalized incompatible trace element variation diagram for the skarn samples taken from the Xitian ore field. Normalizing factors are from [76]. Data for the Xitian Indosinian and Yanshanian granites are from [28,38].

5.5. Wolframite Sr-Nd-Pb Isotopic Composition Analysis

The wolframite Sr-Nd isotopic compositions of the four samples analyzed in this study are given in Table S5 of Supplementary Materials. The initial $^{87}Sr/^{86}Sr$ and $^{143}Nd/^{144}Nd$ ratios were calculated using the muscovite ^{40}Ar-^{39}Ar dating result of 150.0 ± 0.6 Ma. These samples have high initial $(^{87}Sr/^{86}Sr)_i$ ratios (0.71282–0.72003), low initial $(^{143}Nd/^{144}Nd)_i$ ratios (0.511644–0.512155), negative $\varepsilon_{Nd}(t)$ values ranging from −15.6 to −5.6, and old model ages (T_{DM2}) of 2168–1396 Ma.

The Pb isotopic compositions of the samples are shown in Table S5 of Supplementary Materials. All samples are enriched in radiogenic Pb, with $^{206}Pb/^{204}Pb$ ratios of 18.489–18.569, $^{207}Pb/^{204}Pb$ ratios of 15.724–15.877, and $^{208}Pb/^{204}Pb$ ratios of 39.055–39.335, respectively. The Pb isotopic values were calculated with the muscovite ^{40}Ar-^{39}Ar dating result of 150.0 ± 0.6 Ma using single-stage Pb isotopic evolution model [77], with values of Δβ ranging from 26.09 to 36.06, Δγ values from 48.95 to 58.62, and μ values from 9.81 to 9.99, respectively.

6. Discussion

6.1. Timing of Mineralization and Granitic Magmatism

In order to constrain the time interval of mineralization in the Xitian W-Sn deposit, several studies have been carried out using various dating technologies [19,40–42,78]. It was first reported the muscovite ^{40}Ar-^{39}Ar isotopic ages of ore-bearing quartz vein in skarn and greisen in the Longshang deposit, with ages of 155.6 ± 1.3 Ma and 157.2 ± 1.4 Ma, respectively [41]. Then, some authors obtained the metallogenic age of other deposits in the Xitian ore field, such as Heshuxia (molybdenite Re-Os age of 150.0 ± 2.7 Ma; [78]), Shantian (molybdenite Re-Os age of 158.9 ± 2.2 Ma; [40]), and Hejiangkou (molybdenite Re-Os age of 225.5 ± 3.6 Ma; [43]). An overwhelming majority of dating minerals are collected from ore-bearing quartz vein and/or greisen with ages ranging from 159 Ma to 149 Ma which has a congruent relationship with the early Yanshanian granitic magmatism [19,40–42,78]. It seems that the quartz vein and/or greisen type ore bodies in the Xitian W-Sn deposit is close to the early Yanshanian magmatic hydrothermal activities. It was reported a molybdenite Re-Os age of 225.5 ± 3.6 Ma of altered granite in the Hejiangkou deposit, regarded as the proof for the Indosinian mineralization in Xitian ore field [43]. Here, we reported a muscovite ^{40}Ar-^{39}Ar isotopic age of the quartz vein type ore body (156.0 ± 0.7 Ma) in the Hejinagkou deposit of the Xitian ore field, indicating that the quartz vein type W-Sn mineralization in this deposit is closely linked to Yanshanian granitic magmatic activities. However, there is still no direct chronologic evidence to seek an answer for the skarn-type mineralization and restricting the genesis of the skarn-type mineralization in this area. We reported the muscovite ^{40}Ar-^{39}Ar isotopic ages of ore-bearing quartz vein for the Hejiangkou deposit and Heshuxia deposit, and two hydrothermal zircon U-Pb ages for the altered granites are 156.6 ± 0.7 Ma, 149.5 ± 0.8 Ma, 152.8 ± 1.1 Ma, and 226.0 ± 2.8 Ma, respectively. Together with the ages published and obtained in this study, we draw a conclusion that there two phases of metallogenic events have been recorded with the time interval of the quartz vein and greisen type W-Sn mineralization ranging from 159 to 149 Ma and skarn-type W-Sn mineralization ca. 225 Ma in the Xitian ore field, respectively.

The Yanshanian period is a significant time interval of granitic magmatism and W-Sn metallogenesis in the Nanling region, which was regarded as the period of mineralization explosion in South China [79]. However, compared with the Yanshanian W-Sn mineralization event in the Nanling region, the Indosinian W-Sn mineralization event is inconspicuous (Table S6 of Supplementary Materials). Although only a few deposits were determined to be the products of Indosinian magmatic hydrothermal events, such as Shuiyuanshan (220.6 ± 1.1 Ma; [31]), Yeziwo (227.2 ± 1.5; [31]), Hehuaping (224.0 ± 1.9 Ma; [32]), and Limu (214.1 ± 1.9 Ma; [33]), it indicates that the Indosinian is also an important epoch for the W-Sn mineralization in South China.

6.2. Physico-Chemical Conditions of the Ore-Forming Fluids

Recently, several studies proved that chemical compositions of zircons could be used as a valid tracer to reflect the physico-chemical conditions of the magmatic melt and/or hydrothermal fluid [80–84]. Specifically, the positive Ce anomalies in zircon are the result of the oxidation of Ce^{3+} to Ce^{4+}. Ce^{4+} is compatible in zircon and can substitute for Zr^{4+}, Hf^{4+} and other tetravalent elements in zircon lattice. Thus, Ce anomalies can provide information for the oxidation state of magma and/or related fluid, where higher Ce/Ce* ratios are in accordance with the high oxygen fugacity (fO_2; [83–85]. However, zircon Eu anomalies are not the efficient tracer, because Eu anomalies in zircon are controlled not only by the redox state of the fluids but also by the crystallization history of plagioclase [86,87]. Due to the relatively low ratios of Ce/Ce* for the zircons from sample 19-4s1 and 24-15s1, almost all the zircon grains plot in the field of low fO_2 indicating that these formed in a lower fO_2 environment (Figure 11a). Using the model proposed by [83], we also estimate the redox conditions for the samples 19-4s1 and 24-15s1 (Table S2 of Supplementary Materials), yielding the relatively low oxygen fugacities with log fO_2 values ranging from −19 to −15 (mean = −17) and from −19 to −13 (mean = −15), respectively. This data further confirmed that both the Indosinian and Yanshanian hydrothermal fluids are reducing fluids which are favorable for the W-Sn mineralization [88–91].

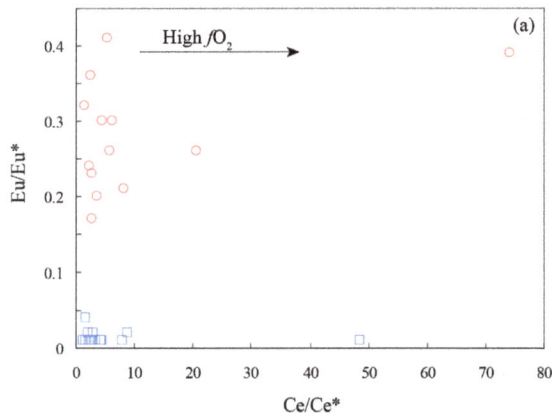

Figure 11. *Cont.*

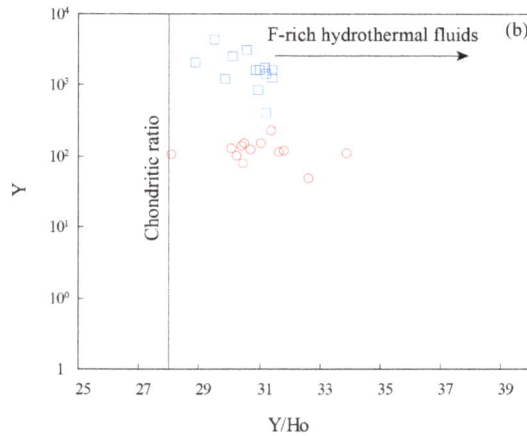

Figure 11. (**a**) Ce/Ce* versus Eu/Eu*; and (**b**) Y/Ho versus Y plots of zircon grains from the Xitian ore field. (**b**) is modified from [82]. Symbols are the same as those in Figure 7.

As an efficient tracer, Y/Ho ratios can provide evidence about the chemical characteristics of source fluids [81]. Fractionation between the Y and Ho occurs in highly evolved granitic melts or hydrothermally altered granites with participation of F-rich fluids which contain high concentrations of Y, Li, B, and/or P [81,82,92,93]. Zircon grains of the sample 19-4s1 and 24-15s1 are ranging from 28 to 34 (mean = 31) and from 29 to 31 (mean = 31), respectively, which are higher than the chondritic value of 28 (Figure 11a; [94]). The high ratios of zircons from these two samples suggest that they were crystallized in F-rich fluids, which are consistent with the existence of abundant fluorite ore bodies (Figure 11b).

6.3. Source of Ore-Forming Metals

On account of the extremely similar geochemical behavior, the REEs are always involved in the geological process in group, and the hydrothermal metamorphism will not change the composition mode and distribution mode of REEs in minerals or rocks, making them efficient tracers for determining the source rocks and element migration mechanism in ore-forming processes [95–98]. According to the chondrite-normalized REE patterns of Xitian skarns, they all show a good consistency, exhibiting right-dipping V-type curves with obvious negative Eu anomalies. Furthermore, the REE patterns of Xitian skarns are extremely similar to those of Xitian Indosinian granites, which are distinctly different to Yanshanian granites (Figure 10b; [28,38]). It indicates that the origin of skarn has a genetic relationship with Indosinian granites rather than Yanshanian granites. The Indosinian granitoids of the Xitian pluton are high-Si, high-K, weakly to strongly peraluminous, and highly fractionated S-type granites with high initial ^{87}Sr/^{86}Sr isotope ratios (0.71397–0.71910), low ε_{Nd}(t) values ranging from -10.1 to -9.4, and old Nd model ages (1858–1764 Ma), indicating that the Xitian Indosinian granites were mainly originated from partial melting of Paleoproterozoic metamorphic basement with small amounts of mantle-derived magma involved [38]. Since the Indosinian granites are closely related to the skarn type W-Sn mineralization, it can be inferred that the source of ore-forming metals from skarn type ore bodies should be mainly originated from a crustal source.

Radiogenic Sr, Nd, and Pb isotopes are powerful tools not only to determine magma sources, but also to determine ore-forming metals in minerals [99–103]. As the most important tungsten-bearing mineral in the tungsten deposit, geochemical compositions and mineralogical features of wolframite can provide abundant metallogenetic information, i.e., fluid, environment, and metal source [1,5,44,104]. The samples of wolframite analyzed in this study have high initial ^{87}Sr/^{86}Sr ratios

(0.71282–0.72003), indicating that the wolframites in quartz vein type ore bodies are characterized by a crustal source. Furthermore, these samples also have negative $\varepsilon_{Nd}(t)$ values ranging from −15.9 to −5.6 and cover a wide range of model ages (T_{DM2} = 2168 − 1399 Ma). In addition, almost all the samples are plotted near the evolution line of the Upper Continental Crust (UCC), and two samples are plotted in the field of the Meso-Paleoproterozoic low mature basement (Figure 12a), indicating that the ore-forming metals are mainly originated from the crustal source, which are proved by the diagram of age versus $\varepsilon_{Nd}(t)$ (Figure 12b). Lead isotopic compositions of wolframites from the Xitian deposit are characterized by high radiogenic Pb isotope values with $^{206}Pb/^{204}Pb$ ratios of 18.489–18.569, $^{207}Pb/^{204}Pb$ ratios of 15.724–15.877, and $^{208}Pb/^{204}Pb$ ratios of 39.055–39.335, and high values of μ ranging from 9.70 to 9.99, indicating that Pb was derived from the upper crust. Additionally, the Pb isotopic compositions of the wolframites are similar to the sulfides of the quartz vein type ore bodies, and almost all the samples are plotted primarily toward the upper crust evolution field and/or line in the $\Delta\beta$ versus $\Delta\gamma$ and $^{206}Pb/^{204}Pb$ versus $^{207}Pb/^{204}Pb$ diagrams (Figure 12c,d). Briefly, based on the diagenetic and metallogenic geochronology, the relationship between the metallogenic and magmatic activities, and the evidence of mineral isotopes, we deduce that the ore-forming metals of quartz vein/greisen type ore bodies were also derived from a crustal source.

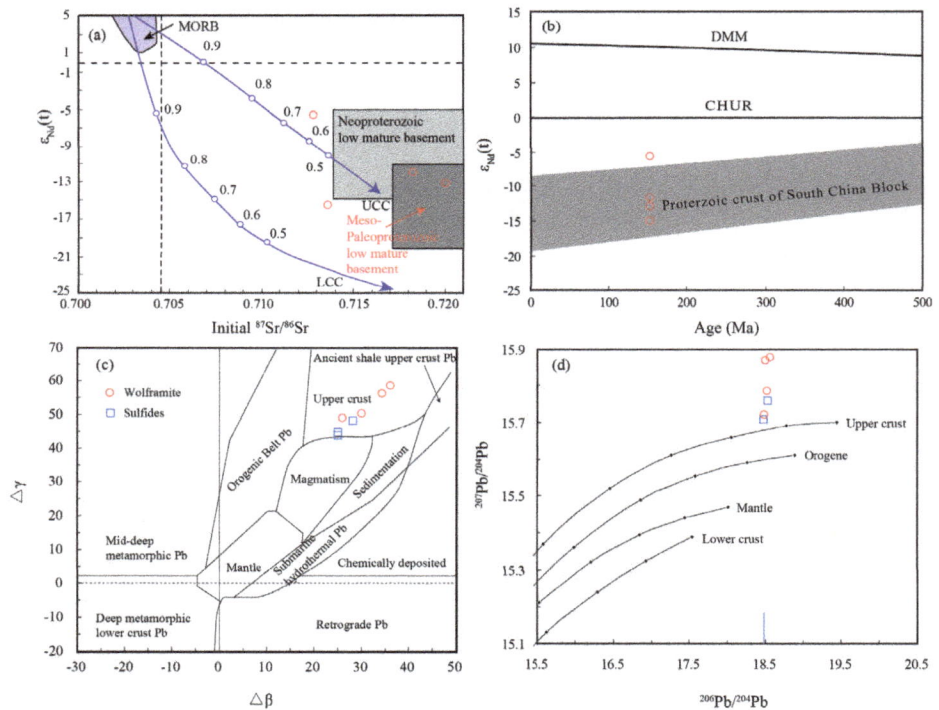

Figure 12. (a) Initial $^{87}Sr/^{86}Sr$ ratios versus $\varepsilon_{Nd}(t)$; and (b) Age versus $\varepsilon_{Nd}(t)$ plots for the wolframite from the Xitian ore field; (c) $\Delta\beta$ versus $\Delta\gamma$; and (d) $^{206}Pb/^{204}Pb$ versus $^{207}Pb/^{204}Pb$ diagrams for the wolframite from the Xitian ore field. (a–d) are modified from [59,78,105,106]. Data of sulfides are from [107]. MORB: Mid Ocean Ridge Basalt; DMM: depleted end-member; CHUR: chondritic uniform reservoir.

7. Conclusions

1. Hydrothermal zircon U-Pb and muscovite ^{40}Ar/^{39}Ar dating suggests that there are two epochs of W-Sn mineralization in the Xitian ore field, with skarn-type W-Sn mineralization at ca. 226 Ma and quartz vein/greisen type W-Sn mineralization at ca. 156 Ma.

2. The ore-forming fluids for the two metallogenic events are both characterized by enrichment in F and low oxygen fugacities.

3. The ore-forming metals for the skarn-type and quartz vein/greisen type W-Sn mineralization are both originated from a crust source.

Supplementary Materials: The following are available online at http://www.mdpi.com/2075-163x/8/3/111/s1, Table S1: LA-ICP-MS zircon U-Pb isotopic compositions of altered granite in Xitian W–Sn ore field, Table S2: LA-ICP-MS zircon trace element compositions of altered granite in Xitian W-Sn ore field (ppm), Table S3: ^{40}Ar/^{39}Ar laser stepwise heating analytical data for two muscovite samples from the Xitian W-Sn ore field, Table S4: Major and trace element compositions of the skarn in Xitian W-Sn ore field, Table S5: Sr-Nd-Pb isotopic compositions of wolframite in Xitian W-Sn ore field, Table S6: Synthesis of the metallogenic ages of the W-Sn deposits associated with the granitic pluton in the Nanling range.

Acknowledgments: This study was financially supported by the China Geological Survey (Grant No. 12120114052101), the National Key R&D Program of China (No. 2016YFC0600404) and the National Natural Science Foundation of China (Nos. 41673040, 41502067). The manuscript has been greatly benefited from constructive comments by Editors-in-Chief Paul Sylvester, and other two anonymous reviewers. It has also been significantly improved in both English and science by Lei Liu of the University of Science and Technology of China. Editor Ms. Queenie Wang is also appreciated for their helpful suggestions.

Author Contributions: Jingya Cao, and Qianhong Wu conceived and designed the experiments; Hua Kong and Xiaoshuang Xi took part to the field campaigns; Huan Li took part in the discussion; Qianfeng Huang and Biao Liu analyzed the data; Jingya Cao, Qianhong Wu and Xiaoyong Yang wrote the paper.

Conflicts of Interest: The authors declare no conflict of interest.

References

1. Kempe, U.; Wolf, D. Anomalously high Sc contents in ore minerals from Sn-W deposits: Possible economic significance and genetic implications. *Ore Geol. Rev.* **2006**, *28*, 103–122. [CrossRef]

2. Mao, J.W.; Xie, G.Q.; Guo, C.L.; Chen, Y.C. Large-scale tungsten-tin mineralization in the Nanling region, South China: Metallogenic ages and corresponding geodynamic processes. *Acta Petrol. Sin.* **2007**, *23*, 2329–2338. (In Chinese)

3. Mao, J.W.; Xie, G.Q.; Guo, C.L.; Yuan, S.D.; Cheng, Y.B.; Chen, Y.C. Spatial-temporal distribution of Mesozoic ore deposits in south China and their metallogenic settings. *Geol. J. Chin. Univ.* **2008**, *14*, 510–526. (In Chinese)

4. Mao, J.; Pirajno, F.; Cook, N. Mesozoic metallogeny in East China and corresponding geodynamic settings-An introduction to the special issue. *Ore Geol. Rev.* **2011**, *43*, 1–7. [CrossRef]

5. Neiva, A.M.R. Geochemistry of cassiterite and wolframite from tin and tungsten quartz veins in Portugal. *Ore Geol. Rev.* **2008**, *33*, 221–238. [CrossRef]

6. Moura, A.; Dória, A.; Neiva, A.M.R.; Leal Gomes, C.; Creaser, R.A. Metallogenesis at the Carris W-Mo-Sn deposit (Gerês, Portugal): Constraints from fluid inclusions, mineral geochemistry, Re-Os and He-Ar isotopes. *Ore Geol. Rev.* **2014**, *56*, 73–93. [CrossRef]

7. Cheng, Y.; Mao, J.; Liu, P. Geodynamic setting of Late Cretaceous Sn-W mineralization in southeastern Yunnan and northeastern Vietnam. *Solid Earth Sci.* **2016**, *1*, 79–88. [CrossRef]

8. Chicharro, E.; Boiron, M.; López-García, J.Á.; Barfod, D.N.; Villaseca, C. Origin, ore forming fluid evolution and timing of the Logrosán Sn-(W) ore deposits (Central Iberian Zone, Spain). *Ore Geol. Rev.* **2016**, *72*, 896–913. [CrossRef]

9. Liu, P.; Mao, J.; Cheng, Y.; Yao, W.; Wang, X.; Hao, D. An Early Cretaceous W-Sn deposit and its implications in southeast coastal metallogenic belt: Constraints from U-Pb, Re-Os, Ar-Ar geochronology at the Feie'shan W-Sn deposit, SE China. *Ore Geol. Rev.* **2017**, *81*, 112–122. [CrossRef]

10. Soloviev, S.G.; Kryazhev, S.G.; Dvurechenskaya, S.S. Geology, mineralization, stable isotope, and fluid inclusion characteristics of the Vostok-2 reduced W-Cu skarn and Au-W-Bi-As stockwork deposit, Sikhote-Alin, Russia. *Ore Geol. Rev.* **2017**, *86*, 338–365. [CrossRef]

11. Soloviev, S.G.; Kryazhev, S.G.; Dvurechenskaya, S.S. Geology, mineralization, and fluid inclusion characteristics of the Lermontovskoe reduced-type tungsten (Cu, Au, Bi) skarn deposit, Sikhote-Alin, Russia. *Ore Geol. Rev.* **2017**, *89*, 15–39. [CrossRef]

12. Zhang, Y.; Yang, J.; Chen, J.; Wang, H.; Xiang, Y. Petrogenesis of Jurassic tungsten-bearing granites in the Nanling Range, South China: Evidence from whole-rock geochemistry and zircon U-Pb and Hf-O isotopes. *Lithos* **2017**, *278*, 166–180. [CrossRef]

13. Zhao, W.W.; Zhou, M.; Li, Y.H.M.; Zhao, Z.; Gao, J. Genetic types, mineralization styles, and geodynamic settings of Mesozoic tungsten deposits in South China. *J. Asian Earth Sci.* **2017**, *137*, 109–140. [CrossRef]

14. U.S. Geological Survey. *Mineral Commodity Summaries 2016*; U.S. Geological Survey: Reston, VA, USA, 2016; p. 202.

15. Wang, D.H.; Chen, Y.C.; Chen, Z.H.; Liu, S.B.; Xu, J.X.; Zhang, J.J.; Zeng, Z.L.; Chen, F.W.; Li, H.Q.; Guo, C.L. Assessment on mineral resource in Nanling region and suggestion for further prospecting. *Acta Geol. Sin.* **2007**, *81*, 882–890. (In Chinese)

16. Chen, J.; Wang, R.; Zhu, J.; Lu, J.; Ma, D. Multiple-aged granitoids and related tungsten-tin mineralization in the Nanling Range, South China. *Sci. China-Earth Sci.* **2013**, *56*, 2045–2055. (In Chinese) [CrossRef]

17. Guo, C.; Mao, J.; Bierlein, F.; Chen, Z.; Chen, Y.; Li, C.; Zeng, Z. SHRIMP U-Pb (zircon), Ar-Ar (muscovite) and Re-Os (molybdenite) isotopic dating of the Taoxikeng tungsten deposit, South China Block. *Ore Geol. Rev.* **2011**, *43*, 26–39. [CrossRef]

18. Hu, R.; Wei, W.; Bi, X.; Peng, J.; Qi, Y.; Wu, L.; Chen, Y. Molybdenite Re-Os and muscovite Ar-40/Ar-39 dating of the Xihuashan tungsten deposit, central Nanling district, South China. *Lithos* **2012**, *150*, 111–118. [CrossRef]

19. Liang, X.; Dong, C.; Jiang, Y.; Wu, S.; Zhou, Y.; Zhu, H.; Fu, J.; Wang, C.; Shan, Y. Zircon U-Pb, molybdenite Re-Os and muscovite Ar-Ar isotopic dating of the Xitian W-Sn polymetallic deposit, eastern Hunan Province, South China and its geological significance. *Ore Geol. Rev.* **2016**, *78*, 85–100. [CrossRef]

20. Mao, J.; Cheng, Y.; Chen, M.; Pirajno, F. Major types and time-space distribution of Mesozoic ore deposits in South China and their geodynamic settings. *Miner. Deposita* **2013**, *48*, 267–294.

21. Peng, J.; Zhou, M.; Hu, R.; Shen, N.; Yuan, S.; Bi, X.; Du, A.; Qu, W. Precise molybdenite Re-Os and mica Ar-Ar dating of the Mesozoic Yaogangxian tungsten deposit, central Nanling district, South China. *Miner. Deposita* **2006**, *41*, 661–669. [CrossRef]

22. Yao, Y.; Chen, J.; Lu, J.; Wang, R.; Zhang, R. Geology and genesis of the Hehuaping magnesian skarn-type cassiterite-sulfide deposit, Hunan Province, Southern China. *Ore Geol. Rev.* **2014**, *58*, 163–184. [CrossRef]

23. Chen, Y.; Li, H.; Sun, W.; Ireland, T.; Tian, X.; Hu, Y.; Yang, W.; Chen, C.; Xu, D. Generation of Late Mesozoic Qianlishan A2-type granite in Nanling Range, South China: Implications for Shizhuyuan W-Sn mineralization and tectonic evolution. *Lithos* **2016**, *266–267*, 435–452. [CrossRef]

24. Yuan, S.; Peng, J.; Hu, R.; Li, H.; Shen, N.; Zhang, D. A precise U-Pb age on cassiterite from the Xianghualing tin-polymetallic deposit (Hunan, South China). *Miner. Deposita* **2008**, *43*, 375–382. [CrossRef]

25. Zhang, W.; Hua, R.; Wang, R.; Chen, P.; Li, H. New dating of Dajishan granite and related tungsten mineralization, South Jiangxi Province, China. *Front. Earth Sci.* **2007**, *1*, 218–225. [CrossRef]

26. Dong, S.H.; Bi, X.W.; Hu, R.Z.; Chen, Y.W. Petrogenesis of Yaogangxian granites and inplications for W mineralization, Hunan province. *Acta Petrol. Sin.* **2014**, *30*, 2749–2765. (In Chinese)

27. Guo, C.; Chen, Y.; Zeng, Z.; Lou, F. Petrogenesis of the Xihuashan granites in southeastern China: Constraints from geochemistry and in-situ analyses of zircon U-Pb-Hf-O isotopes. *Lithos* **2012**, *148*, 209–227. [CrossRef]

28. Zhou, Y.; Liang, X.; Wu, S.; Cai, Y.; Liang, X.; Shao, T.; Wang, C.; Fu, J.; Jiang, Y. Isotopic geochemistry, zircon U-Pb ages and Hf isotopes of A-type granites from the Xitian W-Sn deposit, SE China: Constraints on petrogenesis and tectonic significance. *J. Asian Earth Sci.* **2015**, *105*, 122–139. [CrossRef]

29. Li, X.H.; Liu, D.Y.; Sun, M.; Li, W.X.; Liang, X.R.; Liu, Y. Precise Sm-Nd and U-Pb isotopic dating of the supergiant Shizhuyuan polymetallic deposit and its host granite, SE China. *Geol. Mag.* **2004**, *141*, 225–231. [CrossRef]

30. Zhang, S.M.; Chen, Z.H.; Shi, G.H.; Li, L.X.; Qu, W.J.; Li, C. Re-Os isotopic dating of molybdenite from Dajishan tungsten deposit in Jiangxi Province. *Miner. Depos.* **2011**, *30*, 1113–1121. (In Chinese)

31. Zhang, R.; Lu, J.; Wang, R.; Yang, P.; Zhu, J.; Yao, Y.; Gao, J.; Li, C.; Lei, Z.; Zhang, W. Constraints of in situ zircon and cassiterite U-Pb, molybdenite Re-Os and muscovite Ar-40-Ar-39 ages on multiple generations of granitic magmatism and related W-Sn mineralization in the Wangxianling area, Nanling Range, South China. *Ore Geol. Rev.* **2015**, *65*, 1021–1042. [CrossRef]

32. Cai, M.H.; Chen, K.X.; Chen, K.X.; Liu, G.Q.; Fu, J.M.; Yin, J.P. Geological characteristics and Re-Os dating of molybdenites in Hehuaping tin-polymetallic deposit, Southern Hunan Province. *Miner. Depos.* **2006**, *25*, 263–268. (In Chinese)

33. Yang, F.; Li, X.F.; Feng, Z.H.; Bai, Y. 40Ar/39Ar dating of muscovite from greisenized granite and geological significance in Limu tin deposit. *J. Guilin Univ. Technol.* **2009**, *29*, 21–24. (In Chinese)

34. Cai, Y.; Ma, D.S.; Lu, J.J.; Huang, H.; Zhang, R.Q.; Qu, W.J. Re-Os geochronology and S isotope geochemistry of Dengfuxian tungsten deposit, Hunan Province, China. *Acta Petrol. Sin.* **2012**, *28*, 3798–3808. (In Chinese)

35. Li, X.H.; Chung, S.L.; Zhou, H.W.; Lo, C.H.; Liu, Y.; Chen, C.W. Jurassic intraplate magmatism in southern Hunan-eastern Guangxi: Ar-40/Ar-39 dating, geochemistry, Sr-Nd isotopes and implications for the tectonic evolution of SE China. *Geol. Soc. Spec. Publ.* **2004**, *226*, 193–215. [CrossRef]

36. Li, X.; Xiao, R.; Feng, Z.; Chunxia, W.; Tang, Y.; Bai, Y.; Zhang, M. Ar-Ar ages of hydrothermal muscovite and igneous biotite at the Guposhan-Huashan district, Northeast Guangxi, South China: Implications for mesozoic W-Sn mineralization. *Resour. Geol.* **2015**, *65*, 160–176. [CrossRef]

37. Yuan, S.; Peng, J.; Shen, N.; Hu, R.; Dai, T. Ar-40-Ar-39 isotopic dating of the Xianghualing Sn-polymetallic ore field in southern Hunan, China and its geological implications. *Acta Geol. Sin. Engl.* **2007**, *81*, 278–286.

38. Wu, Q.; Cao, J.; Kong, H.; Shao, Y.; Li, H.; Xi, X.; Deng, X. Petrogenesis and tectonic setting of the early Mesozoic Xitian granitic pluton in the middle Qin-Hang Belt, South China: Constraints from zircon U-Pb ages and bulk-rock trace element and Sr-Nd-Pb isotopic compositions. *J. Asian Earth Sci.* **2016**, *128*, 130–148. [CrossRef]

39. Xiong, Y.; Shao, Y.; Zhou, H.; Wu, Q.; Liu, J.; Wei, H.; Zhao, R.; Cao, J. Ore-forming mechanism of quartz-vein-type W-Sn deposits of the Xitian district in SE China: Implications from the trace element analysis of wolframite and investigation of fluid inclusions. *Ore Geol. Rev.* **2017**, *83*, 152–173. [CrossRef]

40. Guo, C.L.; Li, C.; Wu, S.C.; Xu, Y.M. Molybdenite Re-Os Isotopic Dating of Xitian Deposit in Hunan Province and Its Geological Significance. *Rock Miner. Anal.* **2014**, *33*, 142–152. (In Chinese)

41. Ma, L.Y.; Fu, J.M.; Wu, S.C.; Xu, D.M.; Yang, X.J. 40Ar/39Ar isotopic dating of the Longshang tin-polymetallic deposit, Xitian ore field, eastern Hunan. *Geol. Chin.* **2008**, *35*, 706–713. (In Chinese)

42. Wang, M.; Bai, X.J.; Hu, R.G.; Cheng, S.B.; Pu, Z.P.; Qiu, H.N. Direct dating of cassiterite in Xitian Tungsten-Tin polymetallic deposit, South-East Hunan, by 40Ar/39Ar Progressive Crushing. *Geotecton. Metallog.* **2015**, *39*, 1049–1060. (In Chinese)

43. Deng, X.W.; Liu, J.X.; Dai, X.L. Geological characteristics and molybdenite Re-Os isotopic age of Hejiangkou tungsten and tin polymetallic deposit, East Hunan, China. *Chin. J. Nonferrous Met.* **2015**, *25*, 2883–2897. (In Chinese)

44. Xiong, Y.Q.; Shao, Y.J.; Liu, J.P.; Wei, H.T.; Zhao, R.C. Ore-forming fluid of quartz-vein type tungsten deposits, Xitian ore field, eastern Hunan, China. *Chin. J. Nonferrous Met.* **2016**, *26*, 1107–1119. (In Chinese)

45. Zhao, J.; Zhou, M.; Yan, D.; Zheng, J.; Li, J. Reappraisal of the ages of Neoproterozoic strata in South China: No connection with the Grenvillian orogeny. *Geology* **2011**, *39*, 299–302. [CrossRef]

46. Yu, C. The characteristic target-pattern regional ore zonality of the Nanling region, China (I). *Geosci. Front.* **2011**, *2*, 147–156. [CrossRef]

47. Mao, J.; Takahashi, Y.; Kee, W.; Li, Z.; Ye, H.; Zhao, X.; Liu, K.; Zhou, J. Characteristics and geodynamic evolution of Indosinian magmatism in South China: A case study of the Guikeng pluton. *Lithos* **2011**, *127*, 535–551. [CrossRef]

48. Mao, J.; Ye, H.; Liu, K.; Li, Z.; Takahashi, Y.; Zhao, X.; Kee, W. The Indosinian collision-extension event between the South China Block and the Palaeo-Pacific plate: Evidence from Indosinian alkaline granitic rocks in Dashuang, eastern Zhejiang, South China. *Lithos* **2013**, *172–173*, 81–97. [CrossRef]

49. Niu, Y.; Liu, Y.; Xue, Q.; Shao, F.; Chen, S.; Duan, M.; Guo, P.; Gong, H.; Hu, Y.; Hu, Z. Exotic origin of the Chinese continental shelf: New insights into the tectonic evolution of the western Pacific and eastern China since the Mesozoic. *Sci. Bull.* **2015**, *60*, 1598–1616. [CrossRef]

50. Wang, Y.; Fan, W.; Sun, M.; Liang, X.; Zhang, Y.; Peng, T. Geochronological, geochemical and geothermal constraints on petrogenesis of the Indosinian peraluminous granites in the South China Block: A case study in the Hunan Province. *Lithos* **2007**, *96*, 475–502. [CrossRef]

51. Wang, Y.; Fan, W.; Zhang, G.; Zhang, Y. Phanerozoic tectonics of the South China Block: Key observations and controversies. *Gondwana Res.* **2013**, *23*, 1273–1305. [CrossRef]

52. Yan, Q.; Shi, X.; Castillo, P.R. The late Mesozoic-Cenozoic tectonic evolution of the South China Sea: A petrologic perspective. *J. Asian Earth Sci.* **2014**, *85*, 178–201. [CrossRef]

53. Zhang, G.; Guo, A.; Wang, Y.; Li, S.; Dong, Y.; Liu, S.; He, D.; Cheng, S.; Lu, R.; Yao, A. Tectonics of South China continent and its implications. *Sci. China Earth Sci.* **2013**, *56*, 1804–1828. [CrossRef]

54. Hua, R.M.; Chen, P.R.; Zhang, W.L.; Liu, X.D.; Lu, J.J.; Lin, J.F.; Yao, J.M.; Ql, H.W.; Zhang, Z.; Gu, S.Y. Metallogenic systems related to Mesozoic and Cenozoic granitoids in South China. *Sci. China Earth Sci.* **2003**, *46*, 816–829. [CrossRef]

55. Hua, R.M.; Zhang, W.L.; Gu, S.Y.; Chen, P.R. Comparison between REE granites and W-Sn granite in the Nanling region, South China, and their mineralizations. *Acta Petrol. Sin.* **2007**, *23*, 2321–2328. (In Chinese)

56. Fan, W.M.; Wang, Y.J.; Guo, F.; Peng, T.P. Mesozoic mafic magmatism in Hunan-Jiangxi province and the lithospheric extension. *Earth Sci. Front.* **2003**, *10*, 159–169. (In Chinese)

57. Li, J.G.; Li, J.K.; Wang, D.H.; Liu, J.; He, H.H. The deep tectonic features of Qitianling ore-concentrated area in Southern Hunan province and its contrains to the regional ore-forming process. *Acta Geol. Sin.* **2014**, *88*, 695–703. (In Chinese)

58. Cai, Y.; Lu, J.; Ma, D.; Huang, H.; Zhang, H.; Zhang, R. The Late Triassic Dengfuxian A-type granite, Hunan Province: Age, petrogenesis, and implications for understanding the late Indosinian tectonic transition in South China. *Int. Geol. Rev.* **2015**, *57*, 428–445. [CrossRef]

59. Guo, C.; Zeng, L.; Li, Q.; Fu, J.; Ding, T. Hybrid genesis of Jurassic fayalite-bearing felsic subvolcanic rocks in South China: Inspired by petrography, geochronology, and Sr-Nd-O-Hf isotopes. *Lithos* **2016**, *264*, 175–188. [CrossRef]

60. Wu, S.C.; Long, Z.Q.; Xu, H.H.; Zhou, Y.; Jiang, Y.; Pan, C.C. Structural characteristics and prospecting significance of the Xitian tin-tungsten polymetallic deposit, Hunan province, China. *Geotecton. Metallog.* **2012**, *36*, 217–226. (In Chinese)

61. Bureau of Geology and Mineral Resources of Hunan Province. *Regional Geology of the Hunan Province*; Geological Publishing House: Beijing, China, 1988; p. 507. (In Chinese)

62. Wang, Y.J.; Fan, W.M.; Guo, F.; Peng, T.P.; Li, C.W. Geochemistry of Mesozoic mafic rocks adjacent to the Chenzhou-Linwu fault, South China: Implications for the lithospheric boundary between the Yangtze and Cathaysia blocks. *Int. Geol. Rev.* **2003**, *45*, 263–286. [CrossRef]

63. Chen, D.; Ma, A.J.; Liu, W.; Liu, Y.R.; Ni, Y.J. Research on U-Pb Chronology in Xitian pluton of Hunan province. *Geoscience* **2013**, *27*, 819–830. (In Chinese)

64. Wiedenbeck, M.; Alle, P.; Corfu, F.; Griffin, W.L.; Meier, M.; Ober, F.; Von Quadt, A.; Roddick, J.C.; Speigel, W. Three natural zircon standards for U-Th-Pb, Lu-Hf, trace-element and REE analyses. *Geostand. Geoanal. Res.* **1995**, *19*, 1–23. [CrossRef]

65. Elhlou, S.; Belousova, E.; Griffin, W.L.; Pearson, N.J.; O Reilly, S.Y. Trace element and isotopic composition of GJ-red zircon standard by laser ablation. *Geochim. Cosmochim. Acta* **2006**, *70*, A158. [CrossRef]

66. Liu, Y.; Gao, S.; Hu, Z.; Gao, C.; Zong, K.; Wang, D. Continental and oceanic crust recycling-induced melt-peridotite interactions in the trans-north china orogen: U-Pb dating, Hf isotopes and trace elements in Zircons from mantle xenoliths. *J. Petrol.* **2010**, *51*, 537–571. [CrossRef]

67. Andersen, T. Correction of common lead in U-Pb analyses that do not report 204Pb. *Chem. Geol.* **2002**, *192*, 59–79. [CrossRef]

68. Ludwig, K.R. *ISOPLOT 3.00: A Geochronological Toolkit for Microsoft Excel*; Berkeley Geochronology Center: Berkeley, CA, USA, 2003; p. 39.

69. Wang, S.S. Age determinations of 40Ar-40K, 40Ar-40Ar and radiogenic 40Ar released characteristics on K-Ar geostandards of China. *Chin. J. Geol.* **1983**, *4*, 315–323. (In Chinese)

70. Qiu, H.N.; Bai, X.J.; Liu, W.G.; Mei, L.F. Automatic 40Ar/39Ar dating technique using multicollector ArgusVI MS with home-made apparatus. *Geochimica* **2015**, *44*, 477–484. (In Chinese)

71. Koppers, A. ArArCALC-software for Ar-40/Ar-39 age calculations. *Comput. Geosci.* **2002**, *28*, 605–619. [CrossRef]

72. Chen, F.; Siebel, W.; Satir, M.; Lu, M.T.; Saka, K. Geochronology of the Karadere basement (NW Turkey) and implications for the geological evolution of the Istanbul zone. *Int. J. Earth Sci.* **2002**, *91*, 469–481. [CrossRef]

73. Hoskin, P.; Schaltegger, U. The composition of zircon and igneous and metamorphic petrogenesis. *Rev. Miner. Geochem.* **2003**, *53*, 27–62. [CrossRef]

74. Taylor, S.R.; McLennan, S.M. *Continental Crust: Its Composition and Evolution: An Examination of the Geochemical Record Preserved in Sedimentary Rocks*; Blackwell Scientific: Hoboken, NJ, USA, 1985; p. 312.

75. Hoskin, P. Trace-element composition of hydrothermal zircon and the alteration of Hadean zircon from the Jack Hills, Australia. *Geochim. Cosmochim. Acta* **2005**, *69*, 637–648. [CrossRef]

76. Sun, S.S.; McDonough, W.F. Chemical and isotopic systematics of oceanic basalts: Implications for mantle compositions and processes. In *Magmatism in the Ocean Basins*; Saunders, A.D., Norry, M.J., Eds.; Geological Society, London Special Publication: London, UK, 1989; Volume 32, pp. 313–345.

77. Zartman, R.E.; Doe, B.R. Plumbotectonics-The model. *Tectonophysics* **1981**, *75*, 135–162. [CrossRef]

78. Liu, G.Q.; Wu, S.C.; Du, A.D.; Fu, J.M.; Yang, X.J.; Tang, Z.H.; Wei, J.Q. Metallogenic ages of the Xitian tungsten-tin deposit, Eastern Hunan Province. *Geotecton. Metallog.* **2008**, *32*, 63–71. (In Chinese)

79. Hua, R.M.; Mao, J.W. A preliminary discussion on the Mesozoic metallogenic explosion in East China. *Miner. Depos.* **1999**, *18*, 300–308. (In Chinese)

80. Ferry, J.M.; Watson, E.B. New thermodynamic models and revised calibrations for the Ti-in-zircon and Zr-in-rutile thermometers. *Contrib. Miner. Petrol.* **2007**, *154*, 429–437. [CrossRef]

81. Irber, W. The lanthanide tetrad effect and its correlation with K/Rb, Eu/Eu*, Sr/Eu, Y/Ho, and Zr Hf of evolving peraluminous granite suites. *Geochim. Cosmochim. Acta* **1999**, *63*, 489–508. [CrossRef]

82. Kozlik, M.; Raith, J.G.; Gerdes, A. U-Pb, Lu-Hf and trace element characteristics of zircon from the Felbertal scheelite deposit (Austria): New constraints on timing and source of W mineralization. *Chem. Geol.* **2016**, *421*, 112–126. [CrossRef]

83. Trail, D.; Watson, E.B.; Tailby, N.D. Ce and Eu anomalies in zircon as proxies for the oxidation state of magmas. *Geochim. Cosmochim. Acta* **2012**, *97*, 70–87. [CrossRef]

84. Trail, D.; Watson, E.B.; Tailby, N.D. The oxidation state of Hadean magmas and implications for early Earth's atmosphere. *Nature* **2011**, *480*, 79–238. [CrossRef] [PubMed]

85. Li, H.; Watanabe, K.; Yonezu, K. Zircon morphology, geochronology and trace element geochemistry of the granites from the Huangshaping polymetallic deposit, South China: Implications for the magmatic evolution and mineralization processes. *Ore Geol. Rev.* **2014**, *60*, 14–35. [CrossRef]

86. Burnham, A.D.; Berry, A.J. An experimental study of trace element partitioning between zircon and melt as a function of oxygen fugacity. *Geochim. Cosmochim. Acta* **2012**, *95*, 196–212. [CrossRef]

87. Kong, D.; Xu, J.; Chen, J. Oxygen isotope and trace element geochemistry of zircons from porphyry copper system: Implications for Late Triassic metallogenesis within the Yidun Terrane, southeastern Tibetan Plateau. *Chem. Geol.* **2016**, *441*, 148–161. [CrossRef]

88. Bhalla, P.; Holtz, F.; Linnen, R.L.; Behrens, H. Solubility of cassiterite in evolved granitic melts: Effect of *T*, *f*O_2, and additional volatiles. *Lithos* **2005**, *80*, 387–400. [CrossRef]

89. Lehmann, B. *Metallogeny of Tin*; Lecture Notes in Earth Sciences Berlin Springer Verlag: Berlin, Germany, 1990; Volume 32, p. 211.

90. Li, X.; Chi, G.; Zhou, Y.; Deng, T.; Zhang, J. Oxygen fugacity of Yanshanian granites in South China and implications for metallogeny. *Ore Geol. Rev.* **2017**, *88*, 690–701. [CrossRef]

91. Linnen, R.L.; Pichavant, M.; Holtz, F. The combined effects of *f*O_2 and melt composition on SnO_2 solubility and tin diffusivity in haplo-granitic melts. *Geochim. Cosmochim. Acta* **1996**, *60*, 4965–4976. [CrossRef]

92. Bau, M. Controls on the fractionation of isovalent trace elements in magmatic and aqueous systems: Evidence from Y/Ho, Zr/Hf, and lanthanide tetrad effect. *Contrib. Miner. Petrol.* **1996**, *123*, 323–333. [CrossRef]

93. Veksler, I.V.; Dorfman, A.M.; Kamenetsky, M.; Dulski, P.; Dingwell, D.B. Partitioning of lanthanides and Y between immiscible silicate and fluoride melts, fluorite and cryolite and the origin of the lanthanide tetrad effect in igneous rocks. *Geochim. Cosmochim. Acta* **2005**, *69*, 2847–2860. [CrossRef]

94. Anders, E.; Grevesse, N. Abundances of the elements: Meteoritic and solar. *Geochim. Cosmochim. Acta* **1989**, *53*, 197–214. [CrossRef]

95. Bau, M. Rare-earth element mobility during hydrothermal and metamorphic fluid-rock interaction and the significance of the oxidation state of europium. *Chem. Geol.* **1991**, *93*, 219–230. [CrossRef]

96. Boulvais, P.; Fourcade, S.; Moine, B.; Gruau, G.; Cuney, M. Rare-earth elements distribution in granulite-facies marbles: A witness of fluid-rock interaction. *Lithos* **2000**, *53*, 117–126. [CrossRef]

97. Chen, L.; Qin, K.Z.; Li, G.M.; Xiao, B.; Li, J.X.; Jiang, H.Z.; Chen, J.B.; Zhao, J.X.; Fan, X.; Han, F.J. Geochemical characteristics and origin of skarn rocks in the Nuri Cu-Mo-W deposit, Southern Tibet. *Geol. Explor.* **2011**, *47*, 78–88. (In Chinese)

98. Zamanian, H.; Radmard, K. Geochemistry of rare earth elements in the baba Ali magnetite skarn deposit, western Iran-a key to determine conditions of mineralisation. *Geologos* **2016**, *22*, 33–47. [CrossRef]

99. Jahn, B.; Valui, G.; Kruk, N.; Gonevchuk, V.; Usuki, M.; Wu, J.T.J. Emplacement ages, geochemical and Sr-Nd-Hf isotopic characterization of Mesozoic to early Cenozoic granitoids of the Sikhote-Alin Orogenic Belt, Russian Far East: Crustal growth and regional tectonic evolution. *J. Asian Earth Sci.* **2015**, *111*, 872–918. [CrossRef]

100. Pankhurst, M.J.; Vernon, R.H.; Turner, S.P.; Schaefer, B.F.; Foden, J.D. Contrasting Sr and Nd isotopic behaviour during magma mingling; new insights from the Mannum A-type granite. *Lithos* **2011**, *126*, 135–146. [CrossRef]

101. Volkert, R.A.; Feigenson, M.D.; Mana, S.; Bolge, L. Geochemical and Sr-Nd isotopic constraints on the mantle source of Neoproterozoic mafic dikes of the rifted eastern Laurentian margin, north-central Appalachians, USA. *Lithos* **2015**, *212–215*, 202–213. [CrossRef]

102. Xiong, D.X.; Sun, X.M.; Shi, G.Y.; Wang, S.W.; Gao, J.F.; Xu, T. Trace elements, rare earth elements (REE) and Nd-Sr isotopic compositions in scheelites and their implications for the mineralization in Daping gold mine in Yunnan province, China. *Acta Petrol. Sin.* **2006**, *22*, 733–741. (In Chinese)

103. Zhang, Z.; Zuo, R. Sr-Nd-Pb isotope systematics of magnetite: Implications for the genesis of Makeng Fe deposit, southern China. *Ore Geol. Rev.* **2014**, *57*, 53–60. [CrossRef]

104. Tindle, A.G.; Webb, P.C. Niobian wolframite from Glen Gairn in the Eastern Highland of Scotland-A microprobe investigation. *Geochim. Cosmochim. Acta* **1989**, *53*, 1921–1935. [CrossRef]

105. Sun, T.; Zhou, X.M.; Chen, P.R.; Li, H.M.; Zhou, H.Y.; Wang, Z.C.; Shen, W.Z. Petrogenesis of Mesozoic strongly peraluminous granites in the Nanling, China. *Sci. China Earth Sci.* **2003**, *33*, 1209–1218. (In Chinese)

106. Zhu, B.Q.; Li, X.H.; Dai, T.M. *Theory and Application of Isotope Systematics in Earth Sciences*; Science Press: Beijing, China, 1998; p. 333. (In Chinese)

107. Yao, Y. Magnesian and Calcic Skarn Type Tin-Polymetallic Mineralization in the Nanling Range: Case Study from Hehuaping and Xitian. Ph.D. Thesis, Nanjing University, Nanjing, China, 2012. (In Chinese)

![minerals logo] *minerals*

MDPI

Article

Geochronology, Petrology, and Genesis of Two Granitic Plutons of the Xianghualing Ore Field in South Hunan Province: Constraints from Zircon U–Pb Dating, Geochemistry, and Lu–Hf Isotopic Compositions

Lizhi Yang [1], Xiangbin Wu [1,*], Jingya Cao [2,*], Bin Hu [1], Xiaowen Zhang [3], Yushuang Gong [4] and Weidong Liu [5]

[1] Key Laboratory of Metallogenic Prediction of Nonferrous Metals and Geological Environment Monitoring, Ministry of Education, Central South University, No. 932, Lushan Road, Changsha 410083, China; yanglizhidz@csu.edu.cn (L.Y.); binhu1999@hotmail.com (B.H.)

[2] CAS Key Laboratory of Crust-Mantle Materials and Environments, University of Science and Technology of China, Hefei 230026, China

[3] School of Environment and Safety Engineering, University of South China, Hengyang 421001, China; zhangxiaowen02@sina.com

[4] Sinomine Resource Exploration Co. Ltd., Beijing 100089, China; gys1016@163.com

[5] No. 7 Institute of Geology and Mineral Exploration of Shandong Province, Linyi 276006, China; wolf1210@sina.com

* Correspondence: Wuxb133@126.com (X.W.); jingyacao@csu.edu.cn (J.C.)

Received: 21 April 2018; Accepted: 12 May 2018; Published: 15 May 2018

Abstract: Two small-sized granitic plutons, outcropped in Xianghualing ore field, South Hunan (South China), have a close relationship with the super large-scale Sn–W polymetallic mineralization in this ore field. The Laiziling and Jianfengling plutons are composed of medium- to coarse-grained two-mica and coarse-grained biotite granites, respectively, and have zircon U–Pb ages of 156.4 ± 1.4 Ma and 165.2 ± 1.4 Ma, respectively. Both of the Laiziling and Jianfengling granites are characterized by extremely similar elemental and Lu–Hf isotopic compositions with high contents of SiO_2, Al_2O_3, Na_2O, K_2O, high A/CNK ratios, negative $\varepsilon_{Hf}(t)$ values (ranging from -3.86 to -1.38 and from -5.44 to -3.71, respectively), and old T_{DMC} ages (ranging from 1.30 to 1.47 Ga and from 1.32 to 1.56 Ga, respectively). These features indicate that they both belong to highly fractionated A-type granites, and were formed in an extensional setting and from the same magma chamber originated from the Paleoproterozoic metamorphic basement of South China with a certain amount of mantle-derived magma involved with temperatures of ca. 730 °C and low oxygen fugacity.

Keywords: zircon U–Pb dating; geochemistry; Lu–Hf isotopes; Xianghualing; South Hunan

1. Introduction

South Hunan, located in the central part of the Shi-Hang zone, is well-known for its world-class W–Sn–Pb–Zn polymetallic deposits and reserves (Figure 1a). The Shi-Hang zone, well-known as the collision suture between the Yangtze Block and Cathaysia Block in the Neoproterozoic, is also an important granitic magmatic belt and polymetallic metallogenic belt [1–3]. As a significant part of the Shi-Hang zone, the W–Sn–Pb–Zn mineralization in this South Hunan possesses an obvious zoning feature from east to west: Shizhuyuan and Yaogangxian W deposits in the eastern part, Furong, Xianghualing and Furong Sn deposits in the middle part, and Huangshaping and Baoshan Pb–Zn deposits in the western part (Figure 1b). Previous studies have revealed that these deposits were formed in 165–150 Ma, which were the significant part of the Jurassic metallogenic explosion event of South China [4–10]. In addition, these deposits have a genetic relationship with the granitic magmatic activity in this area, and it has been proved by the geological and geochronological evidences [4–9,11–15]. Due to the large-scale W–Sn–Pb–Zn polymetallic mineralization, the granitic plutons related with these large deposits have been drawn the attention of geologists, and abundant geochronological and geochemical data have been reported recently, such as Qitianling pluton (155.5 ± 1.3 Ma, associated with the Furong Sn deposit [13]), Qianlishan pluton (157 ± 2 Ma, associated with the Shizhuyuan W deposit [15]), Yaogangxian pluton (156.9 ± 0.7 Ma, associated with the Yaogangxian W deposit [11]), Huangshaping pluton (154.3 ± 1.9 Ma, associated with the Huangshaping Pb–Zn deposit [16]), and Baoshan pluton (158 ± 2 Ma, associated with the Baoshan Cu–Mo–Pb–Zn deposit [6]). These coeval granitic plutons in South Hunan, related to different metallic mineralization, have been an ideal place to probe into the magmatism and related mineralization of South China.

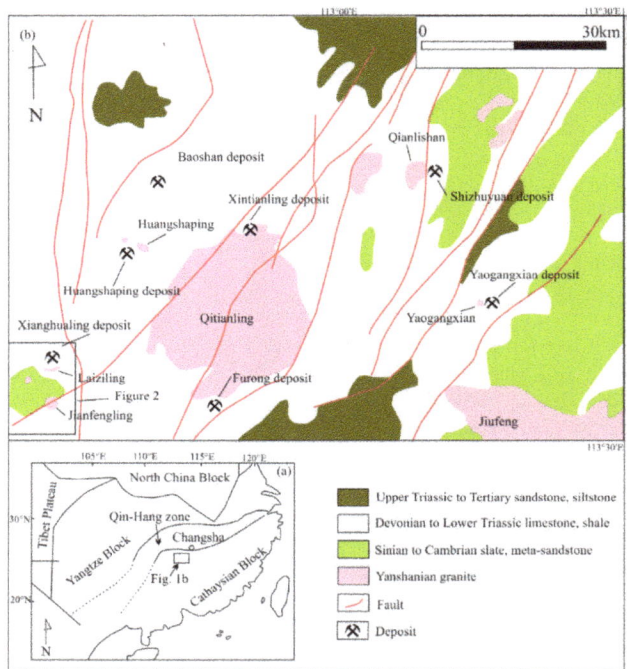

Figure 1. (**a**) Geological sketch map of South China; (**b**) Geological sketch map of the South Hunan province (modified from [8]), showing the distribution of granitic plutons, and related deposits.

The Laiziling and Jianfengling plutons, located in Xianghualing ore field, South Hunan province, are two small-sized granitic plutons, however, they have close relationship with the super-large Xianghualing Sn deposit and large Dongshan W deposit, respectively, both in time and space [7,17,18]. Then, it is the perfect laboratory for studying the theory of little intrusion forming large deposit. However, former studies have been focused on the abundant Sn–W polymetallic mineralization and genesis of the singly pluton. Additionally, a lack of systematic geochronological, geochemical, and isotopic analysis makes it unclear for the genesis and tectonic setting of these granitic plutons. Furthermore, few works have been conducted on the relationship between the Laiziling and Jianfengling plutons. Then, in this paper we report new data of zircon U–Pb dating, bulk-rock geochemical compositions and zircon Lu–Hf isotopes of Laiziling and Jianfengling granites, aiming to outline the petrogenesis of these two plutons, constrain the source and origin of the granitic magmas, discuss the tectonic setting, and clarify the relationship between these two plutons.

2. Geological Background

The Xianghualing ore field, located in the Chenzhou city, South Hunan province, is one of the biggest Sn–W–Pb–Zn ore fields in China, and consists of Xianghualing Sn deposit (a super-large Sn deposit), Dongshan W deposit (a large W deposit) and many small-medium sized deposits (Figure 2).

The strata, outcropped in the Xianghualing ore field, are composed of Quaternary sediments, Jurassic-Cretaceous sandstone and shale, Carboniferous carbonate and clastic rocks, Mid-Upper Devonian limestone and dolomite, and Permian quartz sandstone and shale, however, the Mid-Upper Devonian rocks are dominant in this area (Figure 2). The faults can be subdivided into five groups, based on theirs trend: NE-, NWW-, NNW-, NW-, NNE-, and EW-trending, however, the NE-trending faults are dominant and acted as the passable and ore-hosting structures in this area (Figure 2, [19]). The intrusive rocks consist of Laiziling, Jianfengling, and some little granitic plutons, and are intruded into the Mid-Upper Devonian limestone and dolomite, and Permian quartz sandstone and shale (Figure 2). Previous studies have revealed that these granitoids are emplaced in Late Jurassic [18], indicating that they were the important part of the Jurassic magmatic activity in South China.

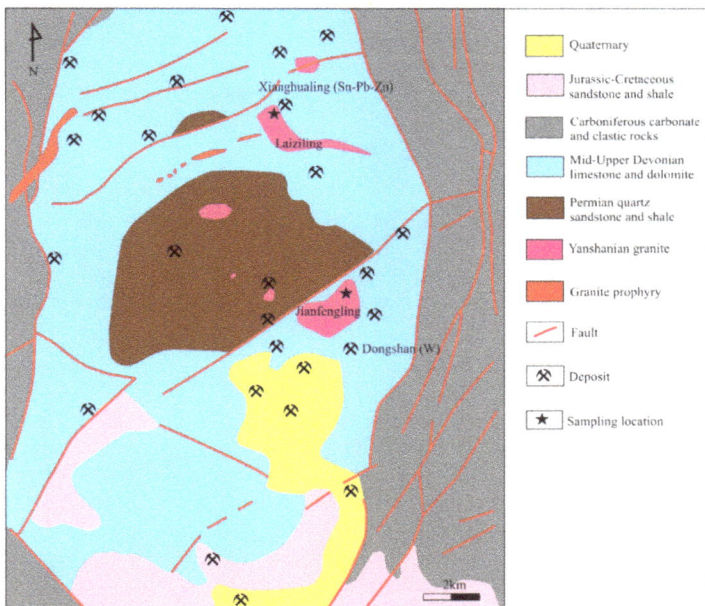

Figure 2. Schematic geological map of the Xianghualing ore field showing the location of samples (modified from [7]).

The Laiziling pluton, occupying an area of 2.2 km^2, is composed of the medium- to coarse-grained two-mica granites. It is characterized by massive-, leucocratic- and porphyroid-texture, and consist of quartz (~40%), K-feldspar (~30%), plagioclase (~20%), biotite (~5%), and muscovite (~5%) (Figure 3a–c). The accessory minerals contain zircon, apatite, sphene, and magnetite. The Jianfengling pluton, occupying an area of 4.4 km^2, is composed of coarse-grained biotite granites. They are also characterized by massive-, leucocratic- and porphyroid-texture, and consists of quartz (~40%),

K-feldspar (~30%), plagioclase (~25%), and biotite (~5%) (Figure 3d–f). The accessory minerals contain zircon, apatite, sphene, and magnetite.

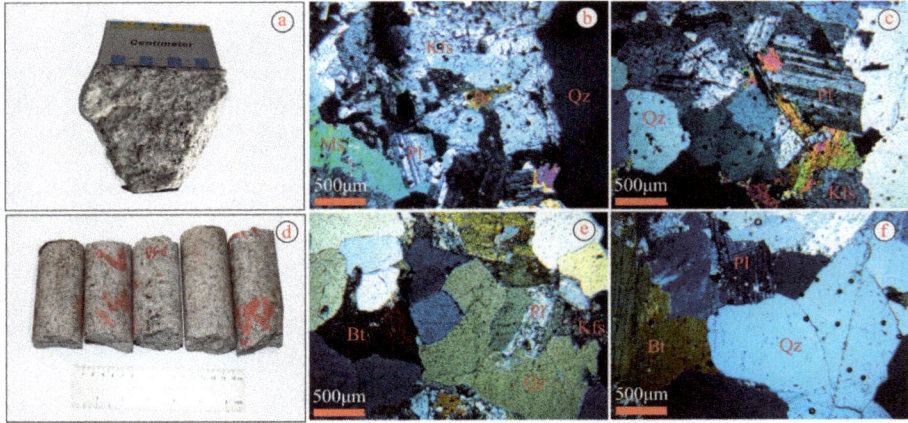

Figure 3. Photos of representative rocks samples (**a,d**) and relevant microphotos (**b,c,e,f**). Photos (**a–c**) refer to medium-to coarse-grained two-mica granite from Laiziling pluton; Photos (**d–f**) refer to coarse-grained biotite granite from Jianfengling pluton. Kfs—K-feldspar; Pl—plagioclase; Qz—quartz; Bt—biotite; Ms—muscovite.

3. Sampling and Analytical Methods

Samples of Laiziling and Jianfengling plutons were collected from drill and underground mine, respectively (Figure 2). Zircon grains used for LA-ICPMS U–Pb dating and Lu–Hf isotopic analyses were separated from a medium- to coarse-grained two-mica granite (sample No. Lzl-1) and a coarse-grained biotite granite (sample No. Ds-6), which were collected from Xianghualing and Dongshan deposits, respectively.

3.1. In Situ LA-ICPMS Zircon U–Pb Dating and Trace Element Compositions

Zircon grains were separated from samples Lzl-1 and Ds-6 using magnetic and heavy liquid separation techniques, and were hand-picked under a binocular microscope before mounted in epoxy resin and polished. Cathodoluminescence (CL) techniques were used to reflect the internal structures of the zircon grains, with a scanning electron microscope (TESCAN MIRA 3 LMH FE-SEM, TESCAN, Brno, Czech Republic) at the Sample Solution Analytical Technology Co., Ltd., Wuhan, China. Zircon grains for U–Pb dating and trace elements analyses were carried out using Laser Ablation Inductively-Coupled Plasma Mass Spectrometry (LA-ICPMS, Agilent, Santa Clara, CA, USA) method at the In situ Mineral Geochemistry Lab, Ore Deposit and Exploration Centre (ODEC), Hefei University of Technology, China. The instrument of an Agilent 7900 Quadrupole ICP-MS coupled to a Photon Machines Analyte HE 193-nm ArF Excimer laser ablation system was used for the analyses. Standard zircon 91500 (1062 ± 4 Ma; [20]) and standard silicate glass (NIST SRM610) was applied to be as external standards for dating and trace element analysis. Quantitative calibration for zircon U–Pb dating and trace elements were performed by ICPMSDataCal 10.7 [21,22], and common Pb was corrected with the model proposed by [23]. Weighted mean age calculation and Concordia diagrams were conducted with the help of an ISOPLOT program from [24].

3.2. Major and Trace Elements Analysis

Bulk-rock major and trace elements analyses were finished at the ALS Geochemistry Laboratory in Guangzhou, China. Before the analyses, samples were crushed in a steel jaw crusher, and then powdered in an agate mill to grain size of 74 μm. The detailed methodology for major element compositions are as follows: Loss of ignition (LOI) was determined after igniting sample powders at 1000 °C for 1 h. A calcined or ignited sample (0.9 g) was added to 9.0 g of Lithium Borate Flux ($Li_2B_4O_7$–$LiBO_2$), mixed well and fused in an auto fluxer between 1050 and 1100 °C. A flat molten glass disk was prepared from the resulting melt. This disk was then analyzed by a Panalytical Axios Max X-ray fluorescence (XRF, Panalytical, Almelo, The Netherlands) instrument, with analytical accuracy of ca. 1–5%.

Trace element compositions were measured using ICP-MS (Perkin Elmer Elan 9000, Perkin, Waltham, MA, USA), after 2-day closed beaker digestion using a mixture of HF and HNO_3 acids in Teflon screw-cap bombs. Detection limits, defined as 3 s of the procedural blank, for some critical elements are as follows (ppm): Th (0.05), Nb (0.2), Hf (0.2), Zr (2), La (0.5) and Ce (0.5). The analytical accuracy is better than 5%.

3.3. Zircon Lu–Hf Isotope Analysis

The zircon Lu–Hf isotopes were conducted on a Neptune Plasma multi-collector inductively coupled plasma mass spectrometer (MC-ICP-MS, NePtune Plus, Thermo Fisher Scientific, Waltham, MA, USA) equipped with New Wave 213 nm FX ArF-excimer laser ablation system, at the laboratory of the Xi'an Institute of Geology and Mineral Resource, Chinese Academy of Geological Sciences, Xi'an, China. Instrumental parameter and data acquisition followed that described by [25,26]. The laser beam diameters were used by 50 μm, 10 Hz repetition rate and 15 J/cm^2 energy density. Helium was used as carrier gas to transport laser eroded matter in Neptune (MC-ICP-MS). Zircon standard GJ-1 was used as external calibration to evaluate the reliability of the analytical data, the recommended $^{176}Hf/^{177}Hf$ ratio of 0.282006 ± 24 (2σ, [26]). Isobaric interference of ^{176}Lu on ^{176}Hf was corrected measuring the intensity of the interference-free ^{175}Lu isotope and using a recommended $^{176}Lu/^{175}Lu$ ratio of 0.02655 (2σ, [27]). Similarly, the isobaric interference of ^{176}Yb on ^{177}Hf was corrected against the $^{176}Yb/^{172}Yb$ ratio of 0.5886 (2σ, [28]) to calculate $^{176}Hf/^{177}Hf$ ratios. In doing so, a normalizing $^{173}Yb/^{171}Yb$ ratio of 1.12346 for the analyzed spot itself was automatically used in the same run to calculate a mean β_{Yb} value, and then the ^{176}Yb signal intensity was calculated from the ^{173}Yb signal intensity and the mean β_{Yb} value [29,30]. In this work, we adopted the decay constant for ^{176}Lu of 1.865×10^{-11} a^{-1} [31], the present-day chondritic ratios of $^{176}Hf/^{177}Hf$ = 0.282772 and $^{176}Lu/^{177}Hf$ = 0.0332 [32], the present-day depleted mantle value of $^{176}Hf/^{177}Hf$ = 0.28325 [33] and $^{176}Lu/^{177}Hf$ = 0.0384 [34]. All the Lu–Hf isotope results are reported in 2σ error. The data processing and related parameters calculation was finished with the help of an Excel program "Hflow".

4. Results

4.1. Zircon U–Pb Dating

Most of the zircons from medium- to coarse-grained two-mica granite (sample No. Lzl-1) of Laiziling pluton are euhedral, with obvious internal oscillatory zoning in CL images (Figure 4a), indicating a magmatic origin of these zircons [35]. The length of these zircons are from 60 to 150 μm with length-to-width ratios of 1:1 to 3:1. The contents of U and Th are 402–4683 ppm (mean = 1209 ppm) and 232–2132 ppm (mean = 615 ppm), with Th/U ratios of 0.44–0.73 (mean = 0.55), which also indicate that they were typical magmatic zircons [35]. The $^{206}Pb/^{238}U$ ages of fourteen zircons vary from 152.5 Ma to 166.4 Ma which plot on or near the concordant curve (Supplementary Materials Table S1), and a weighted mean $^{206}Pb/^{238}U$ age of 156.4 ± 1.4 Ma (MSWD = 1.6) was yielded (Figure 4b).

Most of the zircons from coarse-grained biotite granite (sample No. Ds-6) of Jianfengling pluton are also featured by euhedral and obvious internal oscillatory zoning in CL images (Figure 4c), indicating a magmatic origin of these zircons [35]. The length of these zircons are from 50 to 200 μm with length-to-width ratios of 1:1 to 3:1. The contents of U and Th are 177–2779 ppm (mean = 954 ppm) and 94–1732 ppm (mean = 498 ppm), with Th/U ratios of 0.38–0.76 (mean = 0.55), which also indicate that they were typical magmatic zircons [35]. The $^{206}Pb/^{238}U$ ages of twenty-two zircons vary from 160.1 Ma to 170.7 Ma which plot on or near the concordant curve (Supplementary Materials Table S1), and a weighted mean $^{206}Pb/^{238}U$ age of 165.2 ± 1.4 Ma (MSWD = 0.47) was obtained (Figure 4d).

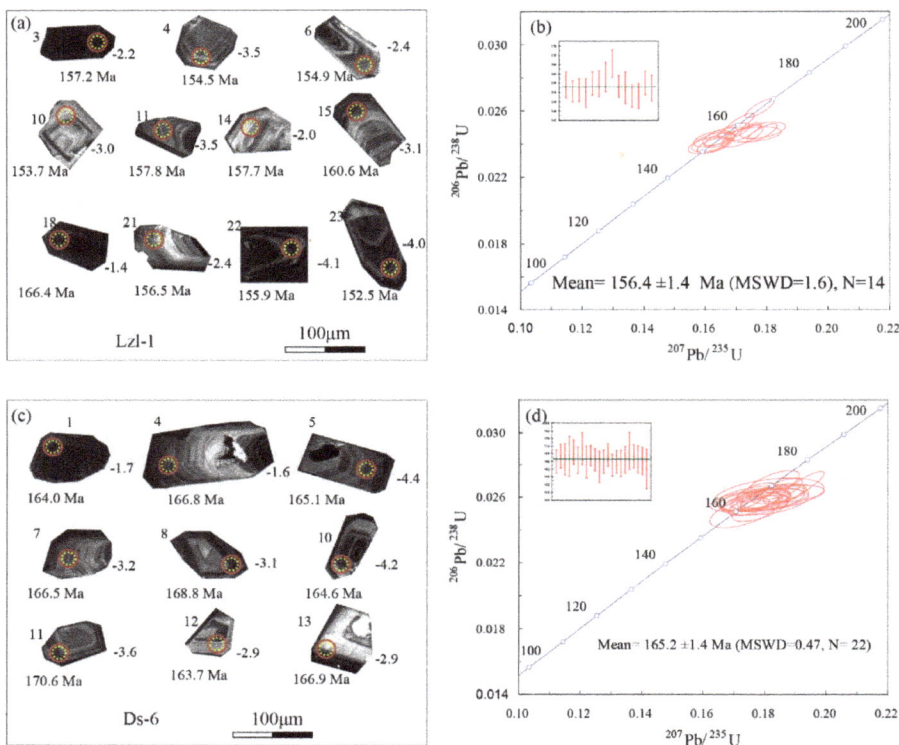

Figure 4. Cathodoluminescence (CL) images of zircon grains (**a**,**c**) and concordant diagrams of zircon U–Pb ages (**b**,**d**) from the Laiziling and Jianfengling granites, respectively. Red and yellow circles are spots for the zircon U–Pb dating and Lu–Hf isotopes analyses, respectively in (**b**,**d**).

4.2. Trace Element Compositions of Zircons

The trace element compositions of zircon grains are listed in Supplementary Materials Table S2. Zircon grains of sample Lzl-1 have relatively high contents of Ti and REEs (rare earth elements), and are from 4.09 to 11.99 ppm (mean = 8.41 ppm) and from 593 to 1440 ppm (mean = 1026 ppm), respectively. They are enriched in HREEs (heavy rare earth elements) and depleted in LREEs (light rare earth elements), with LREE/HREE ratios of 0.02–0.04 (mean = 0.03). The chondrite normalized REE patterns are featured by left-leaning steep slopes, and obvious positive Ce anomalies (Ce/Ce* = 7.03–30.52, mean = 15.57), and negative Eu anomalies (Eu/Eu* = 0.01–0.06, mean = 0.03, Figure 5a).

Zircon grains from sample Ds-6 have a little higher Ti and REE contents than those of sample Lzl-1, with Ti content of 3.71–18.31 (mean = 10.27) and REE content of 479–1691 (mean = 956). They also are enriched in HREEs and depleted in LREEs, with LREE/HREE ratios of 0.02–0.04 (mean = 0.03). The chondrite normalized REE patterns are featured by left-leaning steep slopes, and obvious positive

Ce anomalies (Ce/Ce* = 5.35–34.21, mean = 18.41), and negative Eu anomalies (Eu/Eu* = 0.01–0.11, mean = 0.04, Figure 5b).

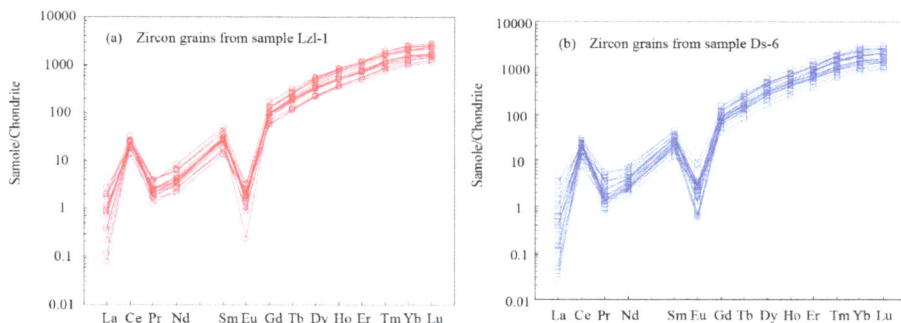

Figure 5. Chondrite-normalized REE (rare earth element) chemistry of zircon grains for the samples taken from the Laiziling (**a**) and Jianfengling (**b**) plutons, with normalizing factors from [36].

4.3. Major and Trace Element Compositions

Major and trace element compositions of the granites from Laiziling and Jianfengling plutons are presented in Supplementary Materials Table S3. The Laiziling granites are characterized by high contents of SiO$_2$ (73.92–74.61%, mean = 74.36%), Al$_2$O$_3$ (13.62–14.26%, mean = 13.90%), Na$_2$O (3.66–3.95%, mean = 3.81%), and K$_2$O (3.61–4.20%, mean = 3.89%) and low contents of TiO$_2$ (0.02–0.03%, mean = 0.03%), MgO (0.06–0.09%, mean = 0.08%), and P$_2$O$_5$ (0.01%). The Jianfengling granites have the similar major element composition to that of the Laiziling granites, characterized by high contents of SiO$_2$ (74.07–75.38%, mean = 74.79%), Al$_2$O$_3$ (13.19–13.86%, mean = 13.46%), Na$_2$O (2.05–5.08%, mean = 3.16%) and K$_2$O (2.38–4.21%, mean = 3.08%), and low contents of TiO$_2$ (0.01–0.04%, mean = 0.03%), MgO (0.01–0.07%, mean = 0.05%) and P$_2$O$_5$ (0.01%). In addition, all the samples are plotted in the field of granite in the SiO$_2$ vs. Na$_2$O + K$_2$O diagram, indicating that these two-rock types are both typical granites (Figure 6). All the samples from Laiziling pluton are plotted in the field of high-K calc-alkaline, however, the samples from Jianfengling pluton are plotted in the field of high-K calc-alkaline and calc-alkaline (Figure 7a). Both of the granites have high A/CNK (molar Al$_2$O$_3$/(CaO + Na$_2$O + K$_2$O)) values, with Laiziling granites of 1.17–1.23 (mean = 1.19) and Jianfengling granites of 1.06–1.64 (mean = 1.36), respectively, indicating that they belong to peraluminous series (Figure 7b). They also have high differentiation index values (DI), ranging from 91 to 93 (mean = 92) and from 84 to 94 (mean = 89) for the Laiziling and Jianfenging granites, respectively.

Both of the granites from Lalziling pluton and Jianfengling pluton have similar trace element contents and primitive-mantle normalized patterns, which are enriched in Rb, U, Nb, and Sm, and depleted in Ba, Sr, P, and Ti (Figure 8a). They also have the similar REE contents and chondrite normalized patterns, with ΣREEs of 341–370 ppm (mean = 358 ppm) and of 297–425 ppm (mean = 329 ppm) for the Laiziling and Jianfengling granites, respectively (Figure 8b). They also have obvious negative Eu anomalies, with Eu/Eu* values of 0.01 for granites from both of the plutons.

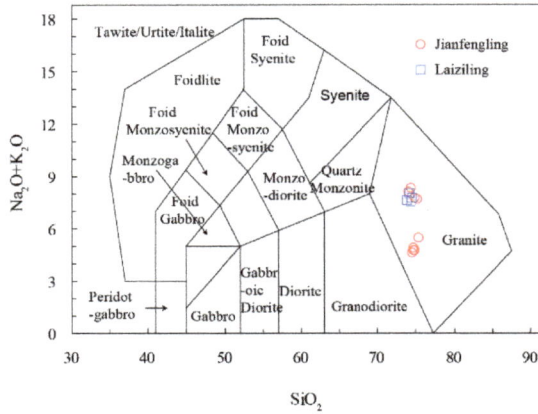

Figure 6. Classification diagram of igneous rocks for the samples from the Laiziling and Jianfengling plutons (modified from [37]).

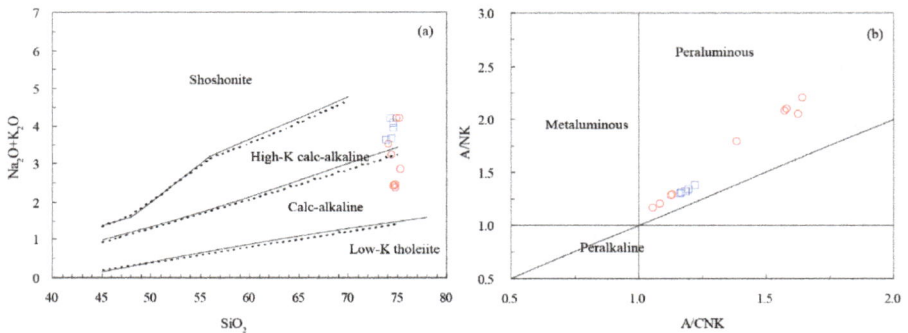

Figure 7. SiO$_2$ versus Na$_2$O + K$_2$O (**a**) and A/CNK versus A/NK (**b**) diagrams for the samples from the Laiziling and Jianfengling plutons (**a** and **b** are modified from [38,39], respectively). Symbols are as in Figure 6. A/CNK = molar Al$_2$O$_3$/(CaO + Na$_2$O + K$_2$O); A/NK = molar Al$_2$O$_3$/(Na$_2$O + K$_2$O).

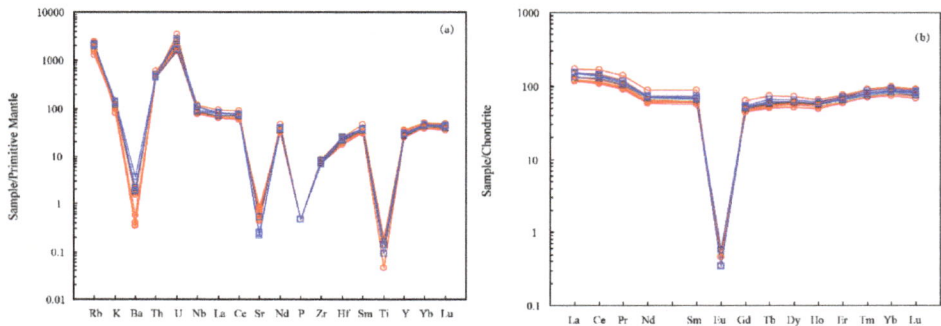

Figure 8. Primitive-mantle-normalized trace element (**a**) and chondrite-normalized REE (**b**) variation diagrams for samples from the Laiziling and Jianfengling plutons. Normalizing factors are from [36,40], respectively. Symbols are as in Figure 6.

4.4. Zircon Lu–Hf Isotopic Compositions

The zircon Lu–Hf isotopic compositions and related parameters for the granites from Laiziling pluton (sample No. Lzl-1) and Jianfengling pluton (sample No. Ds-6) are listed in Supplementary Materials Table S4. Result for the sample Lzl-1 have variable $^{176}Lu/^{177}Hf$ ratios of 0.000572–0.007548, and similar present-day $^{176}Hf/^{177}Hf$ ratios of 0.282562–0.282658. The calculated initial $^{176}Hf/^{177}Hf$ (Hf$_i$) ratios vary from 0.282559 to 0.282636, with $\varepsilon_{Hf}(t)$ values of −3.86 to −1.38 (mean = −2.91) and T$_{DMC}$ages of 1.30 to 1.47 Ga (mean = 1.39 Ga), which were calculated by the zircon U–Pb age of 156.4 Ma (Figure 9a,b).

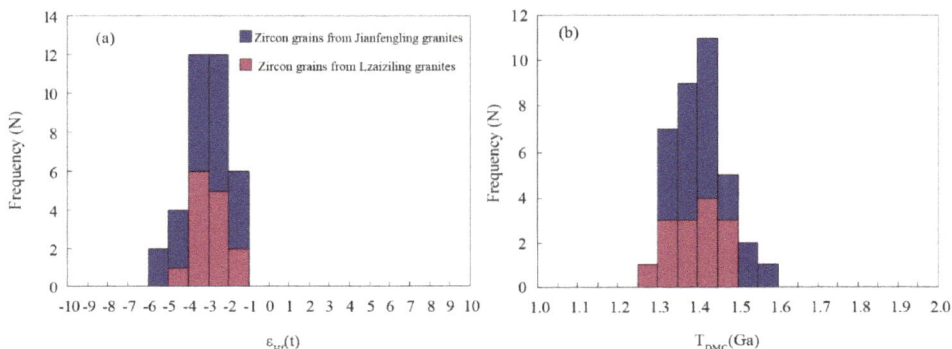

Figure 9. Statistical histograms for $\varepsilon_{Hf}(t)$ (**a**) and T$_{DMC}$ (**b**)values for the zircon grains from the Laiziling and Jianfengling plutons.

Zircon spots from sample Ds-6 also show variable $^{176}Lu/^{177}Hf$ ratios of 0.000429–0.003164 and similar present-day $^{176}Hf/^{177}Hf$ ratios of 0.282520–0.282630. The calculated initial $^{176}Hf/^{177}Hf$ (Hf$_i$) ratios vary from 0.282516 to 0.282625, with $\varepsilon_{Hf}(t)$ values of −5.44 to −3.71 (mean = −3.17) and T$_{DMC}$ ages of 1.32 to 1.56 Ga (mean = 1.42 Ga), which were calculated by the zircon U–Pb age of 165.2 Ma (Figure 9a,b).

5. Discussion

5.1. Genetic Type of the Granitic Rocks: An A-Type Affinity

The issue on the classification of the granitic rocks has been a hot topic for decades, and many types of granitic rocks are proposed based on the different standards, among which the classification of I-, S-, M-, and A-type granite are well accepted all over the world [41–51]. The term of A-type granite was first proposed by [45] and defined by their alkaline, anhydrous and anorogenic nature. Then, many geologists enriched and improved the concept of A-type granite, making it a significant component of the granite series [44,52–55]. Generally, in terms of the elemental compositions, the A-type granites have high contents of SiO$_2$, K$_2$O, Na$_2$O, Zr, Nb, REE, Y, and Ga, and low contents of CaO, Sr, Ba, and so forth, and characterized by high ratios of Ga/Al and (K$_2$O + Na$_2$O)/CaO [43]. The Laiziling and Jianfengling granites are characterized by high contents of SiO$_2$ (average ca. 74%), total alkalis (K$_2$O + Na$_2$O, average ca. 6.9%), total REE, and Ga, with depletion in Sr and Ba, which are similar to the major- and trace-element compositions of A-type granites [43]. Both of the Laiziling and Jianfengling granites have high 10,000 Ga/Al ratios, most of which are higher than 4, and are plotted in the field of A-type granite in the related discrimination diagrams (Figure 10). In addition, the extremely low content of P$_2$O$_5$ (0.01%) for the Laiziling and Jianfengling granites which differs from the typical S-type granites indicates that they might not belong to S-type granite [46]. The peraluminous nature, which most of A/CNK ratios are higher than 1.1, indicates that these granites are unlikely I-type granite [47]. Furthermore, the high content of FeO^t, K$_2$O, and Na$_2$O and low content of MgO also reveal that they might be likely A-type granites, since most of the samples are plotted in the field of A-type granites (Figure 11) [56,57].

Recent studies have revealed that most of the late Mesozoic granitic plutons in Nanling were mainly composed of the A-type granites, forming a NE-trending granite belt [58,59]. Generally, granites of this belt in Nanling were exposed at the central of the Shi-Hang zone proposed by [58] (Figure 1a). In addition, numerous A-type granitic plutons have been identified in the past few decades along the Shi-Hang zone, including Guposhan [60], Xitian [61], Qitianling [62], Laiziling [63], and so on. Consequently, geochemical characters of Laiziling and Jianfengling granites, together with the regional geology of the Jurassic granites along the Shi-Hang zone, reveal that they have an affinity of A-type rather than S- and I-type granite.

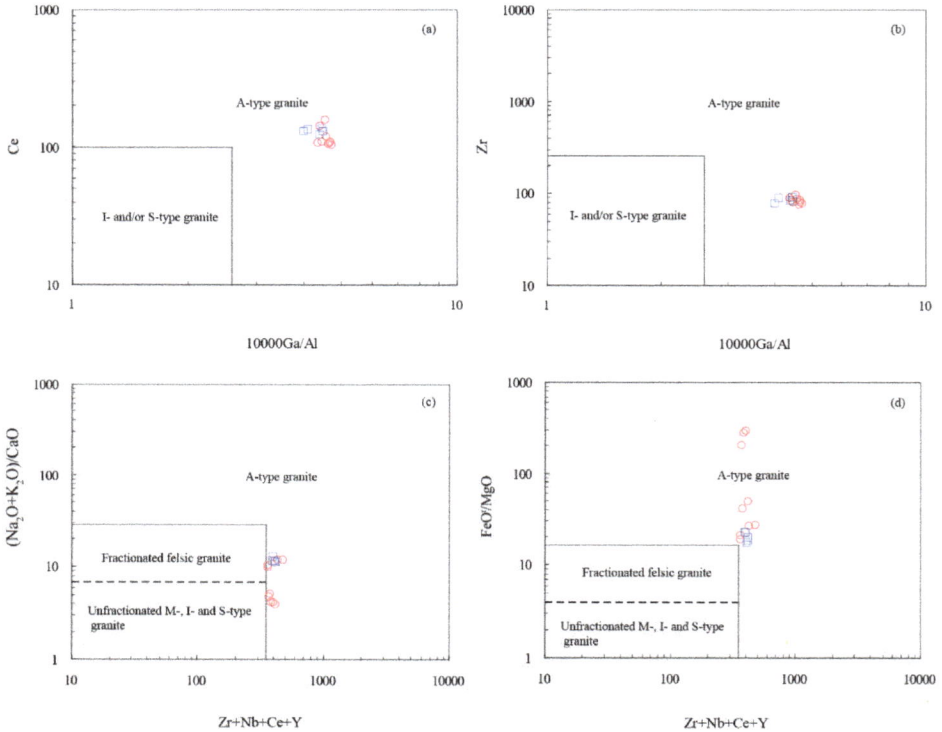

Figure 10. Discrimination diagrams on the types of granites from the Laiziling and Jianfengling plutons (modified from [43]). (**a**) 10000 Ga/Al versus Ce; (**b**) 10000 Ga/Al versus Zr; (**c**) Zr + Nb + Ce + Y versus (Na$_2$O + K$_2$O)/CaO; (**d**) Zr + Nb + Ce + Y versus FeO^t/MgO. Symbols are as in Figure 6.

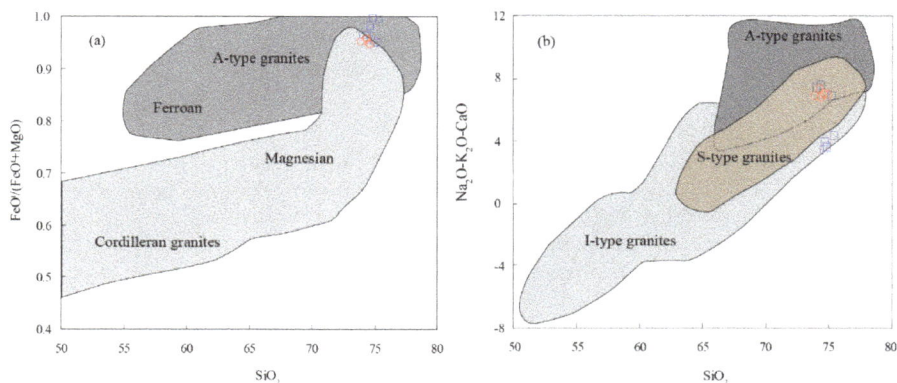

Figure 11. Discrimination diagrams on the types of granites from the Laiziling and Jianfengling plutons (modified from [56]). (**a**) SiO_2 versus $FeO^t/(FeO^t + MgO)$; (**b**) SiO_2 versus $Na_2O + K_2O - CaO$. Symbols are as in Figure 6.

5.2. Genesis of Laiziling and Jianfengling Granites

5.2.1. Temperatures

Temperature is a significant index to reflect the magma process and the genesis of granites [42,64–68]. As one of the most stable minerals in igneous rocks, zircons can be resistant to a certain degree of weathering and alteration in many kinds of geological events. In addition, the Zr partition coefficient and Ti content in zircon is sensitive to the temperature [65,67–69]. Then, based on those theories, [68] conducted an experiment on the solubility of Zr in melt at 860, 930 and 1020 °C and a model of zircon saturation thermometer was proposed to estimate the temperature of magmatic melt. Based on the crystal growth experiments of zircon in siliceous melt at different levels of temperature, the zircon Ti thermometer was first proposed by [69]. Furthermore, Ferry et al. [65] revised and replenished the model, making it an important and useful tool to reflect the temperatures of magmatic melt.

Then, in order to probe into the temperatures of Laiziling and Jianfengling granites, we used these two calculation models to estimate the temperatures of these granites. The results show that the calculated temperatures range from 738 to 751 °C (mean = 743 °C) and from 708 to 749 °C (mean = 725 °C) for the Laiziling and Jianfengling granites, respectively, with the help of zircon saturation thermometer (Supplementary Materials Table S3). The results calculated by zircon Ti thermometer show the similar temperatures for the Laiziling and Jianfengling granites, ranging from 668 to 757 °C (mean = 724 °C) and from 661 to 797 °C (mean = 739 °C), respectively (Supplementary Materials Table S2). The consistent temperatures, based on both of the calculated models, indicate that the both of Laiziling and Jianfengling granites crystalized from magmas with relatively high temperature (ca. 730 °C). Furthermore, the evidence that the zircons from both the Laiziling and Jianfengling granites lack of inherited core reveals that these temperatures can be as the minimum estimation for the magmatic melts.

5.2.2. Oxygen Fugacities and Fractional Crystallization

Similar to the temperature, oxygen fugacity is also a significant index to reflect the redox condition of magma melt, not only for the genesis of granites but also for their close relationship with the mineralization of different metals [60,70–81]. For example, high oxygen fugacity plays an important role in controlling the formation of porphyry Cu–Au and epithermal Au–Cu deposits, whereas, low oxygen fugacity is in favor of the W–Sn–Mo mineralization [75,76,79,82]. Recent studies revealed that some elements (Eu, Ce, and so on) in zircon can be an efficient tracers to reflect the oxidation status of magma [70,80,83]. Since the Eu and Ce are multivalent elements, with Eu^{2+} and Eu^{3+} for Eu, and Ce^{4+} and Ce^{3+} for Ce, respectively. Since valence of Ce and Eu is sensitive to the redox conditions of the melt, then the Ce^{4+}/Ce^{3+} and Eu^{3+}/Eu^{2+} ratios can be a useful parameters to reflect the redox conditions of the melt [84]. Based on the results from an experiment at different levels of temperature and oxygen fugacity, [80] proposed a model to calculate the oxygen fugacity of magma during zircon crystallization. The calculation results show that both of the Laiziling and Jianfengling granites have similar oxygen fugacities, with $\log(fO_2)$ values of -18 to -15.7 (mean = -16.5) and -18.2 to -14.8 (mean = -16.2), respectively. In addition, almost all the samples from both of these two plutons are plotted in the field between the IW (iron-wustite)- and FMQ (fayalite-magnetite-quartz)- buffer, and were close to the IW-buffer in the T versus $\log(fO_2)$, indicating that they have relatively low oxygen fugacities (Figure 12). Then, based on the evidences above, we can conclude that the Laiziling and Jianfengling granites were crystalized from a reducing magma.

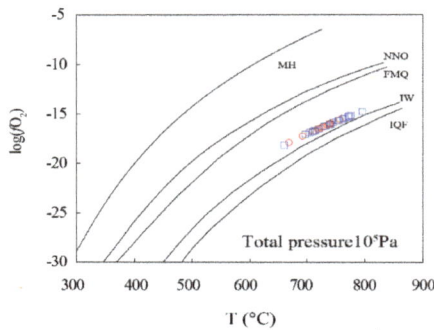

Figure 12. Temperature ($°C$) versus $\log(fO_2)$ diagram for the zircon grains from the Laiziling and Jianfengling granites (modified from [85]). Symbols are as in Figure 6. MH (magnetite-hematite); NNO (Ni-NiO); FMQ (fayalite-magnetite-quartz); IW (iron-wustite); IQF (iron-quartz-fayalite).

The fractional crystallization process has been proved by the depletion of P, Ta, Sr, Ti, Ba, and Eu of these granites, which represents the fractional crystallization of plagioclase, apatite, ilmenite, K-feldspar, and other minerals (Figure 8a,b). In addition, the positive correlation between the Rb and Ba and negative correlation between Rb and Sr suggest that the fractional crystallization of plagioclase and biotite is significant during the evolution of magma process (Figure 13a,b).

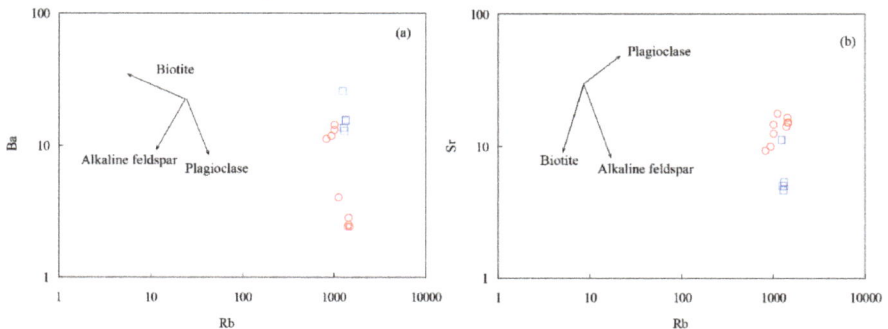

Figure 13. Rb versus Ba (**a**) and Rb versus Sr (**b**) diagrams for the Laiziling and Jianfengling granites. Symbols are as in Figure 6.

5.2.3. Magma Source

The source and genesis of A-type granite have long been a debatable topic for decades, and many models have been proposed to explain that, for example, fractional crystallization of mantle-derived magma [41], partial and/or complete melting of granulite [41], partial and/or complete melting of calc-alkali metasomatized mantle [54], partial melting of old granodiorite [86], partial melting of crust [43,52,87], and magma mixing [88,89].

The elemental compositions of the Laiziling and Jianfengling granites reveal that they were unlikely originated from the fractional crystallization of mafic rocks, and the model of fractional crystallization of mafic magma can rule out. The A-type granite nature of these granites can rule out the model of partial melting of old granodiorite which is mainly I-type granites. In addition, these granites are aluminous A-type granites with high A/CNK ratios, and the aluminous A-type granites could be generated from the partial melting of a felsic infracrustal source [42]. The Lu–Hf compositions of these granites from Laiziling and Jianfengling plutons are characterized by negative $\varepsilon_{Hf}(t)$ values (mean = −2.91 and −3.17, respectively) and old T_{DMC} ages (mean = −1.39 Ga and 1.42 Ga, respectively), indicating that they were likely mainly originated from a crustal source. However, the Lu–Hf isotopic features of Laiziling and Jianfengling granites differ from these coeval granites which were originated from the partial melting of the Proterozoic basement with no and/or few mantle materials involved in the Nanling range, such as Taoxikeng [90], Dengfuxian [91], and Xihuashan plutons [92] (Figure 14a,b). In addition, the Lu–Hf isotopic features of Laiziling and Jianfengling granites are similar to these coeval granites which were originated from the mixing of mantle and crustal materials, such as Jiuyishan [93], Guposhan [94] and Qitianling plutons [62] (Figure 14a,b). Thus, we can conclude that the Laiziling and Jianfengling plutons might likely be originated from the partial melting of Proterozoic basement of South China with a certain amount of mantle-derived magma involved.

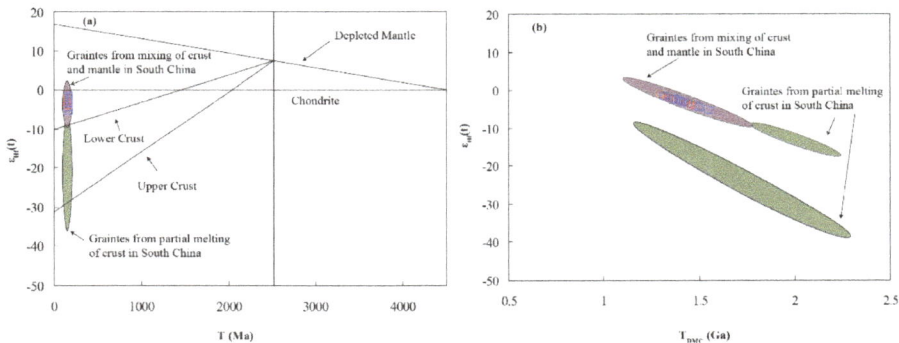

Figure 14. (**a**) Age versus $\varepsilon_{Hf}(t)$ and (**b**) T_{DMC} versus $\varepsilon_{Hf}(t)$ plots for the samples from the Laiziling and Jianfengling plutons. Data of granites from mixing of crust and mantle are from [62,93,94]; Data of granites from partial melting of crust in South China are from [90–92]. Symbols are as in Figure 6.

5.2.4. Relationship between the Two Granitic Plutons and Genesis of Laiziling and Jianfengling Granites

As stated above, we obtain two zircon ages for the Laiziling and Jianfengling granites, which are 156.4 ± 1.4 Ma and 165.2 ± 1.4 Ma, respectively, and these ages are consistent with the former studies within the uncertainty [17,63]. However, the relationship between the two granitic rocks were poorly understood, since they have an age interval of ca. 10 Ma during the emplacement of magma. Then, in order to probe into the relation between the two granitic plutons, some evidence we should ignore includes: (1) the similar major element compositions with high contents of SiO_2,

Al_2O_3, Na_2O, and K_2O, low contents of TiO_2, MgO, and P_2O_5; (2) terrifically similar trace element primitive-mantle normalized patterns and REE chondrite normalized patterns; and (3) nearly parallel zircon Lu–Hf isotopic compositions. These proofs indicate that both of the Laiziling and Jianfengling plutons might be originated from the same magma chamber, although, their emplaced age of Laiziling pluton is ca. 10 Ma after that of Jianfengling pluton. The new evidence was also provided by the mineral compositions with the occurrence of muscovite in Laiziling granites rather than in Jianfengling granites, since the residual magma will be enriched in Al, Si, K, Na, and so on, during the process of fractional crystallization.

Then, together with the evidences above, the genesis of the Laiziling and Jianfengling plutons might be concluded as following: (1) primary magma chamber was formed from mixing of partial melting of Proterozoic basement and a certain amount of mantle-derived magma; (2) the magma uplifted and intruded into the Paleozoic strata in ca. 165 Ma and Jianfengling pluton formed; and (3) during the process of fractional crystallization, the residual magma which was enriched in Al, uplifted and emplaced in ca. 156 Ma leading to the formation of Laiziling pluton.

5.3. Tectonic Settings

The tectonic settings of A-type granite have been a hot spot for decades, however, an overwhelming number of studies have revealed that A-type granites were formed in extensional settings, such as intraplate rift, mantle plume, back-arc extension, post-collisional extension and so on [43,45,54]. In addition, the A-type granite can be subdivided into two types of granites: A1-type granite associated with the intraplate rift and/or mantle plume and A2-type granite associated with back-arc extension, intraplate extension, and/or post-collisional extension [95]. Based on the discrimination diagrams from [95], the Laiziling and Jianfengling granites are all plotted in the field of A2-type granite, indicating that these granites belong to A2-type granite which are likely associated with the back-arc extension, intraplate extension, and/or post-collisional extension (Figure 15a,b). Furthermore, these granites are plotted in the field of within plate granite (WPG) in the diagrams proposed by [96], indicating an intraplate setting for these granites (Figure 15c,d). The results demonstrate that the Laiziling and Jianfengling granites might likely be emplaced in an intraplate extensional setting. As a part of Jurassic tectonic-magmatic activity in South China, the Laiziling and Jianfengling plutons might be formed in the same tectonic setting with other coeval granitic plutons, such as Qitianling, Guposhan, and Jiuyishan plutons [62,93,94]. However, the geodynamic mechanism triggering the extensional setting and magma activity in South China has long been in debate for decades [59,97–107]. Several models have been proposed to illustrate the geodynamic mechanism, for example, westward subduction of the paleo-pacific plate, mantle plume, post-collision, and so on [97–101,104,105,107]. However, these models concede that the tectonic setting of South China in Jurassic is an extensional setting, and the process of lithospheric extension and thinning occurred in that period [108,109]. Furthermore, these two plutons are located near the Shi-hang zone, which was recognized as the collision belt between the Yangtze and Cathaysia Blocks, and some unsubstantial spots can be the tunnel for the upwelling and emplacement of the mantle magma to mix with the crustal melt.

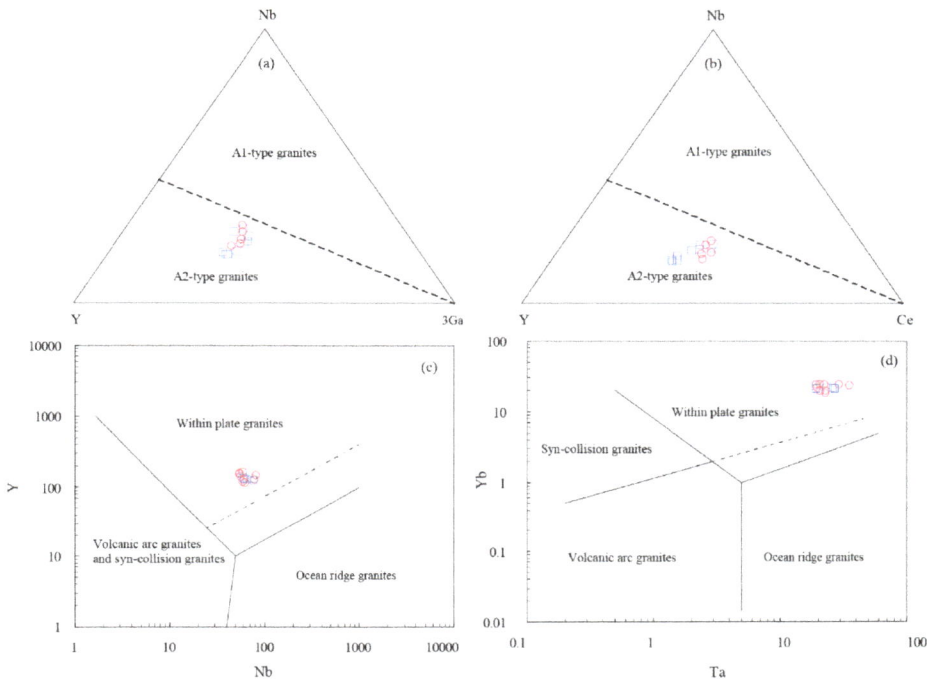

Figure 15. (**a**) Nb–Y–3Ga and (**b**) Nb–Y–Ce triangular diagrams; (**c**) Nb versus Y and (**d**) Ta versus Yb diagrams for the granites from the Laiziling and Jianfengling plutons. (**a**,**b**) are modified from [43]; (**c**,**d**) are modified from [96]. Symbols are as in Figure 6.

6. Conclusions

1. Zircon U–Pb dating yielded precise crystallization ages of 156.4 ± 1.4 Ma and 165.2 ± 1.4 Ma for the Laiziling and Jianfengling plutons in South Hunan, respectively.
2. Both of the Laiziling and Jianfengling granites are high-K, strongly peraluminous, and highly fractionated A-type granites with high temperatures and low oxygen fugacity. They were mainly originated from the Proterozoic basement of South China with a certain amount of mantle-derived magma involved.
3. The Laiziling and Jianfengling plutons were derived from the same magma chamber, and were the products of magma emplacement successively.
4. The granitic magma was emplaced in an extensional setting.

Supplementary Materials: The following are available online at http://www.mdpi.com/2075-163X/8/5/213/s1, Table S1: LA-ICP-MS zircon U–Pb isotopic compositions of granites from Laiziling (No. Lzl-1) and Jianfengling (No. Ds-6) plutons, Table S2: LA-ICP-MS zircon trace element compositions (ppm) of granites from Laiziling (No. Lzl-1) and Jianfengling (No. Ds-6) plutons, Table S3: Major and trace element compositions of the granites from the Jianfengling (sample number titled by DS) and Laiziling (sample number titled by LZL) plutons, Table S4: Zircon Lu–Hf isotopic compositions of granites from Laiziling (No. Lzl-1) and Jianfengling (No. Ds-6) plutons.

Author Contributions: X.W. and J.C. conceived and designed the experiments; B.H. and X.Z. took part in the discussion; Y.G. and W.S. took part in the field campaigns; L.Y., X.W. and J.C. wrote the paper.

Acknowledgments: This study was financially supported by the open fund of state key laboratory of ore deposit geochemistry (grant No. 201509). We also appreciate constructive suggestions and comments by Paul Sylvester, Galina Palyanova, and two anonymous reviewers. We also thank the editor Queenie Wang for her kind help.

Conflicts of Interest: The authors declare no conflict of interest.

References

1. Zhao, J.; Zhou, M.; Yan, D.; Zheng, J.; Li, J. Reappraisal of the ages of Neoproterozoic strata in South China: No connection with the Grenvillian orogeny. *Geology* **2011**, *39*, 299–302. [CrossRef]
2. Cao, J.Y.; Yang, X.Y.; Du, J.G.; Wu, Q.H.; Kong, H.; Li, H.; Wan, Q.; Xi, X.S.; Gong, Y.S.; Zhao, H.R. Formation and geodynamic implication of the Early Yanshanian granites associated with W–Sn mineralization in the Nanling range, South China: An overview. *Int. Geol. Rev.* **2018**. [CrossRef]
3. Yang, M.G.; Mei, Y.W. Characteristics of geology and metallization in the Qinzhou-Hangzhou paleoplate juncture. *Geol. Miner. Resour. South China* **1997**, *3*, 52–59. (In Chinese)
4. Li, H.Y.; Mao, J.W.; Sun, Y.L.; Zou, X.H.; He, H.L.; Du, A.D. Re–Os isotopic chronology of molybdenites in the Shizhuyuan polymetallic tungsten deposit, Southern Hunan. *Geol. Rev.* **1996**, *42*, 261–267.
5. Li, H.; Yonezu, K.; Watanabe, K.; Tindell, T. Fluid origin and migration of the Huangshaping W–Mo polymetallic deposit, South China: Geochemistry and Ar-40/Ar-39 geochronology of hydrothermal K-feldspars. *Ore Geol. Rev.* **2017**, *86*, 117–129. [CrossRef]
6. Lu, Y.; Ma, L.; Qu, W.; Mei, Y.; Chen, X. U–Pb and Re–Os isotope geochronology of Baoshan Cu–Mo polymetallic ore deposit in Hunan province. *Acta Petrol. Sin.* **2006**, *22*, 2483–2492. (In Chinese)
7. Yuan, S.; Peng, J.; Hu, R.; Li, H.; Shen, N.; Zhang, D. A precise U–Pb age on cassiterite from the Xianghualing tin-polymetallic deposit (Hunan, South China). *Miner. Depos.* **2008**, *43*, 375–382. [CrossRef]
8. Peng, J.; Zhou, M.; Hu, R.; Shen, N.; Yuan, S.; Bi, X.; Du, A.; Qu, W. Precise molybdenite Re–Os and mica Ar–Ar dating of the Mesozoic Yaogangxian tungsten deposit, central Nanling district, South China. *Miner. Depos.* **2006**, *41*, 661–669. [CrossRef]
9. Li, S.T.; Wang, J.B.; Zhu, X.Y.; Wang, Y.L.; Han, Y.; Guo, N.N. Chronological characteristics of the Yaogangxian composite pluton in Hunan Province. *Geol. Explor.* **2011**, *47*, 143–150. (In Chinese)
10. Cao, J.Y.; Wu, Q.H.; Yang, X.Y.; Kong, H.; Li, H.; Xi, X.S.; Huang, Q.H.; Liu, B. Geochronology and Genesis of the Xitian W–Sn Polymetallic Deposit in Eastern Hunan Province, South China: Evidence from Zircon U–Pb and Muscovite Ar–Ar Dating, petrochemistry, and Wolframite Sr–Nd–Pb Isotopes. *Minerals* **2018**, *8*, 111. [CrossRef]
11. Dong, S.; Bi, X.; Hu, R.; Chen, Y. Petrogenesis of the Yaogangxian granites and implications for W mineralization, Hunan Province. *Acta Petrol. Sin.* **2014**, *30*, 2749–2765. (In Chinese)
12. Li, H.; Watanabe, K.; Yonezu, K. Geochemistry of A-type granites in the Huangshaping polymetallic deposit (South Hunan, China): Implications for granite evolution and associated mineralization. *J. Asian Earth Sci.* **2014**, *88*, 149–167. [CrossRef]
13. Zhao, K.; Jiang, S.; Jiang, Y.; Liu, D. SHRIMP U–Pb dating of the Furong unit of Qitangling granite from southeast Hunan province and their geological implications. *Acta Petrol. Sin.* **2006**, *22*, 2611–2616. (In Chinese)
14. Li, H.; Watanabe, K.; Yonezu, K. Zircon morphology, geochronology and trace element geochemistry of the granites from the Huangshaping polymetallic deposit, South China: Implications for the magmatic evolution and mineralization processes. *Ore Geol. Rev.* **2014**, *60*, 14–35. [CrossRef]
15. Chen, Y.; Li, H.; Sun, W.; Ireland, T.; Tian, X.; Hu, Y.; Yang, W.; Chen, C.; Xu, D. Generation of Late Mesozoic Qianlishan A2-type granite in Nanling Range, South China: Implications for Shizhuyuan W–Sn mineralization and tectonic evolution. *Lithos* **2016**, *266–267*, 435–452. [CrossRef]
16. Hu, X.; Gong, Y.; Pi, D.; Zhang, Z.; Zeng, G.; Xiong, S.; Yao, S. Jurassic magmatism related Pb–Zn–W–Mo polymetallic mineralization in the central Nanling Range, South China: Geochronologic, geochemical, and isotopic evidence from the Huangshaping deposit. *Ore Geol. Rev.* **2017**, *91*, 877–895. [CrossRef]
17. Xuan, Y.S.; Yuan, S.D.; Yuan, Y.B.; Mi, J.R. Zircon U–Pb age, geochemistry and petrogenesis of Jianfengling plutonin southern Hunan Province. *Miner. Depos.* **2014**, *33*, 1379–1390. (In Chinese)
18. Zhu, J.C.; Wang, R.C.; Lu, J.J.; Zhang, H.; Zhang, W.L.; Xie, L.; Zhang, R.Q. Fractionation, evolution, petrogenesis and mineralization of Laiziling Granite Pluton, Southern Hunan Province. *Geol. J. China Univ.* **2011**, *17*, 381–392. (In Chinese)
19. Xu, Q.D. Identification of the intrusive phases of the composite alkali-feldspathic granite in Xianghualing, Hunan. *Hunan Geol.* **1991**, *10*, 289–294. (In Chinese)

20. Wiedenbeck, M.; Alle, P.; Corfu, F.; Griffin, W.L.; Meier, M.; Ober, F.; Von Quadt, A.; Roddick, J.C.; Speigel, W. Three natural zircon standards for U–Th–Pb, Lu–Hf, trace-element and REE analyses. *Geostand. Geoanal. Res.* **1995**, *19*, 1–23. [CrossRef]

21. Liu, Y.; Gao, S.; Hu, Z.; Gao, C.; Zong, K.; Wang, D. Continental and oceanic crust recycling-induced melt-peridotite interactions in the Trans-North China Orogen: U–Pb Dating, Hf Isotopes and Trace Elements in Zircons from Mantle Xenoliths. *J. Petrol.* **2010**, *51*, 537–571. [CrossRef]

22. Liu, Y.; Hu, Z.; Gao, S.; Guenther, D.; Xu, J.; Gao, C.; Chen, H. In situ analysis of major and trace elements of anhydrous minerals by LA-ICP-MS without applying an internal standard. *Chem. Geol.* **2008**, *257*, 34–43. [CrossRef]

23. Andersen, T. Correction of common lead in U–Pb analyses that do not report 204Pb. *Chem. Geol.* **2002**, *192*, 59–79. [CrossRef]

24. Ludwig, K.R. *ISOPLOT 3.00: A Geochronological Toolkit for Microsoft Excel*; Berkeley Geochronology Center: Berkeley, CA, USA, 2003; p. 39.

25. Wu, F.; Yang, Y.; Xie, L.; Yang, J.; Xu, P. Hf isotopic compositions of the standard zircons and baddeleyites used in U–Pb geochronology. *Chem. Geol.* **2006**, *234*, 105–126. [CrossRef]

26. Geng, J.Z.; Li, H.K.; Zhang, J.; Zhou, H.Y.; Li, H.M. Zircon Hf isotope analysis by means of LA-MC-ICP-MS. *Geol. Bull. China* **2011**, *30*, 1508–1513. (In Chinese)

27. Machado, N.; Simonetti, A. U–Pb dating and Hf isotopic composition of Zircon by Laser-Ablation-MC-ICP-MS. In *Laser-Ablation-ICPMS in the Earth Sciences: Principles and Applications*; Sylvester, P., Ed.; Mineralogical Association of Canada: Québec, QC, Canada, 2001; Volume 29, pp. 121–146.

28. Chu, N.C.; Taylor, R.N.; Chavagnac, V.; Nesbitt, R.W.; Boella, R.M.; Milton, J.A.; German, C.R.; Bayon, G.; Burton, K. Hf isotope ratio analysis using multi-collector inductively coupled plasma mass spectrometry: An evaluation of isobaric interference corrections. *J. Anal. At. Spectrom.* **2002**, *17*, 1567–1574. [CrossRef]

29. Iizuka, T.; Hirata, T. Improvements of precision and accuracy in in situ Hf isotope microanalysis of zircon using the laser ablation-MC-ICPMS technique. *Chem. Geol.* **2005**, *220*, 121–137. [CrossRef]

30. Thirlwall, M.F.; Anczkiewicz, R. Multidynamic isotope ratio analysis using MC-ICP-MS and the causes of secular drift in Hf, Nd and Pb isotope ratios. *Int. J. Mass Spectrom.* **2004**, *235*, 59–81. [CrossRef]

31. Scherer, E.; Munker, C.; Mezger, K. Calibration of the lutetium-hafnium clock. *Science* **2001**, *293*, 683–687. [CrossRef] [PubMed]

32. BlichertToft, J.; Albarede, F. The Lu–Hf isotope geochemistry of chondrites and the evolution of the mantle-crust system. *Earth Planet. Sci. Lett.* **1997**, *148*, 243–258. [CrossRef]

33. Nowell, G.M.; Kempton, P.D.; Noble, S.R.; Fitton, J.G.; Saunders, A.D.; Mahoney, J.J.; Taylor, R.N. High precision Hf isotope measurements of MORB and OIB by thermal ionisation mass spectrometry: Insights into the depleted mantle. *Chem. Geol.* **1998**, *149*, 211–233. [CrossRef]

34. Griffin, W.L.; Pearson, N.J.; Belousova, E.; Jackson, S.E.; van Achterbergh, E.; O'Reilly, S.Y.; Shee, S.R. The Hf isotope composition of cratonic mantle: LAM-MC-ICPMS analysis of zircon megacrysts in kimberlites. *Geochim. Cosmochim. Acta* **2000**, *64*, 133–147. [CrossRef]

35. Hoskin, P.W.O.; Schaltegger, U. The composition of zircon and igneous and metamorphic petrogenesis. *Rev. Miner. Geochem.* **2003**, *53*, 27–62. [CrossRef]

36. Taylor, S.R.; McLennan, S.M. *Continental Crust: Its Composition and Evolution. An Examination of the Geochemical Record Preserved in Sedimentary Rocks*; Blackwell Science Inc.: Boston, MA, USA, 1985; p. 312.

37. Middlemost, E.A.K. Naming materials in the magma/igneous rock system. *Earth Sci. Rev.* **1994**, *37*, 215–224. [CrossRef]

38. Peccerillo, A.; Taylor, S. Geochemistry of Eocene calc–alkaline volcanic rocks from the Kastamonu area, northern Turkey. *Contrib. Mineral. Petrol.* **1976**, *58*, 63–81. [CrossRef]

39. Maniar, P.D.; Piccoli, P.M. Tectonic discrimination of granitoids. *Geol. Soc. Am. Bull.* **1989**, *101*, 635–643. [CrossRef]

40. Sun, S.S.; McDonough, W.F. Chemical and isotopic systematics of oceanic basalts: Implications for mantle compositions and processes. In *Magmatism in the Ocean Basins*; Saunders, A.D., Norry, M.J., Eds.; Geological Society of London Special Paper: London, UK, 1989; Volume 32, pp. 313–345.

41. Collins, W.J.; Beams, S.D.; White, A.J.R.; Chappell, B.W. Nature and origin of A-type granites with particular reference to SE Australia. *Contrib. Mineral. Petrol.* **1982**, *80*, 189–200. [CrossRef]

42. King, P.L.; White, A.J.R.; Chappell, B.W.; Allen, C.M. Characterization and origin of aluminous A-type granites from the Lachlan Fold Belt, Southeastern Australia. *J. Petrol.* **1997**, *38*, 371–391. [CrossRef]

43. Whalen, J.B.; Currie, K.L.; Chappell, B.W. A-type granites: Geochemical characteristics, discrimination and petrogenesis. *Contrib. Mineral. Petrol.* **1987**, *95*, 407–419. [CrossRef]

44. King, P.L.; Chappell, B.W.; Allen, C.M.; White, A.J.R. Are A-type granites the high temperature felsic granites? Evidence from fractionated granites of the Wangrah Suite. *Aust. J. Earth Sci.* **2001**, *48*, 501–514. [CrossRef]

45. Loiselle, M.C.; Wones, D.R. Characteristics and Origin of Anorogenic Granites. *Geochem. Soc. Am.* **1979**, *11*, 468.

46. Chappell, B.W. Aluminium saturation in I- and S-type granites and the characterization of fractionated haplogranites. *Lithos* **1999**, *46*, 535–551. [CrossRef]

47. Chappell, B.W.; White, A.J.R. Two contrasting granite types. *Pac. Geol.* **1974**, *8*, 173–174.

48. Chappell, B.W.; White, A. Two contrasting granite types: 25 years later. *Aust. J. Earth Sci.* **2001**, *48*, 489–499. [CrossRef]

49. Chappell, B.W.; Bryant, C.J.; Wyborn, D. Peraluminous I-type granites. *Lithos* **2012**, *153*, 142–153. [CrossRef]

50. Gao, P.; Zheng, Y.; Zhao, Z. Distinction between S-type and peraluminous I-type granites: Zircon versus whole-rock geochemistry. *Lithos* **2016**, *258–259*, 77–91. [CrossRef]

51. Wu, Q.; Cao, J.; Kong, H.; Shao, Y.; Li, H.; Xi, X.; Deng, X. Petrogenesis and tectonic setting of the early Mesozoic Xitian granitic pluton in the middle Qin-Hang Belt, South China: Constraints from zircon U–Pb ages and bulk-rock trace element and Sr–Nd–Pb isotopic compositions. *J. Asian Earth Sci.* **2016**, *128*, 130–148. [CrossRef]

52. Bonin, B. A-type granites and related rocks: Evolution of a concept, problems and prospects. *Lithos* **2007**, *97*, 1–29. [CrossRef]

53. Grebennikov, A.V. A-type granites and related rocks: Petrogenesis and classification. *Russ. Geol. Geophys.* **2014**, *55*, 1353–1366. [CrossRef]

54. Martin, R.F. A-type granites of crustal origin ultimately result from open-system fenitization-type reactions in an extensional environment. *Lithos* **2006**, *91*, 125–136. [CrossRef]

55. Pankhurst, M.J.; Schaefer, B.F.; Turner, S.P.; Argles, T.; Wade, C.E. The source of A-type magmas in two contrasting settings: U–Pb, Lu–Hf and Re–Os isotopic constraints. *Chem. Geol.* **2013**, *351*, 175–194. [CrossRef]

56. Frost, B.R.; Barnes, C.G.; Collins, W.J.; Arculus, R.J.; Ellis, D.J.; Frost, C.D. A geochemical classification for granitic rocks. *J. Petrol.* **2001**, *42*, 2033–2048. [CrossRef]

57. Frost, C.D.; Frost, B.R. On Ferroan (A-type) Granitoids: Their Compositional Variability and Modes of Origin. *J. Petrol.* **2011**, *52*, 39–53. [CrossRef]

58. Gilder, S.A.; Gill, J.; Coe, R.S.; Zhao, X.X.; Liu, Z.W.; Wang, G.X.; Yuan, K.R.; Liu, W.L.; Kuang, G.D.; Wu, H.R. Isotopic and paleomagnetic constraints on the Mesozoic tectonic evolution of south China. *J. Geophys. Res. Solid Earth* **1996**, *101*, 16137–16154. [CrossRef]

59. Jiang, S.Y.; Zhao, K.D.; Jiang, Y.H.; Dai, B.Z. Characteristics and genesis of Mesozoic A-type granites and associated mineral deposits in the southern Hunan and northern Guangxi provinces along the Shi-Hang belt, South China. *Geol. J. China Univ.* **2008**, *14*, 496–509. (In Chinese)

60. Cao, M.; Qin, K.; Li, G.; Evans, N.J.; McInnes, B.I.A.; Li, J.; Zhao, J. Oxidation state inherited from the magma source and implications for mineralization: Late Jurassic to Early Cretaceous granitoids, Central Lhasa subterrane, Tibet. *Miner. Depos.* **2018**, *53*, 299–309. [CrossRef]

61. Zhou, Y.; Liang, X.; Wu, S.; Cai, Y.; Liang, X.; Shao, T.; Wang, C.; Fu, J.; Jiang, Y. Isotopic geochemistry, zircon U–Pb ages and Hf isotopes of A-type granites from the Xitian W–Sn deposit, SE China: Constraints on petrogenesis and tectonic significance. *J. Asian Earth Sci.* **2015**, *105*, 122–139. [CrossRef]

62. Zhao, K.; Jiang, S.; Yang, S.; Dai, B.; Lu, J. Mineral chemistry, trace elements and Sr–Nd–Hf isotope geochemistry and petrogenesis of Cailing and Furong granites and mafic enclaves from the Qitianling batholith in the Shi-Hang zone, South China. *Gondwana Res.* **2012**, *22*, 310–324. [CrossRef]

63. Yuan, S.D. *Geochronology and Geochemistry of the Xianghualing Tin-Polymetallic Deposit, Hunan Province, China*; Institute of Geochemistry, Chinese Academy of Sciences: Guiyang, China, 2007. (In Chinese)

64. Boehnke, P.; Watson, E.B.; Trail, D.; Harrison, T.M.; Schmitt, A.K. Zircon saturation re-revisited. *Chem. Geol.* **2013**, *351*, 324–334. [CrossRef]

65. Ferry, J.M.; Watson, E.B. New thermodynamic models and revised calibrations for the Ti-in-zircon and Zr-in-rutile thermometers. *Contrib. Mineral. Petrol.* **2007**, *154*, 429–437. [CrossRef]

66. Liu, H.; Xu, Y.; He, B. Implications from zircon-saturation temperatures and lithological assemblages for Early Permian thermal anomaly in northwest China. *Lithos* **2013**, *182*, 125–133. [CrossRef]

67. Miller, C.F.; McDowell, S.M.; Mapes, R.W. Hot and cold granites? Implications of zircon saturation temperatures and preservation of inheritance. *Geology* **2003**, *31*, 529–532. [CrossRef]

68. Watson, E.B.; Harrison, T.M. Zircon saturation revisited: Temperature and composition effects in a variety of crustal magma types. *Earth Planet. Sci. Lett.* **1983**, *64*, 295–304. [CrossRef]

69. Watson, E.B.; Harrison, T.M. Zircon thermometer reveals minimum melting conditions on earliest Earth. *Science* **2005**, *308*, 841–844. [CrossRef] [PubMed]

70. Barth, A.P.; Wooden, J.L. Coupled elemental and isotopic analyses of polygenetic zircons from granitic rocks by ion microprobe, with implications for melt evolution and the sources of granitic magmas. *Chem. Geol.* **2010**, *277*, 149–159. [CrossRef]

71. Brounce, M.; Kelley, K.A.; Cottrell, E.; Reagan, M.K. Temporal evolution of mantle wedge oxygen fugacity during subduction initiation. *Geology* **2015**, *43*, 775–778. [CrossRef]

72. Lee, C.A.; Luffi, P.; Chin, E.J.; Bouchet, R.; Dasgupta, R.; Morton, D.M.; Le Roux, V.; Yin, Q.; Jin, D. Copper systematics in Arc magmas and implications for crust-mantle differentiation. *Science* **2012**, *336*, 64–68. [CrossRef] [PubMed]

73. Lee, C.; Leeman, W.P.; Canil, D.; Li, Z. Similar V/Sc systematics in MORB and arc basalts: Implications for the oxygen fugacities of their mantle source regions. *J. Petrol.* **2005**, *46*, 2313–2336.

74. Qiu, J.; Yu, X.; Santosh, M.; Zhang, D.; Chen, S.; Li, P. Geochronology and magmatic oxygen fugacity of the Tongcun molybdenum deposit, northwest Zhejiang, SE China. *Miner. Depos.* **2013**, *48*, 545–556. [CrossRef]

75. Sun, W.; Huang, R.; Li, H.; Hu, Y.; Zhang, C.; Sun, S.; Zhang, L.; Ding, X.; Li, C.; Zartman, R.E.; et al. Porphyry deposits and oxidized magmas. *Ore Geol. Rev.* **2015**, *65*, 97–131. [CrossRef]

76. Sun, W.; Liang, H.; Ling, M.; Zhan, M.; Ding, X.; Zhang, H.; Yang, X.; Li, Y.; Ireland, T.R.; Wei, Q.; et al. The link between reduced porphyry copper deposits and oxidized magmas. *Geochim. Cosmochim. Acta* **2013**, *103*, 263–275. [CrossRef]

77. Xiao, B.; Qin, K.; Li, G.; Li, J.; Xia, D.; Chen, L.; Zhao, J. Highly oxidized magma and fluid evolution of Miocene Qulong Giant Porphyry Cu-Mo deposit, Southern Tibet, China. *Resour. Geol.* **2012**, *62*, 4–18. [CrossRef]

78. Gao, X.-Q.; He, W.-Y.; Gao, X.; Bao, X.-S.; Yang, Z. Constraints of magmatic oxidation state on mineralization in the Beiya alkali-rich porphyry gold deposit, western Yunnan, China. *Solid Earth Sci.* **2017**, *2*, 65–78. [CrossRef]

79. Yang, Z.; Yang, L.; He, W.; Gao, X.; Liu, X.; Bao, X.; Lu, Y. Control of magmatic oxidation state in intracontinental porphyry mineralization: A case from Cu (Mo–Au) deposits in the Jinshajiang-Red River metallogenic belt, SW China. *Ore Geol. Rev.* **2017**, *90*, 827–846.

80. Trail, D.; Watson, E.B.; Tailby, N.D. Ce and Eu anomalies in zircon as proxies for the oxidation state of magmas. *Geochim. Cosmochim. Acta* **2012**, *97*, 70–87. [CrossRef]

81. Trail, D.; Watson, E.B.; Tailby, N.D. The oxidation state of Hadean magmas and implications for early Earth's atmosphere. *Nature* **2011**, *480*, 79–238. [CrossRef] [PubMed]

82. Sun, Z.L. Geochronology and oxygen fugacity of Mesozoic granites in Nanling area of South China. *J. Earth Sci. Environ.* **2014**, *36*, 141–151. (In Chinese)

83. Burnham, A.D.; Berry, A.J. An experimental study of trace element partitioning between zircon and melt as a function of oxygen fugacity. *Geochim. Cosmochim. Acta* **2012**, *95*, 196–212. [CrossRef]

84. Ballard, J.R.; Palin, J.M.; Campbell, I.H. Relative oxidation states of magmas inferred from Ce(IV)/Ce(III) in zircon: Application to porphyry copper deposits of northern Chile. *Contrib. Mineral. Petrol.* **2002**, *144*, 347–364. [CrossRef]

85. Eugster, H.P.; Wones, D.R. Stability relations of the ferruginous Biotite, Annite. *J. Petrol.* **1962**, *3*, 82–89. [CrossRef]

86. Skjerlie, K.P.; Johnston, A.D. Fluid-Absent Melting Behavior of an F-Rich Tonalitic Gneiss at Mid-Crustal Pressures: Implications for the Generation of Anorogenic Granites. *J. Petrol.* **1993**, *34*, 785–815. [CrossRef]

87. Rutanen, H.; Andersson, U.B.; Vaisanen, M.; Johansson, A.; Frojdo, S.; Lahaye, Y.; Eklund, O. 1.8 Ga magmatism in southern Finland: Strongly enriched mantle and juvenile crustal sources in a post-collisional setting. *Int. Geol. Rev.* **2011**, *53*, 1622–1683. [CrossRef]

88. Villaseca, C.; Orejana, D.; Belousova, E.A. Recycled metaigneous crustal sources for S- and I-type Variscan granitoids from the Spanish Central System batholith: Constraints from Hf isotope zircon composition. *Lithos* **2012**, *153*, 84–93. [CrossRef]

89. Yang, J.H.; Wu, F.Y.; Chung, S.L.; Wilde, S.A.; Chu, M.F. A hybrid origin for the Qianshan, A-type granite, northeast China: Geochemical and Sr–Nd–Hf isotopic evidence. *Lithos* **2006**, *89*, 89–106. [CrossRef]

90. Zhang, Y.; Yang, J.; Chen, J.; Wang, H.; Xiang, Y. Petrogenesis of Jurassic tungsten-bearing granites in the Nanling Range, South China: Evidence from whole-rock geochemistry and zircon U–Pb and Hf–O isotopes. *Lithos* **2017**, *278–281*, 166–180. [CrossRef]

91. Cai, Y. *The Study on Dengfuxian Granite and Its Mineralization in Hunan Province*; Nanjing University: Nanjing, China, 2013. (In Chinese)

92. Guo, C.; Chen, Y.; Zeng, Z.; Lou, F. Petrogenesis of the Xihuashan granites in southeastern China: Constraints from geochemistry and in-situ analyses of zircon U–Pb–Hf–O isotopes. *Lithos* **2012**, *148*, 209–227. [CrossRef]

93. Guo, C.; Zeng, L.; Li, Q.; Fu, J.; Ding, T. Hybrid genesis of Jurassic fayalite-bearing felsic subvolcanic rocks in South China: Inspired by petrography, geochronology, and Sr–Nd–O–Hf isotopes. *Lithos* **2016**, *264*, 175–188. [CrossRef]

94. Gu, S.Y.; Hua, R.M.; Qi, H.W. Zircon LA-ICP-MS U–Pb dating and Sr–Nd isotope study of the Guposhan granite complex, Guangxi, China. *Chin. J. Geochem.* **2006**, *26*, 290–300. [CrossRef]

95. Eby, G.N. Chemical subdivision of the A-type granitoids: Petrogenetic and tectonic implications. *Geology* **1992**, *20*, 641–644. [CrossRef]

96. Pearce, J.A.; Harris, N.B.W.; Tindle, A.G. Trace-element discrimination diagrams for the tectonic interpretation of granitic-rocks. *J. Petrol.* **1984**, *25*, 956–983. [CrossRef]

97. Chen, C.; Lee, C.; Shinjo, R. Was there Jurassic paleo-Pacific subduction in South China? Constraints from (40)Ar/(39)Ar dating, elemental and Sr–Nd–Pb isotopic geochemistry of the Mesozoic basalts. *Lithos* **2008**, *106*, 83–92. [CrossRef]

98. Honza, E.; Fujioka, K. Formation of arcs and backarc basins inferred from the tectonic evolution of Southeast Asia since the Late Cretaceous. *Tectonophysics* **2004**, *384*, 23–53. [CrossRef]

99. Jiang, Y.; Jiang, S.; Dai, B.; Liao, S.; Zhao, K.; Ling, H. Middle to late Jurassic felsic and mafic magmatism in southern Hunan province, southeast China: Implications for a continental arc to rifting. *Lithos* **2009**, *107*, 185–204. [CrossRef]

100. Jiang, Y.; Jiang, S.; Zhao, K.; Ling, H. Petrogenesis of Late Jurassic Qianlishan granites and mafic dykes, Southeast China: Implications for a back-arc extension setting. *Geol. Mag.* **2006**, *143*, 457–474. [CrossRef]

101. Li, Z.; Li, X. Formation of the 1300-km-wide intracontinental orogen and postorogenic magmatic province in Mesozoic South China: A flat-slab subduction model. *Geology* **2007**, *35*, 179–182. [CrossRef]

102. Qiu, Z.; Li, S.; Yan, Q.; Wang, H.; Wei, X.; Li, P.; Wang, L.; Bu, A. Late Jurassic Sn metallogeny in eastern Guangdong, SE China coast: Evidence from geochronology, geochemistry and Sr–Nd–Hf–S isotopes of the Dadaoshan Sn deposit. *Ore Geol. Rev.* **2017**, *83*, 63–83. [CrossRef]

103. Sun, W.; Ding, X.; Hu, Y.; Zartman, R.E.; Arculus, R.J.; Kamenetsky, V.S.; Chen, M. The fate of subducted oceanic crust: A mineral segregation model. *Int. Geol. Rev.* **2011**, *53*, 879–893. [CrossRef]

104. Sun, W.; Ling, M.; Yang, X.; Fan, W.; Ding, X.; Liang, H. Ridge subduction and porphyry copper-gold mineralization: An overview. *Sci. China Earth Sci.* **2010**, *53*, 475–484. [CrossRef]

105. Xie, G.Q.; Hu, R.Z.; Zhao, J.H.; Jiang, G.H. Mantle plume and the relationship between it and Mesozoic large-scale metallogenesis in southeastern China: A preliminary discussion. *Geotecton. Metallog.* **2001**, *25*, 179–186. (In Chinese)

106. Zhao, W.W.; Zhou, M.; Li, Y.H.M.; Zhao, Z.; Gao, J. Genetic types, mineralization styles, and geodynamic settings of Mesozoic tungsten deposits in South China. *J. Asian Earth Sci.* **2017**, *137*, 109–140. [CrossRef]

107. Zhou, X.M.; Li, W.X. Origin of Late Mesozoic igneous rocks in Southeastern China: Implications for lithosphere subduction and underplating of mafic magmas. *Tectonophysics* **2000**, *326*, 269–287. [CrossRef]

108. Fan, W.M.; Wang, Y.; Guo, F.; Peng, T.P. Mezosic mafic magmatism in Hunan-Jiangxi provinces and the lithospheric extension. *Earth Sci. Front.* **2003**, *10*, 159–169. (In Chinese)

109. Wang, Y.J.; Liao, C.L.; Fan, W.M.; Peng, T. Early Mesozoic OIB-type alkaline basalt in central Jiangxi province and its tectonic implications. *Geochimica* **2004**, *33*, 109–117.

![minerals logo] *minerals*

MDPI

Article

Hydrothermal Metasomatism and Gold Mineralization of Porphyritic Granite in the Dongping Deposit, North Hebei, China: Evidence from Zircon Dating

Hao Wei [1], Jiuhua Xu [2,*], Guorui Zhang [1], Xihui Cheng [3], Haixia Chu [4], Chunjing Bian [2] and Zeyang Zhang [2]

1   Geological Testing Center, Hebei GEO University, Shijiazhuang 050031, China; ronghaiwei@163.com (H.W.); zgrsdyney@163.com (G.Z.)
2   School of Civil and Environmental Engineering, University of Science and Technology, Beijing 100083, China; bianjing0314@163.com (C.B.); wagzlxlm@163.com (Z.Z.)
3   MNR Key Laboratory of Metallogeny and Mineral Assessment, Institute of Mineral Resources, Chinese Academy of Geological Sciences, Beijing 100037, China; cheng_xihui@163.com
4   School of Earth Sciences and Resources, China University of Geosciences, Bejiing 100083, China; haixia.chu@cugb.edu.cn
*   Correspondence: jiuhuaxu@ces.ustb.edu.cn

Received: 30 June 2018; Accepted: 15 August 2018; Published: 21 August 2018

Abstract: A porphyritic granite intrusion was recently discovered in the Zhuanzhilian section of the Dongping gold deposit. There is as many as one tonnage of Au in the fractured shear zone within the porphyritic granite intrusion, but no relevant reports concerning the origin and age of the intrusion has been published as yet. In this paper, zircon U-Pb dating is used to study the geochronology of porphyritic granite, in order to find out the evidence of age and the relationship with gold mineralization. There are two groups of zircon $^{207}Pb/^{235}U$-$^{206}Pb/^{238}U$ concordant ages of porphyritic granites: The concordant age of 373.0 ± 3.5 Ma, with the weighted mean age of 373.0 ± 6.4 Ma; and the concordant age of 142.02 ± 1.2 Ma with the weighted mean age of 142.06 ± 0.84 Ma. We believe that the first group might represent the age of residual zircon of alkaline complex, while the second group might be related with main gold mineralization. The obtained results of the petrography and electron probe analysis indicate that the porphyritic quartz and porphyritic granite, as well as gold mineralization, might be products of a late replacement of tectonic-hydrothermal fluid, which was rich in Si, Na and K originally and later yielded gold-forming fluids.

Keywords: porphyritic granite; zircon dating; hydrothermal metasomatism; Dongping gold deposit

1. Introduction

The gold deposits associated with alkaline rocks worldwide are widely distributed and have important economic value. Representative deposits include the US Cripple Crick Gold-Strontium Deposit and Bingham Copper-Gold Deposit; Papua New Guinea Pogel Gold deposits and Radom gold deposits; Glasgow copper-gold deposits in Indonesia and Batu-Haijiao copper-gold deposits; Katia copper-gold deposits in Australia and Skurian copper-gold deposits in Greece. In addition, the Enpaor large gold deposit in Fiji, the large copper-gold deposit in Ollumbrila, Argentina, and the Saskin-Uranium deposit in Canada are also considered to be products of alkaline magma activity [1,2]. The alkaline magmatism and related fluid activities of such deposits are the "carrier" of many large and extra-large gold deposits. For example, the ore-forming rock body of the Kadia Ridgeway copper-gold deposit in Australia is an alkali- and potassium-rich intermediate intrusive rock. Moreover, the mineralization type of the deposit is mainly quartz vein type, and the potassium alteration is closely related to mineralization [3].

The Dongping gold deposit, located in middle northern part of the North China Craton, is the first giant gold deposit discovered among the alkaline complex-hosted in China in the 1980s. The predecessors generally believed that alkaline rock formations in the area formed in the Hercynian period [4,5]. In the 90s, the metallogenic age of the Dongping gold deposit are not uniform, Xiang et al. [6] believed that the normal lead age of the lead isotope of the ore was 127 Ma, which represents the age of gold mineralization. Song et al. [7] believed that metallogenic age was 157–177 Ma, while Li et al. [8] suggested that the age of the main mineralization period in the Dongping gold deposit was 350.9 ± 0.9 Ma. In recent years, Li et al. [5,9] used the hydrothermal zircon U-Pb dating to obtain an altered rock age of 140.3 ± 1.4 Ma, representing the metallogenic age. Since the beginning of this century, the close genetic connection between the gold deposit, and the Yanshanian potassium granite with the age of 135.5 ± 0.4 Ma [10] in the south of Dongping (Figure 1) has gradually attracted the attention of researchers [11,12]. Most people infer that the Yanshanian tectonic-magmatic activity promoted underground hydrothermal fluids to leach and extract gold and other metals from alkaline complexes and Archean Chongli group. Gold and other metals are mineralized in the favorable part of the ductile shear zone, with a suitable physicochemical environment for gold precipitation. The study of isotopic chronology also shows that the gold was mineralized in the Early Cretaceous [4,5].

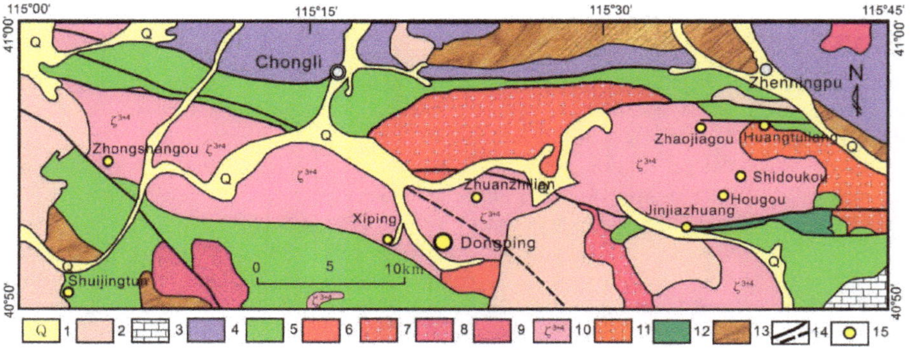

Figure 1. Simplified geological map of Shuiquangou alkaline complex in Zhangjiakou area, Hebei Province China (modified after Li et al. [5]; Song et al. [7]). (1) Quaternary; (2) Yanshanian intermediate-acidic volcaniclastic rocks; (3) Neoproterozoic and Mesoproterozoic cover rocks; (4) Paleo Proterozoic Hongqiyingzi Group; (5) Archean Chongli Group; (6) Shuiquan potassium granite; (7) Honghualiang biotite granite; (8) adamellite; (9) porphyritic granite; (10) Shuiquangou alkaline complex; (11) Hot spring giant porphyritic granite; (12) Ultramafic rocks; (13) Achaean granite gneiss; (14) faults; (15) gold deposits.

In recent years, a porphyritic granite intrusion was discovered in the Zhuanzhilian area of the Dongping Gold deposit during the peripheral exploration. Gold mineralization is intensive within the shear zone of the porphyritic granite intrusion, and Au reserves are estimated to be more than one tonnage; but no relevant research on porphyritic granite was done as yet. The age of gold mineralization in the porphyritic granite and the relations between porphyritic granite and gold mineralization are still questions. Therefore, the age of gold mineralization in porphyritic granite is needed to be clarified. And also, the natures of metasomatic fluids and their relations with porphyritic granite and gold mineralization are significant for further study of ore genesis and regional exploration.

2. Geological Background

The Dongping Gold deposit, located 12 km south to the downtown Chongli of northern Hebei Province, is tectonically in the middle part of the northern margin of the North China Craton (NCC), and at south of Shangyi-Chicheng-Damiao E-W trending deep faults (Figure 1). In the northern block of the fault, the main strata are Paleo-Proterozoic Hongqiyingzi group which is dominated by chlorite-quartz schist and biotitic granulite. While in the southern block, the strata are composed of Archean Chongli Group mainly comprising dihedral granulite and hornblende gneiss, and Neoproterozoic and Mesoproterozoic sedimentary cover rocks, and Mesozoic volcanoclastic rocks (Figure 1). The Chongli group has been affected by regional metamorphism and migmatization. There are three major NWW trending faults in the area, which are secondary faults of Chongli-Chicheng deep fault. They control the distribution of Shuiquangou alkaline complexes as well as Dongping and Hougou gold deposits. Shao et al. [13] suggested that the relatively open structure was formed in Zhangxuan area at 140 Ma or so.

The outcrop strata in the Dongping deposit are mainly the Archean Chongli amphibolite to the granulite facies metamorphic rocks [7]. The deposit is controlled by E-W striking fault and S-N trending fault which are the associated faults of the E-W striking Shangyi-Chongli-Chicheng deep fault. The ore bodies mainly occur at the junction of these two faults within Hercynian Shuiquangou alkaline complex [14]. As for the age of alkaline rock, Luo et al. [4] measured the SHRIMP zircon U-Pb age of 390 ± 6 Ma for the alkaline complex in Dongping gold deposit, and 386 ± 6 Ma for Shuiquangou syenite in Hougou gold deposit. The U-Pb average age of residual magmatic zircon in the deep potassic alteration rock vein of No. 70 is 382.8 ± 3.3 Ma from Li et al. [9], and that of magmatic zircon in K-feldspar-quartz vein of No. 1 is 380.5 ± 2.6 Ma from Li et al. [5]. The Shuiquangou alkaline complex is mainly divided into three parts: The west, the middle, and the east, with an area of about 350 km^2. The western rock section is dominated by the combination of angular syenite. The main rocks are characterized by medium-coarse grain granitic texture, porphyritic texture, massive structure with grain size of 5~15 mm; content of quartz is <3% generally, local up to 8%, K-feldspar accounts for about 50%, plagioclase (albite, oligoclase) accounts for about 20%; dark mineral content is 15~20%, mainly for hornblende and diopside. The middle section of the alkaline complex is dominated by a combination of aegirine-augite syenite. The main rocks are characterized by medium-fine grain granitic texture, massive structure with grain size of 1.5~2 mm, quartz content of 3~10%, K-feldspar (microcline, orthoclase) of 70% to 85%, plagioclase (albite, oligoclase) of 5 to 15%; generally dark minerals account for 5% to 10%, mainly diopside, aegirite, augite, etc. The eastern section of the alkaline complex is dominated by a combination of melanite syenite. The main rocks are characterized by medium-fine grain granitic texture and massive structure with grain size of 1.5~2 mm; quartz content is more than 5%, and K-feldspar accounts for 70–90%, plagioclase accounts for about 5% to 10%; generally dark minerals are 3% to 5%, mainly melanite, diopside, etc.; crushing structure and brittle-ductile shear deformation is common, accompanied by a strong metasomatism [15].

Secondly, there are Yanshanian potassium granites and middle-acidic vein. The age of the zircon LA-ICP-MS U-Pb dating for the volcanic rock of the Zhangjiakou Formation in the Zhangjiakou area is 143.0 ± 3.7 to 136.1 ± 1.4 Ma from Wei et al. [16]. A potash feldspar granite intrusion, namely Shangshuiquan orthoclase granite occurs in the south-eastern part of the mining area, which intruded along the contact zone between the Archean metamorphic rocks, and the Hercynian alkaline complex in the form of small stock. Recently, Shangshuiquan orthoclase granite intrusion has extended to Zhuanzhilian area closely related to the Dongping gold deposit, and has an exposed area of about 8 km^2 and LA-ICP-MS zircon U-Pb age of 142.9 ± 0.8 Ma from Jiang et al. [11]. Mo et al. [10] obtained a zircon U-Pb age of 135.5 ± 0.4 Ma for the single-particle zircon of Shangshuiquan potassium granite. Jiang et al. [17] obtained LA-ICP-MS zircon U-Pb ages of 142.9 ± 0.8 Ma, which are close to the age (140 ± 1.4 Ma) of Dongping gold deposit alteration rock. The dykes are widely distributed in this area, including granite porphyry, monzonite porphyry, quartz syenite porphyry, and hornblende syenite porphyry. The dark gray diorite in the eastern part of the Shuiquangou alkaline complex has an exposed area of about 1 km^2, and the zircon SHRIMP U-Pb age is 139.5 ± 0.9 Ma [18], which is close to the metallogenic age of 140.3 ± 1.4 Ma in the Dongping gold deposit. It is closely related to the formation of the Dongping gold deposit. In summary, the tectonic-hydrothermal fluids caused by the Mesozoic magmatism activities may be the main factor for the gold mineralization.

The mineralization types include potassic altered rock and quartz vein, and the ore belt strikes NNE. The ore is dominated by gold-bearing silica-potassic altered rocks, followed by gold-quartz veins. The ore minerals in the ore are mainly pyrite, pyrrhotite, chalcopyrite, galena, stibnite, sphalerite, native gold and calaverite, and gangue minerals are quartz, K-feldspar, plagioclase, and sericite.

3. Characteristics of Porphyritic Granite

3.1. Occurrence

A porphyritic granite intrusion was discovered recently in the Zhuanzhilian area of the Dongping gold deposit, which occur as ribbon zone striking NEE-SWW with a width of 15–50 m within the Shuiquangou alkaline complex (Figure 2). The porphyritic granite is controlled by the NEE-SWW striking shear zone and intrude into the monzonite of the alkaline rock along the NEE-SWW striking shear zone. Silicification and potassiumization are relatively developed in the shear zone. An outcrop on the surface can be seen (Figure 3C). Exploratory drilling found that gold grades of altered rocks near shear zone in porphyritic granite can be up to 5.96 g/t, and generally reached 0.8~0.5 g/t, suggesting that there is a close relationship between gold mineralization and shear zones in porphyritic granite.

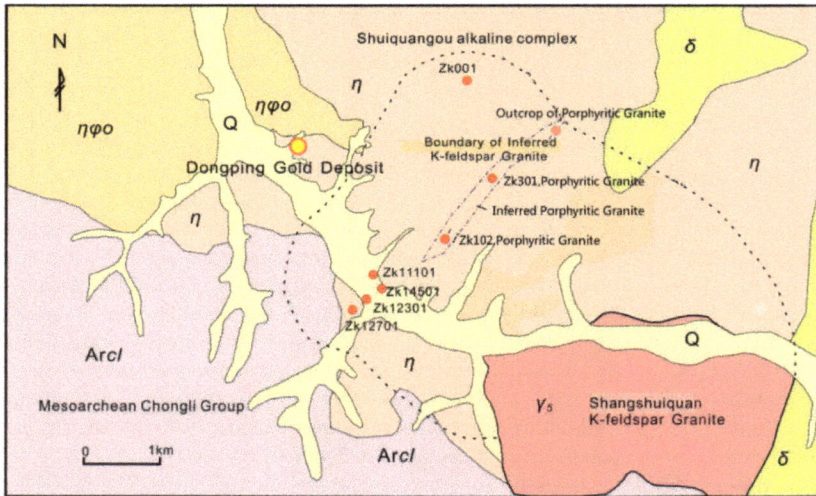

Figure 2. Sketch map of the Dongping gold deposit (after 1:10,000 geological map surveyed by Geological Exploration Report of Dongping Deposit, Chongli Zijin Mining Co., Ltd., Zhangjiakou, China, 2012). Arcl, Archean Chongli group metamorphic rocks; η, Shuiquangou alkaline complex; ηφο, hornblende monzonite; γ$_5$, Shangshuiquan K-feldspar granite; δ, diorite.

Figure 3. Characteristics of porphyritic granite in the Dongping gold deposit. (**A**,**B**) Porphyritic quartz in the porphyritic granite with chlorite veins, ZK301A drilling; (**C**) an outcrop of the porphyritic granite on the surface, Mazhangzi.

3.2. Petrography

The petrological characteristics of porphyritic granite are: Plagioclase (~20%), K-feldspar (~55%), and quartz (~25%). Plagioclase is mainly albite and occurs as fine aggregate (0.05~0.1 mm) (Figure 4C,D). The K-feldspar seems dirty on the surface in thin section, and garnet can be seen between some grains of the K-feldspar. Quartz occurs mainly as phenocrysts or aggregates with sizes ranging from 0.5 to 10 mm. Its characteristics indicate that it is not truly phenocryst. We prefer to call them porphyritic quartz. Lenticular quartz grains with directional distribution can be seen in the fragmented porphyritic granite (Figure 4B). A single porphyritic quartz is composed of one or several quartz crystals surrounded by fine recrystallized quartz (Figure 4A). Tiny feldspar crystals and more fluid inclusions can be seen as inclusions along the growth zone or within the fracture of porphyritic quartz (Figure 4E), including liquid-vapor two-phase aqueous inclusions and, CO_2-H_2O inclusions (Figure 4F). Electron probe analysis on feldspar inclusions within porphyritic quartz was carried out by Xu et al. [19]. The results show two types of feldspar: One is identified as K-feldspar, and the other is identified as albite (Table 1). Albite and K-feldspar typically occur as stripe feldspar in one inclusion.

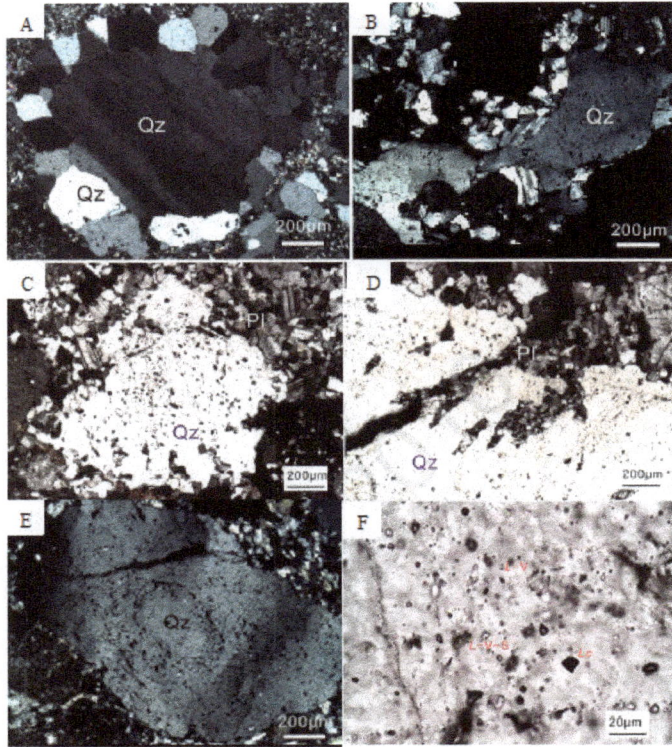

Figure 4. Microscopic features of porphyritic granite and quartz phenocryst in Dongping gold deposit. (**A**) Secondary quartz around the periphery of quartz phenocryst in the strong alteration rocks (sericite-quartz alteration); (**B**) directional distribution of lenticular porphyritic quartz in the fragmented porphyritic granite; (**C**) porphyritic quartz, with its edges surrounded by tiny plagioclase (Pl), containing oligoclase, ZK301A, 283.5 m; (**D**) quartz phenocryst in the porphyritic granite, containing tiny oligoclase aggregates within the fissures, Zhuanzhilian ZK301A, 283.5 m; (**E**) tiny feldspar crystals and fluid inclusions along the growth zone of porphyritic quartz; (**F**) inclusions in the quartz phenocryst containing liquid-vapor two-phase aqueous inclusions and, CO_2-H_2O inclusions.

Table 1. Results of electron probe analysis on feldspar inclusions within porphyritic quartz (quoted from Xu et al. [19]).

Position No.	Results (wt %)											Crystal Chemical Formula	Feldspar Type
	F	SiO$_2$	FeO	K$_2$O	Na$_2$O	MgO	TiO$_2$	CaO	Al$_2$O$_3$	MnO	Total		
MZ-2-1-1	/	68.45	0.39	0.11	11.18	/	/	/	19.70	/	99.88	$(Na_{0.948}K_{0.006})_{0.954}[Al_{1.016}Si_3O_8]$	albite
MZ-2-1-2	/	68.41	0.28	0.09	11.42	/	/	0.04	19.32	/	99.56	$(Na_{0.971}K_{0.005}Ca_{0.002})_{0.978}[Al_{0.999}Si_{3.006}O_8]$	albite
MZ-2-1-3	/	67.88	0.37	0.11	11.54	0.02	/	0.04	19.05	/	99.01	$(Na_{0.989}K_{0.006}Ca_{0.002})_{0.997}[Al_{0.992}Si_{3.006}O_8]$	albite
MZ-2-1-4	/	64.85	0.13	15.78	0.42	/	/	0.02	18.70	0.03	99.97	$(K_{0.929}Na_{0.038}Ca_{0.001})_{0.968}[Al_{1.016}Si_{2.996}O_8]$	K-feldspar
MZ-2-1-5	/	65.31	0.22	15.88	0.39	/	0.05	0.02	18.47	/	100.29	$(K_{0.932}Na_{0.035}Ca_{0.001})_{0.968}[Al_{1.001}Si_{3.007}O_8]$	K-feldspar
MZ-2-1-7	/	68.25	0.25	0.08	11.78	/	/	0.06	19.23	/	99.79	$(Na_{1.002}K_{0.004}Ca_{0.003})_{1.007}[Al_{0.995}Si_{3.001}O_8]$	albite
MZ-2-1-8	/	68.06	0.20	0.10	12.03	/	/	0.03	19.15	/	99.62	$(Na_{1.025}K_{0.006}Ca_{0.003})_{1.034}[Al_{0.992}Si_{2.997}O_8]$	albite
MZ-2-1-9	/	65.23	0.24	15.86	0.52	/	0.04	/	18.43	/	100.32	$(K_{0.931}Na_{0.046})_{0.977}[Al_{0.999}Si_{3.006}O_8]$	K-feldspar
ZK102-1	/	68.03	0.21	0.12	11.24	/	/	0.08	19.50	0.05	99.31	$(Na_{0.959}K_{0.007}Ca_{0.004})_{0.970}[Al_{1.001}Si_{2.998}O_8]$	K-feldspar
ZK102-2	/	64.99	/	16.05	0.23	/	/	/	18.84	/	100.21	$(K_{0.942}Na_{0.021})_{0.963}[Al_{1.021}Si_{2.994}O_8]$	albite
ZK102-3	/	68.22	0.21	0.12	11.68	/	/	/	19.42	/	99.74	$(Na_{0.993}K_{0.007})_{1.00}[Al_{1.004}Si_{2.997}O_8]$	K-feldspar
ZK102-4	/	64.75	0.06	16.02	0.63	/	0.08	/	18.10	/	99.64	$(K_{0.948}Na_{0.057})_{1.005}[Al_{0.989}Si_{3.007}O_8]$	K-feldspar
ZK102-5	/	68.35	0.10	0.13	11.87	/	0.08	0.04	19.36	/	99.93	$(Na_{1.007}K_{0.007}Ca_{0.002})_{1.016}[Al_{0.999}Si_{2.997}O_8]$	albite
ZK102-6	/	65.19	0.06	16.37	0.51	/	/	/	18.46	/	100.59	$(K_{0.959}Na_{0.045})_{1.004}[Al_{0.999}Si_{2.999}O_8]$	K-feldspar
ZK102-7	/	67.92	0.04	0.15	11.80	/	/	0.11	19.59	/	99.65	$(Na_{1.004}K_{0.008}Ca_{0.005})_{1.017}[Al_{1.013}Si_{2.985}O_8]$	albite
ZK102-8	/	64.75	0.04	16.29	0.30	/	/	0.02	18.82	/	100.26	$(K_{0.957}Na_{0.027}Ca_{0.001})_{0.986}[Al_{1.022}Si_{2.985}O_8]$	K-feldspar
ZK102-9	/	65.12	0.16	15.74	1.09	0.02	/	0.02	18.19	/	100.36	$(K_{0.925}Na_{0.097}Ca_{0.001})_{1.023}[Al_{0.986}Si_{3.000}O_8]$	K-feldspar
ZK102-10	/	68.55	0.18	0.16	11.93	/	/	0.03	19.33	/	100.26	$(Na_{1.010}K_{0.009}Ca_{0.001})_{1.020}[Al_{0.995}Si_{2.999}O_8]$	albite
ZK301A-248-1	/	68.24	0.31	0.20	11.31	/	/	0.04	19.69	/	99.84	$(Na_{0.960}K_{0.011}Ca_{0.002})_{0.973}[Al_{1.016}Si_{2.994}O_8]$	albite
ZK301A-248-2	/	64.49	/	16.25	0.33	/	/	/	19.12	/	100.19	$(K_{0.955}Na_{0.029})_{0.984}[Al_{1.038}Si_{2.975}O_8]$	K-feldspar
ZK301A-248-3	/	64.24	0.09	15.93	0.38	/	/	0.03	18.96	/	99.63	$(K_{0.941}Na_{0.034}Ca_{0.001})_{0.976}[Al_{1.035}Si_{2.980}O_8]$	K-feldspar

4. Zircon Dating

4.1. Analytical Method

Zircon particles were separated from samples of porphyritic granite using heavy liquid and magnetic techniques, and then by handpicking under a binocular microscope, at the Hebei Institute of Geological and Mineral Survey (Langfang, China), and targets was made according to Yuan et al. [20,21]. The selected zircon particles are then adhered to a double-sided tape, fixed with an epoxy resin and a curing agent amine, and ground until the zircon particles are maximally polished. To analyze their internal structures, cathodeluminescence (CL) images were obtained at Beijing Zircon Leading Technology Co., Ltd. (Beijing, China). Distinct domains within the zircon were selected for analysis, based on the CL images.

LA-ICP-MS zircon U-Pb dating of sample D35 was performed using an Agilent 7500a ICP-MS (Agilent, Santa Clara, CA, USA) equipped with a 193 nm laser, housed in the Beijing Zircon Leading Technology Co., Ltd. The room temperature was 20 °C and the relative humidity was 30%. NIST610 [22]. A reference material of synthetic silicate glass developed by American National Institute of Standards and Technology, and was used as an external standard material. Si was used as an internal standard element during elemental content analysis. During the test, the selected zircon test area was ablated using a laser beam with helium as a carrier, with a spot diameter of 30 μm and a frequency of 10 Hz. The sampling method is single point erosion, and the data acquisition needs 100s, of which the background measurement time is 40s and the signal measurement time is 60s. A zircon 91500 and a NIST 610 were measured at each of five sample sites. The ablated sample was transported to the ICP-MS using helium as a carrier, mixed with argon gas in a 30 cm³ mixing chamber, and then ionized in a plasma, next, used the mass spectrometer to measure the isotope ratio of the ionized material, then calculated the content of the relevant element and the isotopic age of the measured mineral based on the measured results of the isotope ratio and the corresponding standard mineral. Isotope ratio data processing was performed using GLITTER (ver 4.0, Maequarie University, Sydney, Australia) software. The software in Andersen [23] was used to perform common lead corrections on test data. Age calculation and mapping were performed using ISOPLOT (ver 3.70) software [24].

4.2. Zircon Characteristics

The porphyritic granite sample D35 was used for zircon U-Pb dating. In this study, a total of 245 zircon were selected, with a grain size of 20–150 μm, and 30 sets of zircon data were obtained. Most zircon are subhedral on CL images with irregular contours and are characterized by hydrothermal alteration (Figures 5 and 6). Some zircon have a localized oscillating ring zone (Figure 5A,B and Figure 6A–C), showing magmatic zircon characteristics, which are residual of magmatic zircon. Some zircon are dark black (Figure 6D) (without cathodoluminescence), sheared (Figure 6E,F) or broken into sub-grains (Figure 6E,F), or filled in the edges (Figure 5B,D and Figure 6F) or cracks (Figure 5G,H) of the euhedral magmatic zircon as an irregular gulf-like shape with a rough surface (Figure 5C). This irregular zircon is obviously transformed by shearing and hydrothermal metasomatism, which are defined as the new hydrothermal zircon. Most zircon retain both the residual of magmatic zircon and the characteristics of hydrothermal zircon.

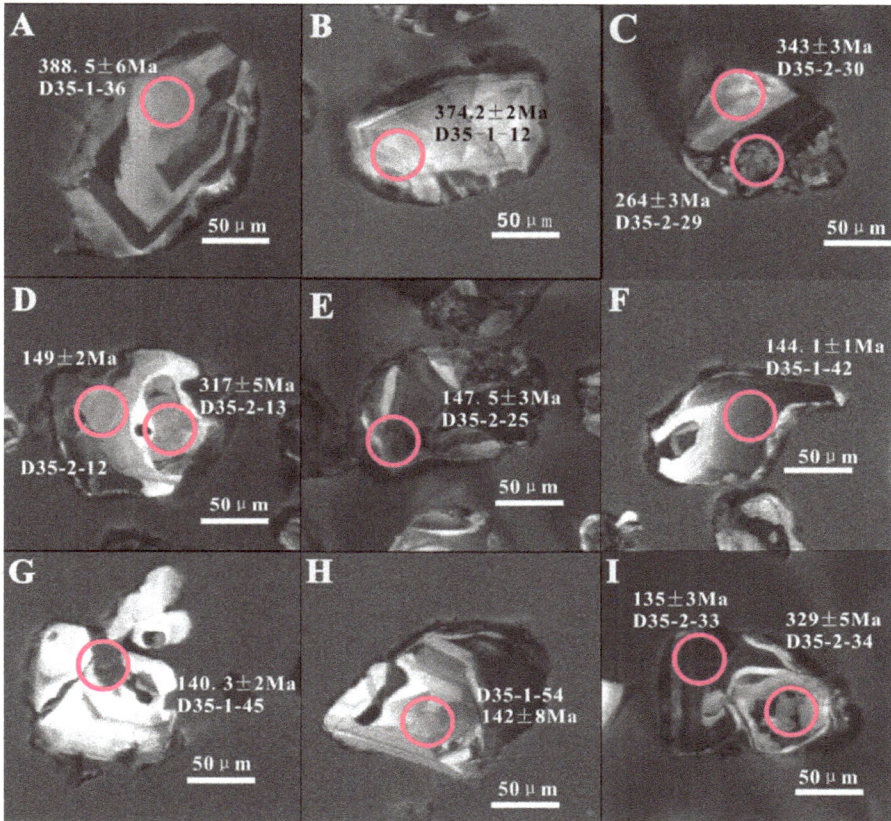

Figure 5. The measurement points on the cathodeluminescence (CL) images of zircon in porphyritic granite. (**A**) Oscillating ring zone of magmatic zircon; (**B**) weak oscillating ring zone of magmatic zircon with dark altered edge; (**C**) bright residual of magmatic zircon with rough surface below; (**D**) the magmatic zircon is transformed by the hydrothermal liquid, showing hydrothermal zircon (149 Ma) on the edge and residual of magmatic zircon (317 Ma) in the center; (**E**) dark hydrothermal zircon; (**F**) dark hydrothermal zircon with bright residual of magmatic zircon; (**G**) dark hydrothermal zircon filling along crack of magmatic zircon; (**H**) the magmatic zircon is transformed by the hydrothermal liquid, showing hydrothermal zircon in the center (142 Ma) and weak oscillating ring zone of magmatic zircon on the edge; (**I**) the magmatic zircon is transformed by the hydrothermal liquid, showing hydrothermal zircon on the edge (135 Ma) and residual of magmatic zircon (329 Ma) in the center.

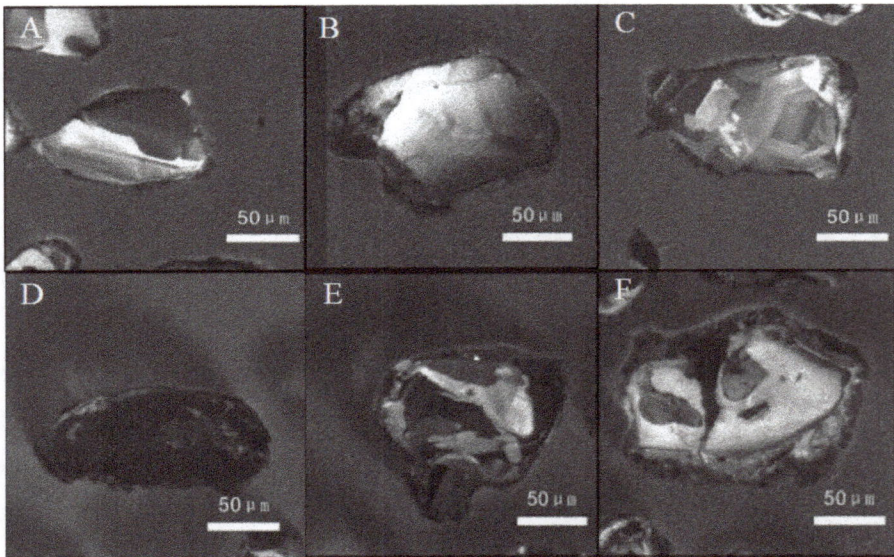

Figure 6. The CL images of zircon in porphyritic granite. (**A–C**) Localized oscillating ring zone of magmatic zircon; (**D**) dark hydrothermal ziron; (**E**) zircon that are sheared or broken into sub-grains; (**F**) broken magmatic zircon with altered edge.

4.3. Results

The typical results of the obtained zircon $^{206}Pb/^{238}U$, $^{207}Pb/^{235}U$, and $^{207}Pb/^{206}Pb$ values and the results of ordinary lead corrected age are shown in Table 2. Most of these measurement points are distributed on or near the U-Pb age concordant curve (Figure 7), showing good concordance. Data of zircon grains from porphyritic granite sample D35 yielded 2 groups of concordant ages at 142.02 ± 1.2 Ma (MSWD = 1.5), and 373.0 ± 3.5 Ma (MSWD = 1.5). The corresponding weighted average ages are 142.06 ± 0.84 Ma (MSWD = 1.13), and 373.0 ± 6.4 Ma (MSWD = 3.3). In general, zircon is geochemically stable by-product mineral. However, the hydrothermal fluids can replace and transform zircon to varying degrees along fractures and lattice defects within the zircon (Figure 5C,D,I and Figure 6A,F). As a result, wider alteration can occur at the edge or fissure of the zircon, resulting in the replacement of the zircon U-Pb system, and the formation of new hydrothermal zircon [25]. In spite of this, due to the stability of zircon itself, some zircon can keep the U-Pb system closed, and the U-Pb dating of zircon can still reflect geological age of primary zircon [4,5,26–28]. In the above two groups of ages, there are 8 measuring points with the old age, and there are obvious magmatic oscillation zone and bright CL images, belonging to the magmatic residual zircon [26,28–31]. This age represents the geological age of the magmatic zircon whose U-Pb system remained unreplaced, that is the age of primary magmatic zircon, which might be the age of residual zircon of Shuiquangou alkaline complex. The young zircon grains have a total of 19 sites with obvious metasomatic structures (Figure 5C,D,I and Figure 6A,F). Some grains of zircon were even converted to complete hydrothermal zircon grains (Figure 6D), which records the age of hydrothermal metasomatism, suggesting that the porphyritic granite suffered metasomatism of later hydrothermal fluid at 142.02 ± 1.2 Ma.

The rare earth element (REE) composition of magmatic and hydrothermal zircons from the porphyritic granites in Dongping Gold deposit are shown in Table 3. Hydrothermal zircon has a significantly higher REE content (Table 3). Magmatic zircon and hydrothermal zircon have distinct characteristics of rare earth patterns (Figure 8). Magma zircon increases rapidly from Pr to Lu.

The hydrothermal zircon is relatively gentle from Pr to Lu. This is consistent with typical magmatic zircon and hydrothermal zircon features.

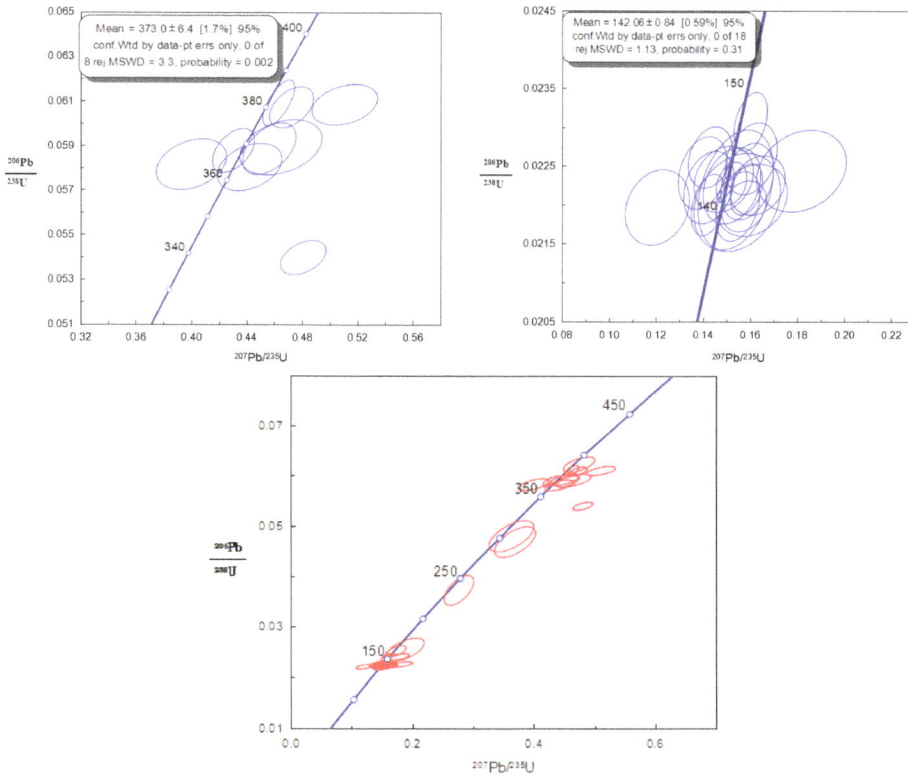

Figure 7. Concordant age diagram and weighted mean age of LA-ICP-MS zircon $^{207}Pb/^{235}U$-$^{206}Pb/^{238}U$ of porphyritic granite from Dongping deposit (D35 sample).

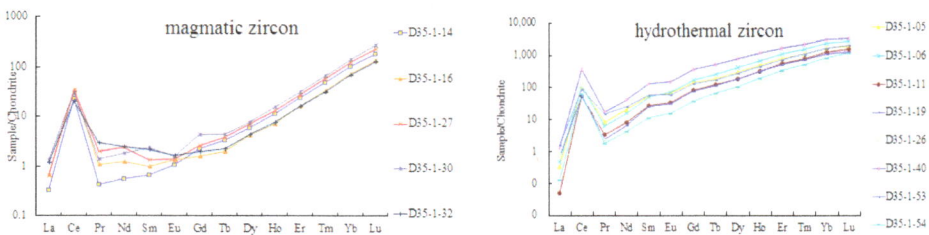

Figure 8. Chondrite-normalized REE patterns of hydrothermal, magmatic zircon.

The porphyritic granite intrudes into the monzonite of the alkaline rock along the NEE-SWW shear zone. The main elements of porphyritic granite and monzonite are shown in Table 4. Both porphyritic granite and monzonite have the characteristics of alkali-rich and high-potassium. However, the SiO$_2$ content of the porphyritic granite is significantly higher than that of the monzonite (Table 4).

Table 2. LA-ICP-MS zircon U-Pb dating results of porphyritic granites in Dongping gold deposit.

Measuring Point Number	Th/U	Isotope Ratio ± 1σ						Age (Ma)					
		207Pb/206Pb		207Pb/235U		206Pb/238U		207Pb/206Pb		207Pb/235U		206Pb/238U	
		Ratio	1σ	Ratio	1σ	Ratio	1σ	Age	1σ	Age	1σ	Age	1σ
D35-1-01	0.26	0.0503	0.0031	0.1538	0.0106	0.0222	0.0005	209.33	146.28	145.26	9.29	141.52	3.23
D35-1-02	0.23	0.0515	0.0040	0.1543	0.0121	0.0220	0.0004	264.88	177.76	145.71	10.64	140.49	2.64
D35-1-05	0.25	0.0534	0.0035	0.1605	0.0100	0.0221	0.0003	346.35	146.28	151.14	8.77	140.66	1.93
D35-1-06	0.28	0.0471	0.0016	0.1421	0.0048	0.0220	0.0003	53.80	77.77	134.95	4.25	140.28	1.66
D35-1-08	0.17	0.0515	0.0022	0.1603	0.0073	0.0226	0.0003	264.88	97.21	150.96	6.35	144.04	1.94
D35-1-11	0.17	0.0501	0.0027	0.1521	0.0081	0.0222	0.0003	198.23	127.76	143.78	7.10	141.85	2.05
D35-1-19	0.12	0.0516	0.0022	0.1546	0.0061	0.0220	0.0003	333.39	98.14	146.00	5.38	140.16	1.87
D35-1-25	0.39	0.0516	0.0014	0.1583	0.0045	0.0222	0.0002	333.39	61.10	149.22	3.96	141.55	1.39
D35-1-26	0.18	0.0467	0.0023	0.1424	0.0071	0.0222	0.0003	35.28	111.10	135.15	6.27	141.69	1.95
D35-1-28	0.34	0.0519	0.0021	0.1573	0.0062	0.0220	0.0002	279.69	90.73	148.36	5.48	140.17	1.37
D35-1-38	0.27	0.0521	0.0025	0.1586	0.0074	0.0223	0.0003	300.06	109.24	149.48	6.48	142.19	2.06
D35-1-41	0.18	0.0485	0.0028	0.1486	0.0083	0.0223	0.0003	124.16	129.61	140.70	7.34	142.37	1.79
D35-1-53	0.11	0.0511	0.0013	0.1558	0.0037	0.0222	0.0002	242.66	57.40	147.04	3.21	141.61	1.34
D35-1-54	0.24	0.0506	0.0033	0.1549	0.0106	0.0222	0.0003	220.44	149.98	146.22	9.28	141.65	2.17
D35-1-55	0.10	0.0531	0.0011	0.4304	0.0101	0.0588	0.0007	331.54	54.63	363.49	7.15	368.18	3.99
D35-1-14	0.02	0.0552	0.0009	0.4631	0.0075	0.0609	0.0007	420.42	35.18	386.40	5.22	381.19	4.12
D35-1-16	0.26	0.0579	0.0026	0.4647	0.0195	0.0589	0.0008	524.11	99.99	387.53	13.54	369.05	5.00
D35-1-27	0.04	0.0567	0.0018	0.4562	0.0127	0.0589	0.0008	479.67	70.36	381.63	8.88	368.65	4.69
D35-1-30	0.05	0.0555	0.0021	0.4419	0.0155	0.0580	0.0007	431.53	86.10	371.62	10.90	363.62	4.31
D35-1-31	0.02	0.0564	0.0012	0.4722	0.0105	0.0607	0.0006	477.82	48.14	392.74	7.22	379.99	3.68
D35-1-32	0.03	0.0610	0.0019	0.5108	0.0157	0.0608	0.0006	638.91	68.51	418.96	10.55	380.74	3.64
D35-1-44	0.07	0.0499	0.0021	0.3999	0.0167	0.0581	0.0008	190.82	91.65	341.55	12.11	364.36	4.62
D35-2-25	0.16	0.0508	0.0025	0.1625	0.0095	0.0231	0.0004	231.55	110.17	152.91	8.32	147.54	3.04
D35-2-30	0.21	0.0586	0.0012	0.4409	0.0209	0.0547	0.0026	553.74	50.91	370.95	14.77	343.65	16.15

Table 3. REE composition of magmatic and hydrothermal zircons from the porphyritic granites in Dongping gold deposit.

No.	D35-1-05	D35-1-06	D35-1-11	D35-1-19	D35-1-25	D35-1-28	D35-1-38	D35-1-41	D35-1-53	D35-1-54	D35-1-14	D35-1-16	D35-1-27	D35-1-30	D35-1-32
La (μg/g)	0.11	0.15	0.02	0.22	0.38	0.17	0.70	0.07	0.50	0.04	0.10	0.21	0.20	0.42	0.37
Ce (μg/g)	88.24	97.08	45.70	45.76	232.90	78.95	41.47	63.14	72.21	52.41	18.35	26.71	28.37	24.34	16.50
Pr (μg/g)	1.02	0.78	0.41	0.31	1.95	1.01	0.19	1.05	1.81	0.23	0.05	0.13	0.25	0.17	0.36
Nd (μg/g)	12.44	9.96	5.00	4.16	24.78	11.72	1.59	11.65	15.81	2.75	0.33	0.74	1.43	1.12	1.48
Sm (μg/g)	10.91	9.92	5.65	5.25	27.54	9.53	1.69	10.12	11.35	2.34	0.13	0.19	0.27	0.46	0.43
Eu (μg/g)	5.17	5.44	2.55	2.45	10.82	4.13	0.97	4.79	4.42	1.23	0.08	0.10	0.10	0.12	0.13
Gd (μg/g)	36.81	45.03	21.53	20.14	85.68	31.87	10.28	32.34	34.24	10.31	0.58	0.41	0.67	1.16	0.53
Tb (μg/g)	9.24	12.53	5.98	5.69	20.90	8.00	3.18	9.00	8.58	3.14	0.16	0.09	0.18	0.21	0.11
Dy (μg/g)	99.57	136.54	61.72	62.83	197.24	85.48	39.34	92.12	89.58	35.33	1.91	1.37	2.33	2.52	1.40
Ho (μg/g)	36.20	50.91	23.53	23.21	65.13	31.25	16.05	34.33	32.84	14.77	0.81	0.52	0.94	1.14	0.57
Er (μg/g)	164.34	232.00	115.69	109.10	255.58	141.89	75.07	153.03	150.94	71.49	5.07	3.37	5.89	6.58	3.47
Tm (μg/g)	36.31	50.06	26.21	24.10	51.07	32.99	17.10	35.44	34.71	17.29	1.59	1.07	1.85	2.14	1.01
Yb (μg/g)	341.27	481.09	263.70	226.83	452.10	316.87	164.90	336.15	357.44	178.51	21.91	15.12	26.10	29.79	14.26
Lu (μg/g)	63.48	87.22	49.54	41.87	75.60	59.09	31.24	63.17	66.33	38.98	5.86	4.36	7.28	8.57	4.03
ΣREE (μg/g)	905.11	1218.68	627.23	571.91	1501.67	812.95	403.78	846.41	880.76	428.81	56.94	54.40	75.87	78.73	44.64
δEu	0.71	0.66	0.62	0.64	0.62	0.65	0.55	0.74	0.63	0.64	0.74	1.07	0.72	0.46	0.80
δCe	24.98	35.19	33.13	35.09	33.57	22.15	27.08	17.65	10.85	63.83	58.43	37.70	26.50	21.86	9.82
Age (Ma)	140 ± 2	140 ± 1	141 ± 2	140 ± 2	141 ± 1	140 ± 1	142 ± 2	142 ± 2	142 ± 1	142 ± 2	381 ± 4	369 ± 5	369 ± 5	364 ± 4	380 ± 4

Table 4. Main element (wt %) analysis result.

Number	ZK102	ZK301A	MZZ-2	HTL202
Discription	Porphyritic Granite	Porphyritic Granite	Porphyritic Granite	Monzonite
SiO_2	71.39	72.40	75.54	65.06
Al_2O_3	14.78	13.75	12.92	15.31
Fe_2O_3	1.95	2.21	1.65	5.88
MgO	0.08	0.12	0.08	0.53
CaO	0.81	0.72	0.12	1.94
Na_2O	5.27	3.89	4.37	3.50
K_2O	5.35	6.51	4.91	5.56
MnO	0.05	0.06	0.02	0.08
TiO_2	0.07	0.07	0.05	0.64
P_2O_5	0.01	0.01	0.01	0.12
Ignition Loss	0.15	0.15	0.22	1.38
FeO	0.95	1.13	0.55	3.04
total	99.90	99.89	99.90	99.99
AR	5.28	6.11	5.93	3.21
A/NK	1.02	1.02	1.03	1.30
A/CNK	0.93	0.93	1.01	1.00
SI	0.56	0.87	0.68	2.85
FL	92.95	93.53	98.70	82.36
MF	97.45	96.53	96.53	94.42

A/CNK = $Al_2O_3/(CaO + Na_2O + K_2O)$; A/NK = $Al_2O_3/(Na_2O + K_2O)$; (AR) = $[Al_2O_3 + CaO + (Na_2O + K_2O)]/[Al_2O_3 + CaO - (Na_2O + K_2O)]$ (wt %); SI = $100 \times MgO/(MgO + Fe_2O_3 + FeO + Na_2O + K_2O)$ (wt %); FL = $100 \times (Na_2O + K_2O)/(CaO + Na_2O + K_2O)$ (wt %); MF = $100 \times (Fe_2O_3 + FeO)/(MgO + Fe_2O_3 + FeO)$ (wt %).

5. Discussion

5.1. Diagenetic Age and Metallogenetic Age of Porphyritic Granite

The gold-bearing porphyritic granite occurs along a NEE-SWW-striking zone with a width of 15–50 m within the Shuiquangou alkaline complex. That is, the mineralized porphyritic granite is controlled by the NEE-SWW shear zone with strong potassium enrichment and silicification. So, the age of porphyritic granite should not be much earlier than 140 Ma according to the age of the relatively open structure [13]. Therefore, the age of 373 ± 3.5 Ma may be for residual zircon of Shuiquangou alkaline complex.

In this paper, although the zircon with an average age of 373.0 ± 3.5 Ma was later replaced by hydrothermal fluids, most of the zircon showed hydrothermal zircon characteristics. However, some zircon may still retain magmatic zircon features. Therefore, it was considered that the residual zircon in the alkaline rock was trapped during emplacement of the porphyritic granite into NE-SW shear zone.

Wei et al. [16] believed that the volcanic activity of the middle-acid volcanic rocks in the Zhangjiakou Formation may provide important heat sources and ore-forming fluids for the Dongping gold deposit. The previous paper introduced there were many medium-acid rock bodies in this area at 140 Ma or so. These rock bodies can provide important heat sources and ore-forming fluids for the Dongping gold deposit. Li et al. [9] selected zircon from the potassium altered rock in the deep vein of No. 70, obtained the zircon age at the Dongping gold deposit as 140.3 ± 1.4 Ma which is the gold metallogenic age. These high-precision isotopic age results are all Early Cretaceous. Over the past decade, there have been many reports of the discovery of the hydrothermal zircon and hydrothermal zircon U-Pb dating in gold deposits [29,32]. Studies have shown that zircon can grow and crystallize directly from medium-low-temperature hydrothermal fluids [33,34]. Dubinska et al. [34] suggested that the zircon grains included two types of primary fluid inclusions containing either liquid CO_2 or aqueous solutions which were used to assign the P–T conditions of the zircon formation (270–300 °C, ca. 1 kbar). The formation temperature and pressure conditions of these hydrothermal zircon are very similar to that of gold-bearing veins. Therefore, the age of gold deposits can be determined by using the hydrothermal zircon U-Pb dating in gold-bearing altered rocks. The hydrothermal zircon age in this study is 142.02 ± 1.2 Ma, which is almost consistent with the mineralization age of 140.3 ± 1.4 Ma obtained by Li et al. [5,9]. It is also consistent with Yanshanian mineralization explosion in the eastern China [35] and the view that there are three major metallogenic stages 200–160, 140 and 120 Ma in the Mesozoic in northern China [36]. According to the above age data, the age of gold mineralization occurred in the porphyritic granite may be 142.02 ± 1.2 Ma.

5.2. Hydrothermal Metasomatism and Gold Mineralization

In the presence of fluids, medium-low metamorphism can also cause changes in zircon structure, composition and age, and the effect of hydrothermal fluids on the zircon U-Pb systems clearly far exceeds the extent of metamorphism [5,28]. The defects formed by radioactive damage inside the zircon provide channels for the hydrothermal reforming. The hydrothermal fluids replace the zircon along the edge of the zircon crystal, cracks or lattice defects, resulting in the destruction of the U-Pb system and forming a complex zircon internal structure. Hydrothermal alteration, especially the hydrothermal action associated with ductile shear, has an important influence on the zircon U-Pb system [37].

The relatively open extensional structure was formed in the Zhangxuan area at 140 Ma or so, where underwent under plating with granulite facies metamorphism at Mesozoic [13]. The gold deposits in the Dongping area are mainly subjected to ductile shearing. Under the conditions of hydrothermal alteration, the zircon grains in the porphyritic granite may be modified. The characteristics and age of zircon in the porphyritic granite show that the formation of porphyritic quartz in porphyritic granites is related to the alteration and metasomatism of late hydrothermal fluids. Generally, chondrite-normalized REE patterns of magmatic zircon maintains a strong Ce anomaly, and the Ce anomaly of hydrothermal zircon is weak [5]. However, both types of zircon have strong Ce anomalies in Figure 8. This may be the fact that some magmatic zircon are not completely replaced by hydrothermal fluids and still retain the residual characteristics of magmatic zircon. Single porphyritic quartz aggregates are composed of one or several quartz crystals surrounded by fine recrystallized quartz, indicating that porphyritic quartz were subjected by later Si-rich hydrothermal fluids. The phenomenon of fine albite grains surrounding and tiny albite veins cutting the porphyritic quartz are also explained as metasomatism of late hydrothermal fluids. However, tiny albite and potassium feldspar occurring as stripe feldspar inclusions along the growth zones in porphyritic quartz may be resulted in trapping from early hydrothermal fluids which were rich in Na and K during porphyritic quartz initially grew. On the other hand, porphyritic granite intrudes into the monzonite of the alkaline rock along the NEE-SWW shear zone. Gold mineralization is intensive within the shear zone of the porphyritic granite intrusion, and Au reserves are estimated to be more than one tonnage. Silicification and potassiumization are relatively developed in the shear zone. The main element (Table 4) reveals that the porphyritic granite has a higher siliceous composition than the surrounding monzonite. These characteristics above suggest that the porphyritic quartz and gold mineralization in the porphyritic granite might be products of a late fluid metasomatism after emplacement of the porphyritic granite into NEE-SWW shear zone.

The porphyritic quartz may be originated from residual Si-rich magma. However, we did not see melt inclusions in porphyritic quartz as yet, indicating that the presence of residual magma could not be possible within the porphyritic granite. The homogenization temperatures of isolated liquid-vapor or vapor-liquid inclusions in porphyritic quartz of mineralizing porphyritic granite are higher than that of inclusions in gold-rich ores, which represents two hydrothermal stages [38]. Xu et al. [19] suggests that isolated liquid-vapor or vapor-liquid inclusions in the porphyritic quartz, might represent the initial fluids that later yielded gold-forming fluids. So, the tiny albite and K-feldspar along the growth belt of porphyritic quartz, might be the initial component of gold-forming fluids.

As for the source of hydrothermal fluid, in combination with age data, it might be associated with volcanic activity during the Late Jurassic—Early Cretaceous period or the evolution of small intrusions, dikes, and hidden rock masses around the ore body during the Yanshanian period. These volcanic activities could provide heat sources and power, also produced later magmatic hydrothermal fluids, which extracted and activated part of the gold from the Shuiquangou alkaline complex. This is an important factor for the enrichment of gold in porphyritic granites, which resulted in precipitation of gold under suitable physical and rock conditions.

It can be seen from the above that the porphyritic granite and gold mineralization in the porphyritic granite might be products of a late fluid metasomatism and are controlled by the shear zone. Therefore, the porphyritic granite and NEE-SWW shear structure can be used as two prospecting indicators for gold.

6. Conclusions

A total of 30 analyses of zircon grains from porphyritic granite sample D35 yielded two groups of concordant ages at 142.0 ± 1.2 Ma (MSWD = 1.5) and 373.0 ± 3.5 Ma (MSWD = 1.5), which might represent the age of residual magma of Shuiquangou alkaline complex and metallogenetic ages of porphyritic granite respectively. Porphyritic quartz in the porphyritic granites might be related to the alteration and metasomatism of Si-rich hydrothermal fluids after emplacement of the porphyritic granite into NEE-SWW shear zone. The porphyritic quartz and gold mineralization might be products of a late fluid metasomatism. The isolated liquid-vapor or vapor-liquid inclusions and tiny albite or K-feldspar along the growth belt of porphyritic quartz, might represent the initial fluids that were rich in Na/K and later yielded gold-forming fluids.

Author Contributions: H.W. wrote the paper and performed data treatment; J.X. formulated the problem, organized the research team, guided the study, participated in writing the manuscript and revised paper; G.Z. provided natural samples and edited paper format; X.C., H.C., C.B., Z.Z. participated in experimental program.

Funding: This work is funded by National Nature Science Foundation of China (grants No. 41672070) and Doctoral Research Foundation of Hebei GEO University (No. BQ201614).

Acknowledgments: We thank the Dongping Gold Mine for its support and Zirconium for the zircon testing service. We are grateful to an anonymous reviewer and Galina Palyanova for helpful comments and suggestions.

Conflicts of Interest: The authors declare no conflict of interest.

References

1. Sillitoe, R.H. Some metallogenic features of gold and copper deposits related to alkaline rocks and consequences for exploration. *Miner. Dépos.* **2002**, *37*, 4–13. [CrossRef]
2. Müller, D. Gold-copper mineralization in alkaline rocks. *Miner. Dépos.* **2002**, *37*, 1–3. [CrossRef]
3. Zhu, Y.P.; Zhao, X.D.; Tan, G.L.; Yao, Z.Y.; Zhao, Y.H. Metallogenetic characteristics of Cadia-Ridgeway Porphyry Gold-Copper Deposit in Australia. *West. China Sci. Technol.* **2015**, *14*, 24–29.
4. Luo, Z.K.; Miao, L.C.; Guan, K. SHRIMP geochronology of the Shuiquangou rock mass and its significance in Zhangjiakou. *Hebei Geochem.* **2001**, *30*, 116–122.
5. Li, C.M.; Deng, J.F.; Chen, L.H. Constraint from two stages of zircon on the metallogenic age in the Dongping gold deposit, Zhangxuan area, northern North China. *Miner. Depos. Geol.* **2010**, *29*, 265–275.
6. Xiang, S.Y.; Ye, J.L.; Liu, J. The origin of alkaline syenite in Hougou gold deposit and its relationship with gold mineralization. *Mod. Geol.* **1992**, *6*, 55–62.
7. Song, G.R.; Zhao, Z.H. *Geological Geology of the Dongping Alkaline Complex in Hebei Province*; Seismological Press: Beijing, China, 1996; pp. 1–181.
8. Li, H.M.; Li, H.K.; Lu, S.N.; Yang, C.L. Using the U-Pb dating of hydrothermal zircon in ore veins to determine the metallogenic age of Dongping gold deposit. *Earth J.* **1997**, *18*, 176–178.
9. Li, C.M.; Deng, J.F.; Su, S.G.; Li, H.M.; Liu, X.M. Two-stage zircon chronology and its significance in potash altered rocks in Dongping Gold deposit, Hebei Province. *J. Earth Sci.* **2010**, *31*, 843–852.
10. Mo, C.H.; Wang, X.Z.; Liang, H.Y. The U-Pb ages and their geological significance of Shangshuiquan granites in the Dongping gold deposit. *J. Mineral.* **1998**, *18*, 298–301.
11. Jiang, S.H.; Nie, F.J. The ^{40}Ar-^{39}Ar isotopic geochronology of the Shuiquangou Complex and its related gold deposits in Northwest Yunnan. *Geol. Forum* **2000**, *46*, 621–627.

12. Nie, F.J.; Jiang, S.H.; Liu, Y. Intrusion-related gold deposits of North China Craton, People's Republic of China. *Resour. Geol.* **2004**, *54*, 299–324. [CrossRef]

13. Shao, J.A.; Wei, C.J.; Zhang, L.Q.; Niu, S.Y.; Mou, B.L. Pyroxene diorite in nuclear department of Zhangxuan Uplift. *Acta Petrol. Sin.* **2004**, *20*, 1389–1396.

14. Liu, B. Metallogenic characteristics of alkaline intrusive complexes in Hougou, Shuiquangou, Northern Hebei Province. *Geol. Resour.* **2001**, *10*, 25–32.

15. Li, C.M. *Study on Chronology and Petrochemistry in Dongping and Hougou Gold Orefields, Northern Hebei Province*; China University of Geoscience: Beijing, China, 2011.

16. Wei, Z.L.; Zhang, H.; Liu, X.M.; Zhang, Y.Q. LA-ICP-MS dating of volcanic rocks in Zhangjiakou Formation, Zhangjiakou area and its geological significance. *Prog. Nat. Sci.* **2008**, *18*, 523–530.

17. Jiang, N.; Zang, S.Q.; Zhou, W.G.; Liu, Y.S. Origin of a Mesozoic granite with A-type characteristics from the North China craton: Highly fractionated from I-type magmas? *Contrib. Mineral. Petrol.* **2009**, *158*, 113–130. [CrossRef]

18. Jiang, N.; Liu, Y.S.; Zhou, W.G.; Yang, J.H.; Zhang, S.Q. Derivation of Mesozoic adakitic magmas from ancient lower crust in the North China craton. *Geochim. Cosmochim. Acta* **2007**, *71*, 2591–2608. [CrossRef]

19. Xu, J.H.; Wand, Y.W.; Bian, C.J.; Zhang, G.R.; Chu, H.X. Unusual quartz phenocrysts in a newly discovered porphyritic granite near the giant Dongping Gold Deposit in Northern Hebei Province, China. *Acta Geol. Sin.* **2018**, *92*, 398–399. [CrossRef]

20. Yuan, H.L.; Gao, S.; Dai, M.N.; Zong, C.L.; Detlef, G.; Gisela, H.F.; Liu, X.M.; Diwu, C.R. Simultaneous determinations of U-Pb age, Hf isotopes and trace element compositions of zircon by excimer laser-ablation quadrupole and multiple-collector ICP-MS. *Chem. Geol.* **2008**, *247*, 100–118. [CrossRef]

21. Yuan, H.; Wu, F.; Gao, S.; Liu, X.; Xu, P.; Sun, D. Determination of U-Pb age and rare earth element concentrations of zircons from Cenozoic intrusions in northeastern China by laser ablation ICP-MS. *Chin. Sci. Bull.* **2003**, *48*, 2411–2421.

22. Pearce NJ, G.; Perkins, W.T.; Westgate, J.A.; Gorton, M.P.; Jackson, S.E.; Neal, C.R.; Chenery, S.P. A compilation of new and published major and trace element data for NIST SRM 610 and NIST SRM 612 glass reference materials. *Geostand. Newsl.* **1997**, *21*, 115–144. [CrossRef]

23. Andersen, T. Correction of common Pb in U-Pb analyses that do not report [204]Pb. *Chem. Geol.* **2002**, *192*, 59–79. [CrossRef]

24. Ludwig, K.R. *User's Manual for Isoplot 3.70: A Geochronological Toolkit for Microsoft Excel*; Special Publication No. 4; Berkeley Geochronology Center: Berkeley, CA, USA, 2008; Volume 76.

25. Kroner, A.; Jaeckel, P.; Williams, I.S. Pb-loss patterns in zircons from a high-grade metamorphic terrain as revealed by different dating methods: U-Pb and Pb-Pb ages of igneous and metamorphic zircon from northern Sri Lanka. *Precambrian Res.* **1994**, *66*, 151–181. [CrossRef]

26. Liati, A.; Gebauer, D.; Wysoczanski, R. U-Pb SHRIMP dating of zircon domains from UHP garnet-rich mafic rocks and late pegmatoids in the Rhodope zone (N Greece): Evidence for Early Cretaceous crystallization and Late Cretaceous metamorphism. *Chem. Geol.* **2002**, *184*, 281–299. [CrossRef]

27. Tomaschek, F.; Kennedy, A.K.; Villa, I.M. Zircons from Syros, Cyclades, Greece-recrystallization and mobilization of zircon during high-pressure metamorphism. *J. Petrol.* **2003**, *44*, 1977–2002. [CrossRef]

28. Li, C.M. Review on zircon genetic mineralogy and zircon micro area dating. *Geol. Surv. Res.* **2009**, *33*, 161–174.

29. Hu, F.F.; Fan, H.R.; Yang, J.H.; Wan, Y.S.; Liu, D.Y.; Zhai, M.G.; Jin, C.W. Mineralizing age of the Rushan lode gold deposit in the Jiaodong Peninsula: SHRIMP U-Pb dating on hydrothermal zircon. *Chin. Sci. Bull.* **2004**, *49*, 1629–1636. [CrossRef]

30. Liu, F.; Li, Y.H.; Mao, J.W.; Yand, F.Q.; Chai, F.M.; Mou, X.X.; Yang, Z.X. Zircon SHRIMP ages and their geological significance in the Abba granites of the Altai Orogenic Belt. *Earth J.* **2008**, *29*, 795–804.

31. Nie, F.J.; Xu, D.Q.; Jiang, S.H.; Hu, P. SHRIMP age of the zircons from the K-feldstolite in the Su-Cha fluorite deposit and its geological significance. *J. Earth Sci.* **2009**, *30*, 803–811.

32. Fu, B.; Terrence, P.M.; Noriko, T.K.; Anthony IS, K.; John, W.V. Distinguishing magmatic zircon from hydrothermal zircon: A case study from the Gidginbung high-sulphidation Au-Ag-(Cu) deposit, SE Australia. *Chem. Geol.* **2009**, *259*, 131–142. [CrossRef]

33. Cherniak, D.J.; Watson, E.B. Diffusion in zircon. *Rev. Mineral. Geochem.* **2003**, *53*, 113–143. [CrossRef]

34. Dubinska, E.; Bylina, P.; Kozlowski, A. U-Pb dating of serpentinization: Hydrothermal zircon from a metasomatic rodingite shell (Sudetic ophiolite, SW Poland). *Chem. Geol.* **2004**, *203*, 183–203. [CrossRef]

35. Deng, J.F.; Mo, X.X.; Zhao, H.L.; Luo, Z.H.; Zhao, G.C.; Dai, S.Q. Catastrophic lithosphere and asthenosphere system in eastern China during the Yanshanian period. *Miner. Depos. Geol.* **1999**, *18*, 309–315.
36. Mao, J.W.; Xie, G.Q.; Zhang, Z.H.; Li, X.F.; Wang, Y.T.; Zhang, C.Q.; Li, Y.F. Mesozoic large-scale mineralization and its geodynamic setting in northern China. *Acta Petrol. Sin.* **2005**, *21*, 170–188.
37. Bao, Z.W.; Zhao, Z.H. Effect of Hydrothermal Alteration on Zircon, U-Pb Dating. In Proceedings of the National Symposium on Petrology and Geodynamic, Nanjing, China, 1–6 November 2006.
38. Zhang, G.R.; Xu, J.H.; Wei, H.; Song, G.C.; Zhang, Y.B.; Zhao, J.K.; He, B.; Chen, D.L. Structure, alteration, and fluid inclusion study on deep and surrounding area of the Dongping gold deposit, Northern Hebei, China. *Acta Petrol. Sin.* **2012**, *28*, 637–651.

MDPI

St. Alban-Anlage 66

4052 Basel

Switzerland

Tel. +41 61 683 77 34

Fax +41 61 302 89 18

www.mdpi.com

Minerals Editorial Office

E-mail: minerals@mdpi.com

www.mdpi.com/journal/minerals

www.ingramcontent.com/pod-product-compliance
Lightning Source LLC
Chambersburg PA
CBHW051845210326
41597CB00033B/5776